WHEN DO THE WAGES GO IN?

JACK STANLEY

First published in Great Britain as a softback original in 2014

Typeset in Mercury

Editing, design and publishing by UK Book Publishing

UK Book Publishing is a trading name of Consilience Media

www.ukbookpublishing.com

ISBN: 978-1-910223-11-6

Cover photos

British money – carlsilver @ freeimages.com

American money – hoboton @ freeimages.com

"Us tradesmen are always chasing the ace, that
one last jolly for those mega bucks.

Along the way we meet people who are golden, people
who for one reason or another we will never forget.

Past and present, wherever they are 'When do
the wages go in' is dedicated to them all."

WHEN DO THE WAGES GO IN?

Sat in the departure lounge at Newcastle airport in the North-east of England, Davey, Calla, Kris, Jonny and Mark were a group of welders and pipe fitters who had worked on and off jobs with each other all over the world for the last 10 years. They were seated around a table full of drinks waiting for a flight to Sanford airport in Orlando, where Davey had landed a supervisor's role on a big process job. Normal practice in the contracting game is getting the good lads he knew and trusted a start – and he did.

Jonny strolls from the busy bar and sits down with his pint of Kronenbourg, "Hey lads, did you see Stevie Williamson in check-in? I did hear the sparky bastard is heading our way; he's going to price up the install side for CJK Electrical."

Kris puts his pint down and wipes the frothy lager from his mouth. "Should be good crack if there's a decent squad of us owa then, instead of listening to them fucking yanks chirp on! By the way is Calla still in the bookies?"

Davey smiles and tells the table: "Lads, he is supposed to be heading owa here to make a few quid. And let me tell you now: whatever he does have the silly twat will be skint before we do a hand's turn on that job!"

Just then Calla strolls over to the table. "Just won a fucking bull's-eye on a hoss – two fences out I rolled the slip up and bounced it off the telly! The favourite fell at the last and my horse ran on to win! I was like a fucking tramp raking through the bits of paper until I found it."

Mark shakes his head. “You are one jammy bastard, Calla, go on get the slurps in, hamma, we don’t board for another two hours,” waving his glass as Calla smiles.

“No problem, Marko, you just make sure you don’t piss yourself on the flight again.” As Calla takes the rest of the lads’ orders Stevie the sparky bastard walks over to the table and puts his laptop bag down.

“Alreet Stevie, lend us your face so we can catch a rat!” Jonny quips as Davey has a dig at him as well. “That’s a posh bag, Stevie – what’s in there? Your fucking ham and cheese sarnies?” The lads all laugh.

***

Jimmy Hunter was sat in the portacabin overlooking the empty shell of a huge building soon to become one of the leading suppliers of technology to best parts of the communication world. The list of names in front of him were not just men who had worked for him but those who had put him right to the top of the ladder in construction. Now Jimmy had a tough choice to make.

The bursting of the door shook Jimmy out of his ponder. “Jim, I can’t fucking believe the squad you have coming over for this first phase!” Malcolm Simmons was a typical manager who forgot the lads he worked with and where he came from! “Davey Ramsey! Johnny Smith! And fucking Andrew Callaghan! It’s not Las fucking Vegas Boulevard we are building – it’s a 100 million dollar state of the art facility in Orlando,” he bellowed to Jimmy.

Jimmy leant back in the chair and lit up a Marlboro light. “Malcolm, these shite Yankee tabs are spoiling my day already and you are now making it worse with your shite! Do you really expect the local lads to throw pipe in like the squad we have coming over!?”

There was a silence as Malcolm sat down and took a cigarette out of Jimmy’s box and lit it up. “You’re right but we better phone Mulder and Scully to warn the fuckers of the invasion that is on its way!!”

The airport lounge at Newcastle is full of typical hustle and bustle – people heading off on their holidays, people heading off to see loved ones and people heading off to make a few honest bob.

"Just had a telly text from Jimmy owa in the US," Dave announces to the table, which was full of half empty glasses of lager and bottles of cider.

"And what did the fat bastard say?" chirps Calla as the rest laugh. Davey holds his phone close to his glasses and squints. "For fuck sake nee wonder them butts you set up are all owa the place, ya blind twat!" The table was always going to be a nest of banter, even more so with the drink flowing so freely!

"Jimmy says here that he's sorting the digs out for us although would it be possible for Andrew Callaghan to run over to Boots before he boards and get some Pampers to save him on fines for pissing the bed!"

"Erc a wet bed's a happy bed, and that fat sackless twat needn't go on – my old man reckons he used to get called Jimmy the sailor coz he used to wake up soaking every morning." Calla was a tall, thin lad, around 6' 3", who was dragged up in the back streets of Redcar – sound lad but wanted to fight the world after the smell of even a barman's fart. He was 29 and already divorced with a 3 year old son whose mother was hell bent on making sure Andrew didn't have a pot to piss in; along with his gambling habits, love of ladies and booze, this wasn't a difficult task!

"Might have something to do with the 20 pints every neet," Dave quips, rocking back on his chair to reminisce. "I remember one neet we got sent to Hull on some job and we had to share, black crashed a boozer after graft and had a good swally, gets to bed and I woke up to the sound of running water – that bastard was only pissing on me. 'Here, Jimmy!' I shouts, smacking him one in the ribs! Dirty bastard finished it off in the wardrobe and I got the fine for the wet bed! He said he wasn't going to pay coz I'd broke his rib! We had to move digs, ended the rest of the stay in a reet shit hole...he is some cunt!"

Dave was the oldest out of the lot, an average built guy, 64 and still well for his age. This was his last stint away then it was feet up with his beloved wife Margaret, who was a right battle axe but the only woman to keep Dave in toe. Dave had served his time in the yards were he had been nicknamed Tungsten due to the fact he was a right hard bastard, a nickname Davey hated! While in the yards Jimmy had been his apprentice learning a lot from Dave – mainly not to back answer! Dave couldn't believe Hunter had got where he was as he had more cheek than an elephant's arse and most of the lads said he wouldn't make a pipe fitter as long as he had a hole in his arse! Nevertheless he had done very well and Dave was pleased as Jimmy often gave him a leg up when work was down.

As the laughter was settling down Jonny pipes up: "Is it reet that Mally noodle spine is in charge?"

"Afraid so," Davey says, batting Calla's feet off the side of his chair.

"Haweh!" snaps Calla.

Jonny was a middle aged bloke who was the best hand in the squad; a jack the lad, Davey said – his only downfall was the fact he was a mackem bastard! Jonny hated Malcolm as he had been in charge of the last job and wouldn't let Jonny get leave over the Christmas – the final nail in the coffin with his marriage, according to Jonny, who shortly after ended up divorced. "That bastard says two words to me about Norway and I'll rip his fucking head off; our lass won't let me see the bairn now, she's being a right cunt!"

"In fairness, Jonny, your lass was like a burst sofa! Not a scratch on the bird you were banging on Saturday!"

"Shut your stupid smoggy mouth, Calla," Jonny snarls, as Mark returns with another round of drinks. Mark was the youngest of the lot. "How lads, you're never ganna guess what just happened!"

All the lads look up puzzled at Mark. "Never mind what happened – get the drinks on the table, son." Davey was anxious to get as many pints of normal beer into him before the three months of American lager which was waiting.

"What's the news then, numb nuts?" asks Calla.

"Well I just got cracking on to this bird at the bar. Don't look but she's over there with the blonde hair! She's heading to the States only to work as a hooters girl!" All the table's heads turn 180 degrees, Calla kicking the leg of Jonny's chair, sending him and his pint backwards crashing to the floor! All the lads burst out laughing as Jonny climbs to his feet. “Ta daaaa! Never spilt a drop!”

With that Calla gets up and walks over to the table with the two girls and sits himself down. Mark sits down next to Davey and sulks into his pint.

“Look son, there’s three months to go yet and we aren’t even out of Blighty so don’t be sulking that Calla is cutting your grass, they’re all the same anyways! Snakes with tits!” Davey’s old words of wisdom brought another laugh to the table.

Kris was the quiet lad in the group, an excellent welder, probably the best argon hand in the game and he knew it. He had worked with the lads on many occasions, like a dream team that were on all the big money jobs. It was a tight circle and apart from the fact all the lads loved the crack and the beer it was all about the Wonga – you had to be top of your trade though otherwise the circle wouldn’t entertain anyone. Kris was working offshore when he got the text off Dave saying there was three months of big dollar; it was good timing as the shutdown on the Piper B was nearly complete so instead of his two weeks off fishing, golfing and shooting, he was sitting drinking Magners with the motley crew! “I bet anyone a tenner Calla gets her number and rides her when he gets to the States,” Kris offers the bet to the table.

“So ladies, a dickie bird tells me you are heading out to the States to fall in love with a big handsome pipe guru!?” Calla smiles over at one of the blondes and winks back at the onlooking table of his pals.

“One hour to boarding, boys.” Stevie Williamson pulls up a chair and places his white wine on the table next to the lads.

“What’s that you’re drinking, gay boy?” Davey asks.

“Oh it’s an acquired taste that you rough piping hands wouldn’t know about!” Stevie quips.

“An acquired taste similar to that of bullshit I get when you have owt to say,” Jonny quickly replies.

“Hawldy hawldy, lads – like I just said to little Mark, we’ve three months to go yet so let’s all play nice.”

Calla returns to his seat with a big smile on his face. “I’m going to be wedged right up that in a few days!” he boasts.

“I bet she has got a fanny like a bill boarder’s bucket!” says Stevie as they all laughed.

"She fucking will have in a few days, hammer!" Calla gets out of his seat, pulling out his wallet. "One for the ditch, girls?!"

"Let it off, Calla, we will be boarding soon, the gate's open and we still have a beer to drink," Jonny says, looking at the display board.

"Eer you're a right raving homo, you are," Calla barks, swigging down the last of his pint, spilling most of it down his shirt. "Al get the nips in then?"

"None of that shite for me, lad," Davey says standing up. "I'm going to splash me shoes and head to the gate."

"Aye, me too." Stevie sips the last of his wine off, following Davey.

"Wasn't offering you one, ya sparky twat!" Calla shouts towards Stevie then staggers off to the bar, bumping into a table full of drinks, causing a stir as Calla, true to drunken form, wanders off without the raise of a hand or an apology.

"He gets right on my tits when he's drunk!" Mark whispers to Kris.

"Gets on my tits sober. Don't worry about him; he'll spoil himself, you wait and see and them Yankee police aren't as forgiving as the Boro plod!" Kris, standing up and tucking his shirt in, pours the last of his cider down his neck. "Haway lad, let's get a roll on!" signalling Mark to his feet as both walk past Calla who is awkwardly carrying three large Jack and cokes back!

"Hold on, I've got yaz a drink."

"Just leave em, don't want to miss the flight!" Mark gives a shy smile to the blonde bombshell he had stolen from him.

Calla flopping back in his chair. "Eighteen fucking quid a round – a will fuck leave em! Wankers last call hasn't went yet!" (Just as the announcement for the passengers on flight to Orlando to proceed to Gate 3) "Bollocks! They'll wait," he says, sipping another Jack down, stretching out.

The lads are forming a queue at the departure gate as Mark and Kris walk in. "Where's silly bollocks?" Davey quips looking puzzled.

"He got another round in. He's a nightmare, Davey. We're going on a flight to the States not a taxi to the Empire!" Kris shakes his head.

"Leave him, he'll be alright!" Davey always stuck up for Calla and had his back, much to the surprise of all the lads.

"Calla can be the 53rd state in the US!" Jonny snides, laughing back in his chair.

The gate opens and the lads begin to shuffle towards the exit holding out their passport and boarding card with still no sign of Calla as the last passenger call is anounced. All the lads shake their heads and step forward onto the air walk towards the aircraft. Looking out the windows – sure enough it is pissing down. Once on the plane it is the typical slow annoying squeeze as passengers mill around looking for their seat and carelessly stopping to throw bags into the overhead lockers, standing on toes, catching people with stray elbows then remembering their iPod and book are in the bag, so repeating the exercise two and three times.

"Gets right on my tits this," Stevie mutters in Jonny's ear who is paying little attention as he is worrying about Calla sitting melted in the airport missing the plane. Just about all passengers are on board as the stewards and stewardesses carry out their final checks on seat belts, blinds and begin doing the head count, re-checking notes as Calla steps onto the plane bold as brass, catching a man at the front with his bag as he heaves it through the aircraft to a wave of tuts and exhaled murmurs.

"Areet, ya pack of twats," he shouts while stuffing his bag in the overhead locker, slamming it shut, flopping down next to Jonny with a huge grin and the sound of his mobile phone ringing out Fools Gold by the Stone Roses.

The stewardess taps him on the shoulder. "Can you please turn that off for the flight."

He replies with a patronizing grin: "Anything for you, beautiful and while we are on the subject of cans, when does the bar open..."

"Once the seatbelt light goes off and we have reached our altitude, sir, we will pass through the cabin with the drinks trolley," again with the professional patronizing smile but with a look of disgust and hatred as well.

"Sound, mark me down for two cold Heineys and a pack of peanuts and whatever you're having," with a wink and a smile, receiving a blushed smile and a flirty laugh from the stewardess – a total turnaround in attitude.

"She wants me." He nudges Jonny as he shuffles down comfortably into his chair

"Ay, wants you to shut the fuck up!" laughs Jonny, giving Calla a playful elbow back.

"Are we there yet?" Calla nudges the back of Davey's chair as the plane taxis along the runway. Davey shakes his head and ignores him, knowing the slightest bit of response will bring more aggravation. Calla leans over the aisle and taps an older lady on the shoulder as the plane gathers momentum along the runway. "The bloody turbulence is bad, isn't it!"

The lady turning smartly replies in a posh north east accent which has the lads all bad laughing: "We haven't taken off yet, you silly man!"

The aircraft begins to ramp up the engines pushing the lads back in their seats as it ascends briefly then banks left and levels out.

'PING' goes the seat belt signs, up jumps Kris like a rat up a shutter. "I've got 10 bar on my piss valve," he says to Marky as he dives into the toilet even before the stewardess is out of her seat.

"Excuse me, gorgeous could you please tell the captain no loop the loops while I'm drinking Heineken!" Calla is hanging out of his seat. Jonny leans in between seats asking Davey or Mark if they want to swap seats for twenty quid.

"Bonny lad, I wouldn't swap seats even if I was sitting next to that cunt out of Silence of the Lambs! What's his name again? Anthony Hopkirk or summit!"

An hour in and the meals arrive for the lads; Calla is already out for the count asleep. "Make sure you leave two Heinekens for the bairn when he wakes up or he'll cry," Jonny asks the gay looking bloke on the cart.

"Bottle of Calpol for me, young'un," Davey asks the steward, pointing towards the red wine. "In fact, give us two – that will get me another few air miles unconscious!"

The plane quietens down as the lads chill with their headphones and iPods on just as Jonny pours half a bottle of water on Calla's pants...

Calla stirs, confused.

"What's up with you?" asks Jonny. Calla, looking weary eyed, dazed and shocked, the corners of his mouth with blobs of dried saliva, says, "I'm soaked."

"You must have pissed yourself, Calla!" Jonny claims.

"Arrr!" replies an unfazed Calla getting to his feet and zigzagging to the bathroom to dry his pants. On his way back he stops at Davey's and puts a cold tin on his forehead! Slightly startled and agitated at being woken, Davey sits up. "Daft arse, might have known it would be you. Get yasel back to sleep and give owa playing the goat!"

Calla squats down and leans into Davey. "Ere a totally forgot it's the bairn's birthday on Sunday. Will you wire me owa a fifty spot into me ma's account, and al square you up when I see Jimmy for an advance?"

Squinting Dave goes to reply but with a frown he sniffs up. "Calla, have you farted?"

"No mate, it's me breath!" He wipes his mouth.

"Oh reet!" replies Davey with a slight nod of the head. "Ay al put some money in, if ya piss off back to ya seat and I don't hear your irritating smoggy yack the rest of the flight?" Davey puts his earphones back in and drops his chair back to recline. Calla, squeezing him on the shoulder and mouthing the words cheers, slowly swings back into his seat, foraging in the seat's magazine holder for his iPod to find the two cans of Heineken Jonny had got him earlier. He starts pressing the bell for attendance.

"What do you want now? Jonny leans in to whisper in Calla's ear.

"Me fucking dinner. What did you have? Was it nice" Calla asks a puzzled looking Jonny

"No it was fucking shite, as always, but you have missed it anyhow. Just get some crisps and fucking suck them – yav woke every cunt up ya mong!"

He adjusts himself for another power nap as Calla taps him on the shoulder. "Eer dya fancy an icy cold Heineken? Hate drinking on me own."

"Or, Calla man!" he frowns then lets out a sigh. "A could dee with an hour, a hate feeling like shite when a land!" Sighing again to Calla's upturned mouth and tilted head, he adds, "Gan on then." Cracking a smile, pulling off his earphones and beginning to wrap them around his iPod he brings his chair upright to the sight of the stewardess frowning slightly at Calla's order for two more Heineys and two jack n cokes.

"Or eer can ya swap these warm tinnies for two cold ones, ta love." Calla rubs his hands like a man who has just landed the jackpot.

"Let me up, am busting for a slash again, av bust the seal." Jonny gets up and wanders forward to the bathrooms as the stewardess squeezes past with Calla's drink hamper. Jonny notices the middle bathrooms are occupied and slips into the business class area to use their toilets. Once in takes full advantage of the complimentary toothbrush kits and sets about scrubbing his railings and daffying his hair up. On his exit he bumps into Stevie whose eyes looked like pin holes without his glasses, stretching and yawning, waiting for the toilet.

"What you doing up here?" Stevie quizzes with a smile squinting towards Jonny.

"Piss off, crab eyes." Jonny gives Stevie a playful dig, pushes past and wanders down the aisle stopping at the only unoccupied seat with Stevie's glasses and book 'Assegai' by Wilbur Smith'. "African adventures, my back eye, bet they're not half as good as my African adventures," he says sniggering to himself, glancing round, jumping into Stevie's chair, fully reclining it, pulling on his sleeping mask and dragging the blanket up to his neck as Stevie gets back.

"Ay very funny, Jonny, haway out me chair." Stevie gently shakes Jonny just as the stewardess arrives.

"Everything ok here?"

"No, this man is disturbing me. Can you please send the peasant back to economy?" barks Jonny.

"Please come with me, sir, I'm afraid you shouldn't be up here without a ticket!" The stewardess turns sideways to nudge Stevie down the aisle.

"Haway Jonny, stop messing around, tell her it's my seat!"

"Miss please! He is keeping me awake!" Jonny quips, snuggling down on the pillow, pulling the blanket over his head.

"Sir, this way please and keep your voice down!"

Flustered and slightly embarrassed at being nudged out of Business and into Economy with just his flight socks on to where Calla had moved across to Jonny's seat, Stevie flops down and responds to a puzzled Calla. "That arsehole Jonny's in my seat, with me mask on, pretending he sits there!"

Calla coughs a laugh spraying Heineken everywhere. "Al sort that out for ya, Stevie, had on." Calla leans up to press the call bell.

"Cheers mate," Stevie says smiling.

"Only coz your crack's shite and your squinty mince pies are spoiling my can," Calla says with a laugh, adjusting himself to speak to the stewardess.

"Excuse me, miss, there seems to be a mix up here." The stewardess and another steward gather around the seats of Calla and Stevie. "About five minutes ago this man appeared and is as annoying as that ginger bird on that quiz show 15-1." Calla nudges Davey's chair and asks him who she is.

"For fuck's sake, Andrew, what did you say?"

The air stewardess chips in and reassures Davey to go back to sleep.

"Ask me again, Calla," says Davey, always mad for quizzes and crosswords.

"The annoying ginger bird on the telly."

Davey sits up in his chair using one of his big bear hands, the other is wrapped around his red wine glass. "Well Calla son, that would be easy. My Margy when she's dusting the ornaments on the telly on a Friday!" His big deep laugh brings a giggle out of both the stewards.

Calla takes his can and clashes it with Davey's wine glass. "Ha-ha you mad old twat!"

Stevie stands up, explaining to the crew that he is business class and shouldn't even be in here let alone be being abused!

Mark leans over the seat, nudging Calla. "He's being saying that since he was 10, Calla!"

Five minutes later Jonny returns to his seat with two bottles of champagne and a big smile on his face. "What time do we land in Orlando, stroker?" he questions Kris who is standing talking to the lads in the aisle.

"Did you not ask the captain when you were up at the front like, you balloon?"

Jonny pulls the champagne cork out with his teeth and pours it into a plastic empty cup that was used for tea earlier, holding his little finger out, taking a sip. "Well I did, my old bean but the cunt was asleep so I pinched his champagne."

PING goes the seatbelt sign signalling the top of the descent and all passengers to return to their seats for landing.

Davey gets up and stretches. "That's just a message from the captain to say 'righto Davey, gan get a piss, son, we'll be doon in 15 minutes'!" He scratches his arse and walks down the aisle smelling his finger.

"Touchdown!" shouts Calla as the Boeing 747 tyres release a huge blast of black rubber with a screech. "You have to get into the American mindset now, boys!" Calla ruffles Mark's hair in front of him.

Kris, leaning over, throws a can top at Calla. "Well that will suit you right down to the ground, empty heed, because they haven't got a scooby about owt"

The plane taxis to the air bridge and Calla starts opening and shutting his belt. "Watch, Jonny, every cunt will start opening their belt – it's like coughing in the doctor's waiting room! One cough then every fucker's at it!"

Standing up and waiting to get off, Mark sends a wink over to the girls he was chatting to at the bar without realizing Calla is watching.

"Aye that's right, ladies, Mark has a glass eye and it twitches like a rabbit's nose when he realizes he has crashed and burned!"

Mark not very happy looks over at Calla. "My fist will crash and burn into your mouth in a minute!"

All the lads in perfect choir tone let out with an "Oooooooooooo oooooooooooooooooooo!!"

The usual hustle and bustle commences to the wave of welcome text message alerts as the plane's seatbelt sign is switched off.

"Wrecks my head this," Kris barks, pulling his bag out of the locker while easing an elbow into the top of some bloke's head. "Sorry, pal," Kris says apologetically as he hears the booming voice of Calla: "Piss off Mark, you little shit, wait your turn!"

"Just pass me bag, Calla – stop being a tit!" Mark back answers trying to squeeze his arm up past Calla to get his bag.

"Both you girlies pipe down will ya!" Jonny barks, sliding past the two of them, grabbing his bag from the overhead locker and catching up with Dave who had already started making his way towards the open door.

Once inside the airport the lads head towards Customs, Davey hanging back to catch a word in the lads' ears, rounding them all up. "Right lads, nee pissing around here, these lads will have you back on British soil gagged, possessions bagged and arseholes ragged before you can say Budweiser!" He gives Calla a stern look and heads towards the shortest customs line, the rest of the lads in tow. A good hour passes before all the lads are stood at the carousel waiting for their cases, Calla's the first off: a burgundy case that looked like it had been washed up with FUCK OFF scribbled in black permanent marker.

"Eyarr there's Calla's fighting rags ere in one piece."

Calla barges through the mass of people all jockeying for position to lift off their cases.

"Calla, who wrote that on your case? Your mam?" chirps Mark to sniggers off the rest of the lads.

"Do yourself a favour, Mark and keep your horrible mouth shut!"

"No actually me and my mate Jamma wrote on each other's case before we went to Magaluf three years ago – his said cunt! Ha-ha, reet laugh!"

An unimpressed Davey pipes up: "Good luck getting that one unopened through this place – hope you've nowt silly in it?"

A worried Calla steps up to Davey's ear. "Eer av brought a box of Kams over! Shite, you think they will tak them off me?"

"Jail for that, Calla, do not pass go, do not collect $200," smirks Kris as he strides towards the carousel to drag his case off!

"Bollocks, Davey I never thought about that, what am I gunna do?"

Davey frowns. "A fucking box? How many in there like, floppy bollocks?"

"Fifty, they are the jellies, I fucking love them and I was hoping to sort these American blurtas out Ron Jeremy style," beams Calla proud as punch.

"Tell them your mam packed it for you and they are multivitamins!" says Jonny striding past heading towards the final checkpoint in the airport.

"Jonny, you think that will work?" shouts Calla as he tries to catch up alongside Jonny.

"Will it fuck, ya thick cunt!" laughs Jonny.

"Don't sweat Cal, play the SOFT lad and say they were prescribed because of your floppy cock!' says Kris, wheeling his case past.

"Or ere, couldn't give a fuck – if they stop me they stop me!" He looks across at the rest of the lads while volleying the back of Mark's case, twisting his wrist and sending it on its side. "Baby arms!" Calla barks, walking off, leaving Mark to pick his case up. "Prick!"

A big sign in the arrivals hall is held high by Jimmy Hunter: T WATTS. Davey is the first through and waits a few minutes for the rest of the lads to catch up with him. "Well we need to look for probably my name," Davey is telling Kris and the boys as they walk into the huge arrivals hall, scanning the boards being held up.

All the lads walk right up to the exit before Calla spots the sign. "Here, Davey, there's your sign!"

Davey bursts out laughing and walks up to Jimmy. "You're not wrong there, bigman, I have a right bunch of twats with me!"

A few handshakes then a sprint for the door for Calla and Jonny to have a well-deserved tab. Jimmy and Davey are the last out with Jimmy pointing towards the car park. All aboard the suburban and the typical questions start firing Jimmy's way. "Digs sorted? We get paid weekly? How long's the trips?"

Jimmy answers the questions with ease until Calla shouts from the back: "Where's the nearest boozer from the digs?!"

Davey was riding shotgun and turns round, giving Calla a stare of 'shut the fuck up'. Jimmy continues to tell the lads who are all ears about the chance to make around two grand a week doing seven days and the hotel-come-apartments were all looked after as well as free dinner every night up at the main hotel. Davey could see out of the corner of his eye Calla leaning forward to ask another question.

"Can I trade my dinner in for five pints?!" Calla seriously rubs his chin at his question. Kris and Jonny both nod in agreement with Calla and wait for Jimmy to finish the long draw on his cigarette.

"You three aren't piss wise! So you don't get the 25 dollar dinner but you get the 25 dollars' worth of drink!"

Jonny shrugs his shoulders and leans right into the driver's seat. "Aye Jimmy but at least we won't be wasting our 25 dollar allowance on non-essentials like food!"

The van pulls into a building site with portacabins and a shell and core of a huge impressive looking facility.

"Extension or bodywork completed in time for the engine to go in, boys," Jimmy proudly states to the lads.

Young Mark who slept all the way sits up, not looking very happy with the fact that it looks like he may have to get togged out and do a day's graft.

Calla flicks his tab out of the window and shouts down to the pilots: "I tell you what, there is absolutely no way I am going on that job today! I'm fucking wrecked!"

Jimmy turns round, opens the door and gets out, leaning back in to say: "Calla, you just sit there, son, I'll bring the rest of the lads into Michelle for their $500 expenses." Jimmy walks away smiling.

"Fucking hold on, wait for me!" Calla climbs over the seats like he is in the 100 metre hurdles.

The offices were large double-stacked portacabins with upstairs and downstairs offices which looked quite small and shabby from the outside but once in they were all brand new and decked with all the latest printers, computers, desks and leather chairs.

"This way, lads," Jimmy hollers to the lads who are walking in single file behind him, Calla bringing up the rear and stopping at a desk to steal an Oreo then fucking it straight in the bin after one bite. "In here," Jimmy says, signalling to the lads, holding open a door entitled Admin & Accounts. Inside was the usual fluster of bodies taking photocopies, filing and typing then into an office out the back where the back of a tall slender dark-haired woman in her early thirties wearing a tight pencil skirt and a white blouse was peering into the top filing cabinet. "Michelle, meet the A-team; lads, this is Michelle Moors, our senior accounts manager!" booms Jimmy.

Davey is the first to step forward and slightly hesitates as Michelle spins round, her dark wavy hair swishing over her beautiful dark skin and the waft of her perfume catching his nose. Davey ponders, looking at her deep blue eyes in awe of her beauty, before holding his hand out. "Pleasure to meet you, bonny lass. Call me Davey!"

"Likewise, I'm Michelle."

All the lads lurch forward into a slight tussle to be the first to introduce themselves to her, Davey holding his big arm across them to introduce them himself, pointing. "Michelle this is Kris, Jonny, Mark and....." as Calla bursts into the office oblivious to the polite introductions– "Eer the biscuits taste like shit!"

"Aye and our Andrew potty mouth!" Davey gives Calla a stern look.

"Sorry love!" Calla responds apologetically, holding his hand up.

"Right Michelle, down to business, the boys here are all snapped off so we'll get them booked onto the site induction tomorrow. I'll get their passports for admin to copy today. If you have their expense packets I'll let them off to get their head down!" speaks Jimmy stepping forward as Michelle leans into the desk drawer to pull out a small metal deposit box. All the lads peer forward to the box like it's a treasure chest, all but Calla who is trying to get an angle to look down her blouse for a look at Michelle's chest. Davey catches him and gives him a cheeky one in the rib. "Behave!" he whispers as Michelle begins calling names to collect the packets, the rustle of thanks. 'Ta love, cheers and thank yous' fill the place, all discreetly sliding them in their pockets – all but Calla who tears his open with his teeth like a Rottweiler at a pork chop, crumpling the envelope up and throwing it in the bin as he thumbs through the hundred and fifty dollar bills fanning them out in excitement. "Steak tonight, Michelle?" He drops her a wink and a smile.

A slightly embarrassed Michelle apologetically replies, "Sorry I've plans," before turning away, catching Kris's eye before busying herself again. "See you later boys!"

Jimmy nudges Calla and the rest of the lads out the office. "Thanks, Michelle. See you outside, lads, I'll be two mins." Jimmy wanders into another office.

Once in the corridor the lads all whip out their packets except Davey who snaps: "Calla man, will you kop on, this isn't a bastard holiday camp! You're lucky to be here!"

Calla knows he's pissed Davey off – Davey was the one bloke who could control Calla, who looked at Davey in better light than his old man who had been in Durham nick since Calla was five for armed robbery of a sunbed shop, thinking because it was chocker on a Friday night around nine it would be minted, only to find his heist got him 630 bed tokens and 37 quid – oh and 12 years in the can. He swallows down. "Sorry Dave, just excited to be eer, al cool me jets!"

Davey, turning his back and heading for a piss, leaves the rest of the lads to wander outside into the sun. The lads begin to stretch and talk among themselves, mainly about Michelle and if she was up for a podging when they hear someone shout in a jock accent, "How's it going, ya bunch of fannies?"

Jonny spinning round scans the area as little William McLaughlin wanders over. "Areet hens, what you knackers deein oot here, weez arse you been tickling?" He shakes Jonny, Kris and Mark's hands.

"Areet, Willy," the lads all chirp. Willy then booms: "Arrr naaaa the job's reet fucked anoo!" laughing as he steps towards Calla. "Areet big man, good to see ya, that's the entertainment sorted!"

Calla smiles as he shakes Willys hand. "Areet ya jock twat, see ya mince pies are getting better, ha-ha. Who you buy ya glegs off, Budget Windows?"

Willy's eyesight had never been his forte and from the age of nine he had worn inch thick lenses. "Ay good one, Big man, still out-run, out-graft, out-drink and out-fuck you, ya English prick!" snides Willy, giving Calla a jab in the arm and taking a boxing stance. "Where's the bigger man at?" Willy asks just as Davey puts his arm around Willy's neck and in a choke hold lifts him from behind and off his feet, Willy's hard hat falling to the ground as he struggles. Willy in a wheezy laugh tries to shout out "FREEEEDOMM". Davey, laughing, drops Willy who dusts himself down and gives Davey a big hug.

"Now then lad, what's this screw like?" Davey quizzes...

Willy pulls out a box of Benson and Hedges and offers the lads an old favourite! "Well me and Hags are sharing an apartment and Keegs and Old Screamer are in the other billet!" Willy continues to tell them that the apartments were out the back of a nice hotel and they were sound, Irish boozer just over the road for a game of pool and a nice pint.

"Gordon Haggerty, Trevor Keegan and Brian Johnson are here!!? Happy days!" says Davey putting his arm around Willy.

"Aye we send in the Scottish lads first to do all the recon, get the wages sorted, then we drop in ready for the assault," Jonny explains.

"Aye summit like that, Jonny, you Hun bastards would get behind wallpaper once the word is out the monies areet!" Willy loving the banter starting already.

Jimmy follows Willy into the hotel car park with all the lads starting to talk about who's sharing with who.

"Just before this gets out of hand lads, me and young Calla will share." Davey, with his arm out of the window, calms the lads in the back.

"Davey, the Queen will fucking mention that on Christmas Day as an act of selfless bravery!" Kris smiles knowing he is happy sharing with either Mark or Jonny.

Jimmy interrupts the plot and puts them all wise. "Lads, everyone knows Davey would be the man for Andrew so that's buttoned up. Jonny, you and young Mark are in together. Kris sonner, you are going to have to share with a lad that's flying tomorrow, a lad called Andy Jones."

Kris slumps back in his chair. "Who the fuck's Andy Jones? I haven't heard of him so he can't be any good!" he rants.

Jimmy leans over the chair and says, "He's the only lad I have ever seen weld as good as you, hammer!"

"Wa hey!" Calla shouts. "Golden balls mark two on the way eh! Where's he from, Fatty?" Calla smiles at Jimmy's stern eyes in the mirror.

"He's from Manchester, a blue not a red, and the only reason I am this fat is because every time I ride your mother she gives me a chocolate biscuit!"

All the lads burst out laughing as the van pulls to a halt. Calla opens the door and throws Jimmy a tab. "The only chocolate thing you would get off my ma is a chocolate finger! She's fucking filthy!"

The 30 minute drive to the digs seemed to fly by as the lads stretched out and enjoyed the scenery.

"Fucking knocks the bollocks off Whitby does this!" Calla shouts.

The minivan swings a right into the road that leads up to a gate house and eases to a stop with Jimmy lowering the window to speak to the security guard named Earl who was around 6" 5', around the same wide and was African American.

"Hey Y'all," his deep gravelly southern voice booms.

"How's it going, Earl? Here is the list of the lads' names who'll be staying in apartments 404, 405 and 417." He hands Earl the list as the big man reads through it, walking back to the gate house, opening the gate and waving the minivan through.

"Fuck, he's a big bastard!" Jonny says looking over his shoulder back to the gate house.

"Wouldn't like a smack in the mouth off him!" Mark says, tapping Jonny.

"Wouldn't like a smack in the mouth off anyone!" barks Jimmy giving Davey a nod.

"Eer did you say 404, 405 and 417? We want 417 – sounds out the way!" Calla shakes Davey's shoulder.

"Will you fuck off, Calla!"

Davey takes the unstable cigarette out of his mouth as the minivan pulls onto the side of the small winding road where on either side are these large town houses made into an apartment upstairs and an apartment downstairs. "These look canny digs, Jim," Dave says as he gets out of the minivan.

"Ay only the best for our lads!" smirks Jimmy as he stands with the three door keys, Calla snatching 417 leaving Jimmy to throw a key at Jonny and one at Kris. "Reet lads, I'm off, get yasel settled in ready for the early rise the morn. Davey, here's a taxi number and the address of this place and there is a Seven Eleven at the top of the road if you need owt, milk, bread! Al give you a bell when I got done, take you all out for a bite to eat about 8-ish!" He climbs back into the van.

"Cheers, Jimmy." Davey waves Jimmy off as the lads study the place before heading off to their apartment, Jonny and Mark getting the upstairs apartment and Kris going it alone in the downstairs, as Davey and Call cross the road to apartment 417.

Inside 417 Calla barges through the door into the hall area where the living room/kitchen is in front of them with a door to his right and a door at the other side of the kitchen. He stops to ponder, knowing he has a decision to make! Whichever room he chose Davey would be in the other one before he had time to change his mind! Before he can even decide Davey throws open the door on the right and then throws Calla into it. "You're in there! I'm out the back away from you – not having you waking me every time you have a daft one!"

Davey picks his case up and wanders to the other room, admiring the size of the living room and kitchen and the 42" plasma TV on the wall! Calla, knowing there was no arguing, scans the room to see a huge double bedroom with an en suite. "Fucking bingo, this will do Cally." He throws his case on the bed then takes a tour into the living area. "Fuck me–" spotting the TV. "Fuck me!" –wandering to the window to see the view. "Fuck me–" looking onto tennis courts and a shared pool, calling out for Davey with excitement. Calla wanders into Davey's room to find him not there and his room bigger. "Fat jammy cunt!" he sneers. "DAVEY?" Calla calls.

"What's that, lad?" Davey stepping into the room from the balcony.

"Or eer yav a fucking balcony aswel, that's shit crack that!" Calla sulks.

"Get ya chin up before a wobble it, bet your room's nee broom cupboard!" barks Davey as Calla sulks off to the living room to see what channels there are.

In 405 the place is upside down where Mark and Jonny have wrestled each other all over the place for the bigger room. Jonny begins to unpack and opens his patio door onto the balcony for some fresh air with a smile as Mark sulks in his room with a fat lip.

Two miles down the road, Willy and the lads were sitting in Dirty Nellys having a few pints and a game of pool. All the lads had worked with Davey and the boys on a good few jobs, always making a good few bob with tough shifts and long stints away from home but always having a good laugh even though it was kind of a split camp.

"Is Davey and the boys coming for a pint or what, Willy?" says Keegs as he pots the black leaving Hags with 6 balls. Trevor was a 6ft 2, good looking lad who did all the inspection and quality work on most of the good jobs, gifted at his precise workmanship, being a good pipefitter himself but also gifted in the ladies department.

"Are you going to fuck off potting balls like that, you big good looking fucker!" Old Screamer had been around the block at 56 but was still a brilliant pipefitter, Keegs being his apprentice in Govan shipyard until the high purity big money jobs took them both on their travels.

Willy puts the big pitcher of Coors on the bar and a couple of pots of hot nuts. "The lads are supposed to be going to JW's with Jimmy for some bait around 8 o'clock but I told Davey to come here for a pint first."

The lads had been there three weeks and were only setting all the kit up for the pipe being delivered on Monday. Gordon Haggerty was on the phone at the door.

"Look at fanny balls out there, he's probably spent a week's wages on the phone to their lass already." Keegs, setting the balls up, asks the barmaid to turn the jukebox up.

"Surprised you haven't took Jemma out for a nice night." Willy smiles as he sits back on the bar stool.

"Too early in the game for that bawbag, three months with her on the arm." Keegs smashes the balls and walks over to the jukebox.

Screamer, ever the dark cloud with a dark and sarcastic sense of humour, adds: "Aye bigman, that's all well and good but I don't think Andy Callaghan will have that outlook at 7 o'clock."

Keegs continues to pot balls at speed, singing out to 'Spit in the rain' an old Del Amitri classic."I don't give a fuck what Calla does – he'd ride a barber's floor, the dirty bastard."

"WELL! It doesn't make a bad person does it, big man!" The boys burst into the bar, Calla ruffling Keegs' blonde hair!

"Areet, Keegs lad, good to see you!"

Keegs straightens up his hair. "Touch my didgery again and all rip ye balls off!"

"Areet lads, was a fucking mess anyway," spits Calla, wandering to the barmaid. "Oy there, darling, can I get some beers?"

Keegs turns to shake the rest of the lads' hands as they all meet and greet. "Now it's a party!"

Jimmy sidesteps the lads and heads to the bar beside Calla who is in good flow with the lines to Jemma as she is pulling the pints. "... Seriously you're beautiful – out with the lads tonight but tomorrow I'm free if you fancy a drink?"

"Oy Romeo, keep it buttoned. Our Jemma's got standards, haven't you love? Put these on my tab." Jimmy hands over his card.

"Hey Jim, ha-ha he's funny, can't understand him – is he English?" Jemma replies beaming with a smile.

"Course am English love, thought you American birds loved the English accent?" Leaning in towards her with a wink.

"Yes Andrew, ENGLISH! You smogs are from another planet," sneers Davey.

"You call it Heaven, Jimmy son...Heaven." Calla, taking two pitchers, winks at Jimmy who is picking the other two pitchers up shaking his head, Jemma following with the glasses. The pool area is jumping as the lads are all trading stories and catching up.

"Jim, sack the steak and stop on the swally with us?" Willy shouts as he leans in to take a shot which doubles in off the back cushion, glancing the 8 ball into place and then dropping in the pocket. "Trick I learned in 'Nam." He stands up, chalking his cue and smiles to Jonny.

"Ah come on, Willy, took you lot for some grub when you got here, few scoops, nice rib-eye, early night, ready for the new job, just what they need, plenty of time for getting mash potatoed!" replies

Davey.

"English Poofs!" Keegs sneers, turning his eye to Jemma who is laughing at Calla who has just whispered something in her ear as she collects the glasses.

"We're happy out with a burger from the bar like, Jim." Jonny takes the cue from Willy who has just wobbled the black.

"What do you think, Davey? I'm not staying late anyhow, heading yem to Skype the missus at 11." Jimmy pours more beer into his glass.

"Ay we'll do that, let the lads have a catch up, saves the lazy cunts doing it the morra on site! Get the burgers in!"

"Ay Jim, get the burgers in. We're fucking hank!" Willy shouts.

"Thought you tight arse jocks wad wangle it so you got something!" Jimmy spits, wandering to the bar

"Bowt time! Years you English rifled our country of its goodness!" hollers Willy, standing on his tip toes.

Calla, snatching the cue from him, says: "See it still hasn't grown back, ya stumpy little mong!"

The lads are spraying beer in laughter just as Hags storms in red in the face and pouring himself a beer as he starts to rant. "Yee naa that witch is on about me wages already! Nee how's the job? How are the digs? Kiss me arse or nowt, just where's my money? Her money? Her fucking money? Lazy bitch hasn't worked in a day in her life! Skin wouldn't graft on her! Tell ya!" He shakes his head, swigging half a pint down in one go.

"Areet Hags, still happily married?" Davey jibes, stepping forward and shaking Hags' hand with a smile.

"Aye Davey, couldn't be better. I hope her fanny heals up and she drowns in her own piss!"

"Might as well heal up, Hags lad, the amount of times you've been in it!" Screamer taps him on the back, heading outside for a smoke!

The bar starts to fill up with regulars and soon the lads are doing shots of all kinds! New barmaids and doormen come to work as the lads play pool and pump money into the jukebox. Mally Simmons wanders into the bar with a young well built lad from Manchester, Andy Jones.

"For fucks sake, Mally is in! Must be checking up to see who is in to be in," Jonny says nudging Keegs.

"Aye he's a snake but he used to be canny crack; who's with him?"

Looking over at the two lads being served and shaking hands with the rest of the lads, Calla turns back from the jukebox and shouts over, "I think that's Andy the golden hand from Manchester, is that right Krissy?"

Kriss was putting his quarters on the table. "Calla, if he can weld your set ups he wants the title of golden hand! Stevie fucking Wonder can set them up better!"

Calla walks over mimicking Shaun Ryder singing 'you're twisting my melon man' as all the lads laugh. "I wouldn't set anything up I couldn't weld myself, and I am not black either!"

Davey sitting half cut on his stool shouts back at Calla: "Good lad, Calla son, these prima donna welders, a fuckin tell ya, they couldn't weld my arse!!"

Mally halters the banter and introduces Andy to all the lads. After a few handshakes Andy pipes up and announces in his thick Wythenshawe accent: "Lads, I am a good welder but I don't think I could weld Davey's crack! Last time I seen an arse that big there was a policeman riding the fucker!!"

The lads playing pool come over and shake Andy's hand as if he is one of their own. Andy leans over to get his pint at the bar and Davey passes it over, spilling a small bit on a fresh looking stocky American lad. "Hey buddy, be careful or I will take you outside and kick your ass!" The statement quietens the banter.

Andy steps back a little shocked as Jimmy Hunter puts his arm on Davey's shoulder. Calla leans his cue against the chair and sticks his head in the medley. "Who the fuck d'ya think you're talking to daft arse!?" standing about 6 inches from the young yank. Gemma the barmaid puts the tray of glasses down and leans into Davey and Jimmy saying this young guy Brad is the town's best street fighter. "I know who my dollars are on, hinny," says Davey taking a swig of his beer.

"Why don't we go outside and I will show you!" the young street fighter threatens Calla. "Sound, haway then daftarse, after you!" Calla follows the young lad towards the door.

"Will he be alright?"Andy quizzes Davey and the lads. Young Mark laughs and says, "If Calla was worried he'd've brayed that cue over his head on the way out!!' Five minutes pass and the two lads come in, arms around each other and no marks of any fighting. Andy looks a bit surprised as he turns to Davey. "I thought he was the town's best street fighter!?"

Davey stands up to go for a piss and as he waddles off his stool he winks at Andy and quips, "Aye he probably is, Andy son, but I don't think he does car parks on a Saturday neet!!"

"Areet lads, this is Hamma, Bradley 'the Hammer' Reid – he is trying to get into that UFC carry on, eer says his old man has a bookies and his sister runs a boozer Roxy's, said its right swank, you can go upstairs to a bit call the sky bar, nee roof on chocka with fanny!" Calla excitedly nudges Brad towards the rest of the lad's handshakes.

"Fuck me, Calla, you went to rip his head off – sounds like you took him on a date!" Kris chirps.

"Anar but there was a squad car outside so a said 'hawld on til he fucks off, Hamma' and he said 'who told you I was called that!' We got talking from there, he's sound!"

Brad steps back from the lads. "Hey you guys should come down to Roxy's, it's ladies' night, fills up around 10 open til 2?"

"Or fuck this!" Jimmy says sliding off his bar stool and finishing his pint signalling the bar girl for his bill. "Not heading down there, it's a reet trek, I'm off to Skype wor lass. You lot watch what ya deein and al pencil you all into the 11 o'clock safety induction – give your nappas time to heal." (A wave of thanks, appreciation and thanks head towards Jimmy.)

"Stop out, Jimmy man, stop being soft? Were guna have another one in here," Calla barks putting his arm round Jimmy.

"Nor Calla, never spoke to her yestda so I'm away. Enjoy your neet!" Jimmy eases past everyone squeezing Dave's knee. "11 o'clock the morn, mak sure these are there!"

"A round of tequilas!" Calla shouts at the barman, who begins racking up the shots, Brad the first to slam his down. "Thanks dude, hey you lot heading down Roxy's? I'm going now, could squeeze a few in my truck?"

"Fucking truck?" Calla quizzes Davey, leaning in, taking hold of Calla's arm.

"Truck is pick up out here, thick shit!"

"Or reet, nah Hamma, you fuck off, we'll follow you down there in a taxi. Hags is on the phone and fuck narz where Willy and Keegs av gone, see ya in a bit!"

Brad waving the lads away as he walks off a bit puzzled at being told to fuck off (sure gunna be fun with these guys he thinks) shaking his head leaving the bar.

Malcolm orders another round of pitchers for the lads as he pushes in towards Davey who is sitting with his back to the bar on a stool. "Tungsten! How the hell are you!"

"Nee botha, Mally, am nee botha, give owa calling me that though, there's a good lad! So what's the story with Jimmy? It's not like him to miss a good swally," Davey asks.

"Yes him and their lass are at it, she's pissed off he's been out here 18 weeks now and it doesn't look like he'll be yem anytime soon!" Mally replies handing over his card to the barman. "Ay keep it open please!"

Callas ears pricking up – "Eer and the tequilas, barman!" with a nod to Mally as the barman racks up the shots again.

Screamer from nowhere pops his head amongst it and lifts his tequila then pours it down his neck. "Tastes like shite and starts a fight!" then picking up a fresh pint shuffles back to his chair leaving the lads all throwing their shots down laughing.

A few more games of pool and a few more drinks and the bar is a lot louder than it was an hour ago. Andy and Kris finish off beating Calla and Keegs at pool. "You're fucking yepless, Calla! Is there anything you are good at?" Keegs is devastated as usual to be beaten at pool.

"There is Trevor, there is: drinking, bucking and putting pipes in."

Davey shouts over, "Well said, Calla son, well said, although I would hold back on the drinking patter – I've spilt more down my green tie!"

Jonny shouts over two taxis will be here in 5 minutes so the lads all start drinking up!! Davey asks the barmaid for another pint and Mally does the same as does Screamer. "We are staying here for one more lads, you go and have a good night and for fucks sake stay out of bother."

Keegs, Calla, Mark and Jonny squeeze into one cab while Andy, Hags, Willy and Kris jump in the other. "Roxy's bar, stroker," Calla gives the cabbie orders.

"How much is it down there pal?" Keegs ever the careful one to have everything in order.

The taxi driver tells them it's about a 20 minute spin costing them around 30 dollars. Both taxis pull up outside a nice looking trendy bar, two doormen outside with little of a queue.

"Hurry the fuck up, I'm bursting for a piss!" Calla quick stepping to the door, the doorman greets the lads and asks them if they have any id.

"Just our driving licence and shit like that, big man. Is that braw or wha?" Willy looks at the bouncer's chest as he pulls out his wallet.

Calla was nipping his cock through his pocket by this stage. "There's mine, Ham, thanks, I need to run to the bogs – where they are!?" The bouncer gives him his card back and asks him politely what a bog is.

"It's a restroom for a Mongol child," says Mark to the bouncer as Calla snatches the licence back and runs off into the corridor.

"There's no charge as you are in before 11 – have a good night," the bouncer growls as the rest of the lads troop in through the corridor into a very plush bar with a spiral staircase up to what looks like another bar. The place is quite full with a good atmosphere, a guy up singing and playing the guitar.

"Two pitchers and 8 glasses please," Willy shouts the drinks in for the lads.

Calla comes back with all his hair done and smelling like a hoower's knickers. "Fucking good bogs them, lads, have a piss, gel, deodorant, aftershave and a bit of chut for a dollar!! I'm coming here every neet to get ready before we gan oot!!" All the lads bursting their holes laughing.

"Where's Keegs at, Willy?" Hags standing there with his glass of lager and phone in the other.

"How the fuck do I know, Gordon? anyway can't you track him down with your phone? The bastard's never out your hand!"

Mark comes back from the toilets pointing up at the balcony "Didn't take Keegs long to sniff out some clunge!" All the lads looking up seeing Trev standing chatting to a lovely blonde lass as her pals were all around him like he was sent from heaven.

"That's it, hammer, chock away, she's getting it!" as Calla picks up his glass and legs it up the spiral staircase.

Hags quizzes Willy, "Back at the other bar, where the fuck did you an Keegs disappear to?"

Willy laughs into his pint. "I bet Keegs $20 he couldn't eat 5 jam doughnuts from Dunking Doughnuts without licking his lips!"

He shakes his head with a smile. "Well, you 20 bucks up?"

Willy, looking up towards Keegs spewing charm all over the group of ladies up the stairs, replies: "Naa the charming bastard did it, then went double or squits he could get the bird who served him to lick it off!"

Hags nearly chokes. "For fuck sake he is some unit!"

"Sure is. I'm $40 down, the cunt!" He taps Hags on the shoulder and heads up the dancers to join the rest of the lads that were like flies around shite. Willy walks straight through the lads –"Move, yee wankers, get by, watch oot–" straight to the tall blonde who is having her ear warmed with some big yank's lies – "Hold this, love and can you get me another!" putting his empty glass in the girl's hand and walks towards an un-occupied table.

"EXCUSE ME, HEY EXCUSE ME!" the blonde cries, leaving the puzzled American and storming after Willy. "HEY YOU! I'M NOT A BARMAID, HEY!" as she turns Willy around and confronts him slamming the empty pot on the table looking for an explanation.

Willy with a wry smile replies: "Of course you're not, Beautiful, but how else was I gunna get you away from Billy boring features? How about joining me for a mojito and I'll tell yee all about the bonny highlands!" putting his arm around a very flattered American.

"Sure, that accent, where are you from?" she asks as they walk towards the bar. The American looking very embarrassed and angry slams his bottle of Bud down and is just about to go after Willy when Keegs, without taking his eyes off the lady, grabs his shoulder. "Think again big man!"

The American, turning around, easing the lady out of the way and squaring up to Keegs: "Say what?"– just as the hairy side of Calla's hand crashes against the side of his face.

"Jog on, petal, your mam's calling!" As Calla turns to face him, the American, rubbing his face, steps back, weighing up his options just as Brad grabs his arm and twists it up his back using his other hand to squeeze his neck. "Time to leave." He frog marches the guy to the stairs. "Jack n coke, Calla, back in 5!"

Calla sarcastically waves the American out, blowing him a kiss. "Nee botha, Hamma–" turning to Keegs, nodding in approval. "That's one cool cunt, pleased a didn't rive on with him now, you ladies want anything to drink?"

"Yeh, get us all a shot of patron silver, with a cranberry back," one replies unfazed by the slight fracas.

"Nee botha love!" Calla strides towards the bar where Willy is well on the gas with the blonde who is now stood with her hand on Willy's chest laughing uncontrollably.

Calla squeezes into the bar beside him. "Eer Willy, that bloke was foaming you took his bird. Had to give him a slap and Hammer's away throwing him out!"

Willy turning slightly– "See how nice I am, big lad? I only did it to give you an excuse to give him a slap, could see he was getting on yee tits!" –winking at Calla, who sarcastically replies: "Jeeez, thanks Willy how ever could I repay you?" rolling his eyes– "Please let me buy you both a drink!"

Willy turning his back says: "Cheers, big man, that'll dee – 2 large mojitos, light on the lime, heavy on the Bacardi!"

"Eer a was...fuck n hell! Joking! Eer barman, can a get 2 large mojitos, 3 bud and 9 patron things with cranberry carry on backs or summit?" Calla asks confused.

"Sure man, where you all at and I'll send them over. You want to pay for them now?" the barman replies.

"Ay you're a legend, just owa there" (pointing) "and ay al pay now – how much?"

"That's $125 sir!"

Calla, flicking through loose ten dollar bills, looks twice at the barman. "Say what? $125 bucks? Ya having a Steffy, aren't ya?"

"No sir, 9 Patrons, $10 each, Mojitos are $10 and the Buds $5! And then a tip for the best barman of course!" the barman leaning back with a smile on his face continues to pour the Patron tequila and the shot of Cranberry juice into shot glasses.

Calla, flicking a hundred, 2 tens & 5 dollar bill: "Your tip is 'Wear a balaclava next time so I feel like I'm being robbed properly'!" Calla turns round and walks away from the bar shaking his head.

Downstairs Kris and Jonny are stood laughing at Screamer who is leaning pissed against the centre table amongst the busy crowds knocking back pint after pint and squeezing birds' arses as they wandered past.

Andy is at the bar being served and his small talk with a lovely American filly in her late 30s leads to an invite to hers and her sister's place just opposite where he is staying. "My sister's away but I will share a cab fare home and even invite you in for a coffee if you like."

Without a second thought Andy asks the barman to drop the pitcher over to the lads and he shuffles his way out the door and straight into the car park where they both headed for the 2 taxis sitting opposite. "Oh no, that looks like my kid brother and his buddies outside of KFC. Don't worry: he's only 18 and will be a bit drunk I'd say."

Andy marches on, his big arm wrapped around his lady friend's waist. Just as they get level the young brother waddles over and asks who his sister's date is.

"Well, this is Andrew and he's from England and he is dropping me home like a gentleman."

The young lad weighs Andy up and down and holds his hand out high and shouts out: "That's cool! Gimmee five, brother!!"

Andy leans into the young lad and says: "If you don't beat it I'll give you more than five, you little shite!!"

Things were going from bad to worse for the lads downstairs.

Screamer was in the process of taking his shirt off and swinging it around his head when a full tray of drinks went over and a few unhappy clubbers reared up. "Calla, Keegs, enough of your bullshit. Let's away doon stairs and get old Screamer yem before we get barred from the best boozer in town!!'

Calla explains to the girl on his arm he has to leave and if she fancies a drink in the week to call in Nelly's. "I'm in every night to practise pool," he explains as he walks off with Willy.

"Practising pots of lager mare like, bigman!" Keegs catches up with the lads as they bool down the stairs to push past the crowd and hook Screamer arm in arm and carry him backwards with him waving his shirt around out the bar. "Thanks lads, saves me a job."

The doorman and Calla's new buddy are smiling as the lads whistle for a cab!

"Aye us British are used to pulling you yanks oot the shite now and again!!" Gordon shouts a bit of abuse at the lads.

"Don't mind him lads, he hasn't gotten over being captured in the First World War!" Kris nudges the doorman as the full squad wait for the 2 cabs to come and collect them.

"Here, where's the new lad?" Mark questions Kris as they get in the cab.

"He left with some dark haired bird about an hour ago – he must have thought he was in that invisible bubble you can climb in after 8 pints!"

The doorman waves the lads off as the cabs head for the garage hoping it is still open with all the boys having a right munch on. Only Calla, Keegs and Mark go into the shop with the cabs waiting! Screamer and Willy are asleep with Kris getting a piss around the back.

"Evening pal, can I have 2 hot dogs, 20 Marlboro lights and 12 bottles of bud please?" Calla is first in the queue!

Keegs and Mark are waiting behind with milk and bread, cereal and a few packs of ham. "Where you two cunts gannin like? A fucking teddy bears picnic!!"Calla ribs the lads.

Mark shouts Calla back as he gets to the door: "You might want to buy a bottle opener for them, daft arse!"

Calla never breaking stride: “Nee need, young flanja, I will use the cheeks of Davey’s arse if I have too!!”

***

The peace and tranquility of Davey’s sleep is woken by a strange sound.

The TV is buzzing with no picture as Calla is half on and half off the sofa with empty bottles and a full ashtray on the floor when Davey enters the sitting room. “Haway knackers, get up, it’s 6 o clock!”

Calla pulls the cushion over his head. “Ah for fucks sake, I’m in fucking bits, I’m staying in the neet.”

Davey goes back into his room as Calla heads into the bathroom, turning the shower on and after 15 minutes he was ready, drinking a cold pint of milk out of the fridge. “Are you ready, Davey or what?”

There’s no answer and Davey’s door is locked. Calla bangs on the door again and he gets a little chuckle from inside with Davey sniggering: “Check your watch, daft arse and I’ll see you in a few hours!” Calla goes into his room and digs his watch out of his hold-all: 3 am in the morning! “Bastard! The fat bastard!!” Calla kicks his sambas off and jumps under the duvet.

Andy Jones however was slipping his trainers on as he tip toed down the hall of Tracey’s apartment and slid out of the door, whistling down the road he went as he could see in the distance the hotel sign of his apartment. Ten minutes and he was at the gate where the big figure of Earl was standing against the wall. “Looks like one of you guys got lucky!” Earl walked over to Andy as he walked through the lifted gate. “Gentlemen don’t disclose any secrets of that nature,” Andy quipped back to Earl as he started to laugh.

“There’s no secrets in this part of town, buddy, and please don’t tell me you never walked out of Tracey Hill’s street!” Andy’s redness gave the game away instantly. “Ah we just kissed and cuddled bigman, no harm in that is there!”

Earl put his huge hand on Andy’s shoulder. “Nope there certainly isn’t but if what she has had in cock was handrail there would be enough to go around that QE2 of yours!" Andy laughed as he said goodnight to Earl and walked off to his apartment scratching his cock.

***

The minivan rolls up to the apartments around 10.30. Davey and Calla are already outside having a smoke as Jimmy jumps out. "Good neet then, where did you finish up?"

"These all went to Roxy's. I stayed at Nelly's with Mally," Davey replies, flicking his cigarette towards the drain.

"Was it any good, Calla?" Jimmy quizzes.

"Or eer Jim, was fucking jam packed with blurt, should've seen these two me n Keegs were cracking onto!"

Jimmy butting in: "Did you shag them?"

"Eer it was nailed on reet but we had to carry Screamer out, he was fucking blotto!"

"So that's a no! Cracking onto them doesn't empty the bags, lad and don't blame Screamer for not getting ya hole!" Jimmy replies looking at Davey with a smile as Mark, Johnny, Kris and Andy head out their apartments, Mark looking like he had got dressed in a Banardos bin, eyes like pissholes in the snow.

"Rough neet, youngen?" Jimmy hollers, lighting up a cigarette and laughing out the smoke.

"Nah Jim, it's hay fever, pollen's shite here!" Mark sniffles out.

"Not as shite as your lies, silly bollocks. Don't worry it's fun and games the day. I will get you shown round the site and then get you lot inducted and lads, be gentle with the H&S officer, he is a fucking vomit and will go into the far end of a fart, but once it's done it's done. Reet haway there's away"

All the lads climb into the minivan. "Eer Davey gan t McDonalds, am fucking hank!" barks Calla.

"Teddy bears picnic, Calla eh, what you think about these apples?" Mark says waving his ham sandwich towards Calla who snatches it from him and fucks it straight out the window, the rest of the lads laughing at Mark who slumps back in his chair arms folded

"There is a canteen on site lads, it's fucking top banana. You can get something there!" Jimmy replies doing a U-turn onto the interstate and gunning the big V8 van down the on ramp.

Jimmy shows the lads around and after a quick tour of the facility and their breakfast the lads are led into a small classroom with a large plasma TV and a laptop set up at the front. There is a jostle for the back seats like kids at school, Calla settling in and swinging his feet onto the next chair and rocking his back to recline, folding his arms and closing his eyes as the rest all find themselves a seat.

Jimmy still at the door shouts back to Calla, "You're gonna have to get forward on that chair and open your eyes for this presentation mind!" Calla not opening his eyes or moving an inch shouts back, "I listen with my ears not my eyes, Jimmy, you daft cunt!" all the lads breaking out laughing as Jimmy shakes his head and walks out.

A middle aged guy tall at 6'5 but thin weighing around 9 stone with short gingery hair, and round spectacles walks in carrying a box. "Good morning guys, my name is Warwick."

"HUNT?" Calla jibes still rocking back on his chair semi asleep.

"Mathews," Warwick replies, Calla's comments going straight over his head as he closes the door and scribes his name on the whiteboard next to the TV then flicking on the laptop and the TV which prompted a power point titled 'Health and Safety in the workplace'. Reaching into the box he begins to unload items.

"A hands on approach to safety with a video will keep you lot awake." The safety manager slams a harness down on the table.

Andy Jones is already in noddy land at the back with Kris as Calla chirps in: "Unless Pamela Anderson is on this video or even better she's here and ganna climb in that harness I think we have seen most of this before, so give us a Blue Peter badge and we can all go and put your job in!"

The safety manager smiles and rubs his wispy ginger goatee. "This job is my baby and believe you me I will be responsible for every single one of you guys going home alive!"

Davey sits up in his chair and reassures Warwick: "Son, you crack on with the safety induction and I will mak sure all these lads leave alive in all, although I can't vouch for the pack of fuckers getting into work alive!"

Two hours and 80 slides later the lights go on. "Any questions, gentlemen?" Warwick smiles down at the weary eyed lads.

Mark puts his hand in the air as Andy wakes up and shouts, "No questions, your honour! Can we be excused!?"

All the lads sign their names and receive a sticker for their hard hat and a safety booklet for their troubles.

Outside there is a load of white hats on their way down to site. Jimmy and Willy march over to the lads and show them the way to the site stores and portacabins. Jimmy opens the stores door and there is an old fella in, around 60, called Charlie, from upstate New York.

"Right Charlie, give the boys their 5 t-shirts each and toolbox and make sure they all sign for it," Jimmy instructs the old timer.

Willy lights up the tab and whispers to Davey: "Jobs a fucking Glenn Hoddle pal, there's 10 of us and 100 yanks! We have the east plant, they have the west!"

Davey trying his t shirt on over his head: "What, just the 10 of us like? This top isn't fat bastard size either!"

Jimmy reaches over and gives him a bigger size and confirms, "There's 10 of you lot who are as good as 100 but there are 4 lads flying in, 2 from Texas, 2 from Ireland. What with Keegs doing the QA/QC and one of the Texans supposed to be shit hot at drawings we'll be sound."

All the lads have the t shirts and hard hats and wait for Charlie to give them their boots. "Caterpillars, I'll wear these bastards the neet." Calla is excited about his new daisy roots!

"I thought you were staying in tonight, son," Davey straightens his t shirt.

Calla puts his arm around Dave and squeezes him. "If you want me to stay in, father, I will!! Till 7 o'clock!!!"

The lads all laugh and head towards the job with Kris shouting back to Jimmy, "Haway Jimbo, show us the next rodeo!!"

Jimmy takes the lads down to the bottom of the site where there is a large wooden facility erected kicking through the two wooden doors. Jimmy signals the lads inside. "Woo hoo, this is Wembley boys, this is where the magic happens," Jonny shouts excitedly wandering over to a bench and laying his tool box down and inspecting the welding kits.

"All fucking top gear here, lads!" Calla throws a bag off the bench sifting through the tools he has tipped out of his box and starts to etch his name onto the bench. "This is Calla's desk, boys!"

Jimmy "Well it was Hags!" signalling to Davey as the rest of the lads throw their boxes down and begin looking around the facility which was littered with all top equipment, showing him an office. "Here you can use this for all the paperwork, it's got a PC in it hooked up to the network and internet so you can get hold of any drawing you need and plot them out. Keep Calla off the porn, Davey!" Jimmy sniggers.

"Steady lad, I've just fathomed that teletext carry on with my mobile, computers wayy lad, fuck off!" Davey shaking his head and nosing around the office before sitting in the large leather chair and rocking it back and taking out his cigarettes. "Or now, this is the touch, mind it's a smoking house, Jimmy!" he says lighting up his cigarette.

"I'm sure Warwick will put pay to that!" Jimmy walks back out to see the lads.

"Make it your home, lads. Davey, haway with me, we'll head up my office, go through the job!" He flicks the AC unit on and walks out, Davey following glancing at the lads on the way out. "Dinnit break owt!"

Calla is straight into Davey's office "Let's get the Frankie fired up!"

Mark and Jonny follow him in leaving Kris and Andy.

"What time did you get in?" Kris asks.

"Got back this morning, lad. Went to this bird's house for a coffee, ended up nudging one into err! Cock's in fucking bits!" Andy replies, sticking his hand down his strides for a shuffle. "Arr you didn't bag up? You daft cunt! If it was that one I saw you with she looked dirtier than a dog's arse, probs got all sorts!" Frowning and nipping his nose.

"Ha, ay lad, wore a blob and she was a reet dirt bag and I was sawing away for ages, fucking ale n blobs have the old lad number than Calla!" Both lads laugh. "Nor she wanted to finish me off with her hand, used this fucking tingly lube! Honest lad, it was like a slippery wank with deep heat, feels like me bastard cock's been nettled!" Kris laughs as Davey's office door opens and Calla barges out with an A2 plot of the St George's flag. "God save our Queen, lads – get this cunt up, sicken the sweatys!"

Davey goes through all the plan drawings with Jimmy and Willy and they look through the schedule to which and where to start first. "All done, piece of piss, Jimmy! Any bonuses?"

Jimmy blows the cig smoke out. "Aye there is a bonus of a next job in either Malaysia or Shanghai if we get this done!"

"A few hard months ahead, lads, 7 till 11. The lads will think we are buying them doughnuts when I tell the bastards that is the hours!"

All 3 lads burst out laughing as Mally comes in the office with 4 other lads. "Oy oy Anto, ya little bollix!!"

Willy is shaking Anthony Nicholson's hand as Anto introduces his pal Jacko Doyle. Mally sits down and puts his boots on. "Well lads, there's plenty of jokes to be had, English Scotsmen and Irish!"

Jimmy tells the lads to sit down and introduces Chuck Weaver and Tommy Jones. "These 2 lads worked with me in Boston, lads – 2 of Texas' finest!" Willy shakes hands with the lads. "Well that's not an unusual name, Tommy, is it!?"

All the office bursts into laughter as Tommy looks down and in his thick Texan accent replies, "I never heard that one before, dipshit, why Chucky I guess I really might have to hit a guy with glasses after all." Tommy being a stocky well put-together bloke with pit shovel hands had Willy on the back foot.

"Ah come on, Tom, we aren't going outside on the green green grass of home already are we!" Jimmy gives Davey the bundle of drawings and tells the 4 lads to go in and get their expenses.

Jacko was a real charmer with the ladies so it didn't take long for Michelle to flutter her eyes and tell Jacko her grandmother was Irish. "So you have a bit of Irish in ya, pet? How would you fancy some more once we get to know each other!?" Jacko sits on her desk running his hand through his thick black hair.

Michelle turns scarlet red. "Maybe Jack, you never know your luck," she says quietly as she gives him his expenses.

"You're some operator, Jacko, I bet the lads before us tried all the moves with your one!" Jack and Anto waited outside for the 2 Texan lads to come out.

"We are being taken back to the ranch to throw in our bags. How you guys fancy a few scoops and shoot some pool at this Irish bar Malcolm mentioned?" Chuck asked the lads, a big 6ft, 16 stone unit of a man but a gentleman all the same.

Anto and Jack looked at each other and Anto offered his hand out to Chuck. "Tell you what, big un, that sounds like a plan to me!!"

***

Jimmy and Keegs were just out of a meeting and had to meet Joe Butler, the American general foreman, for the west plant. "Joe is a good lad, Keegs and we need to get ahead of the game with our lads first because Joe has already started throwing pipe into the racks."

Keegs walked like John Wayne and carried it well. "Nee probs pal, I guarantee Hags has Kris or Andy already welding so I will have all the paperwork in place tomorrow."

With the American install lads loading all areas up with pipe with the overhead cranes, the lads were already dropping it in place with all the kit ready for welding in place. Hags and Screamer were miles ahead with all the brackets and open clips, string lines all in place and drawings on the wall – it looked like the lads had been there a month when Jimmy and Keegs walked in.

"Where's Calla?" Jimmy shouts up to young Mark.

"He's in the fabrication shop, or that tent thing over there."

Jimmy walks in to find Calla and Jonny already about 10 pipes set up ready for Andy.

"Fair play to you, Andy son, I knew you would jump straight in the fucking fab shop for the easy life!" Keegs shouts over at the lads.

Calla pulls his brown leather gloves off and puts the grinder down. "I'd love to talk to you but I'm very busy here and the nerve centre is only on half gas yet!!"

Jimmy turns around to walk out as Calla darts over and puts a foil taped shape knob on Jimmy's hard hat as he slips through the plastic curtain.

Jimmy walks away back to his office, beaming and rubbing his hands with delight. Mally is sat in his chair on the phone.

"Get your arse out of there!" Jimmy barks, waving Mally off. Mally, putting the phone down, jumps out of his chair. "Them lads might be as daft as a ship's cat, but they know how to gan on setting up and clagging metal together!" He squints towards his PC at his emails.

"Ay Jim, we'll still need an eye on them, it's just a matter of time before they start swinging it and blowing shifts!" Mally replies sitting opposite Jimmy who had turned back to Mally and slipped his glasses off and leaning in. "Malcolm lad, you're good at your job, I can't fault you for that but these lads are golden and they are my choice so before you set away with your petty bollock swipes... back off! The daft carry on can be overlooked so long as the job goes in on time with no faults, so don't go upsetting them...that clear?"

Mally stands up with a nod, heading to the door and half turning as he opens it, says: "Ay Jim you're the boss! You might want to take that ball of lag tape off your helmet though! You look a bit of a cunt!"

With a wink Mally leaves, Jimmy whipping his helmet off and riving the ball off and stotting it at the bin. "Calla, you cunt!"

The general buzz is rattling around the fab shop as the lads are backwards and forwards with drawings, lengths of pipes and other materials as Stevie Williamson wanders in. Calla flicks up his mask. "Fucking hell where you been hiding?"

"Just been getting all the wiring diagrams and schematics sorted ready for the..." Calla buts in. "Yawn, fuck off, Stevie you're boring me now–" flicking his mask back down.

Stevie walks into Jim's office who is running through the drawings with Willy. "Areet Dave!" Stevie swings his laptop and tube of drawings at the far end of the Cabin on the desk opposite the big square table the drawings are sprawled over. "Ay sound, you're not telling me your lot are in here?" Davey shouts, standing up and taking off his glasses.

"No, don't worry, he is organizing us an office up in the main facility. He did say we all have to share this fab shop and lay down area," Stevie replies.

"Just wanted to run through these coordination drawings with you."

"It's snug already and there's 4 more lads started!"

Willy chirps in, "Our lads will be on site, maybe pop in to knock up a small panel but we will need to keep our site box here and our gear out back!"

"Ay reet o, when's your lot getting over and who's coming?" Davey asks walking over to his chair and rocking back scratching his head.

"Ah they've nipped our scope back, we are only putting in the scada system for them now, cabling, containment and termination all the kits free issued and they have sorted themselves their own guy to do the software, its gunna be a nightmare, all sorts of problems!"

"Ay bit of back scratching maybe, coz they're not tight with the coin, it's all top kit out here!" Davey replies.

"Getting Mick Taggart, Ben Gibson, Bobby Harris, Owld Peter Thompson Stuey Brown and Rob Brown."

"Ahh for fuck sake! The Browns are owa coming. Willy, you have to see these two twats, smashing lads, great crack and Bobby Harris that cunt can't graft sober... looking forward to seeing Bobby!" Davey barks.

Smirking, Stevie replies, "Mickey gets in in the morn, so I'll have a couple of days with him, then I'm flying yem, gotta job to look at in Consett! It's a shit farm, bit like the Stadium of Light, Davey!"

Davey not looking up and replying: "Aye reeto, like electrical another subject you know nowt about."

Willy chirps in, "Aye enjoy Consett – it's really like Orlando!! In fact I have a bad memory from Consett! I was oot in Whitley Bay this neet, the hotel down there I cannie remember the name but I pulled this bird she was a cracker!!"

Davey lights up a cigarette and stops Willy's tale: "A cracker? What like Robbie Coltrane!?"

Willy continues: "Aye aye bigman, keep them coming, basically she was another English slut! Anyway I gets her home to her place in Consett. 'Be quiet,' she said, 'the dog isn't very well!' The fucking dog a thought?! Anyway this spaniel came in with a gimpy leg and sat looking at me as I tried to get into his mammy's drawers! Every time I got my hand up her skirt the fucking dog would come owa and put its gammy paw on my fucking knee!!"

All the lads are smiling away listening to Willy tale!!

"So to cut a long story short I said to this bird, look I'm very good with animals, I think it just needs a walk. You go upstairs and slip into something like nowt on and I'll take lassie the gimp oot for a wee stroll while I have a tab. Well she was delighted, so me and lassie went oot the back where I walked doon the back lane and had a tab. At the bottom I tied lassie to the lamppost and went back to pump the bird, telling her it was doon the stair asleep!! All was going well until I was taking her from behind standing up bent owa the dresser when she looked oot the curtains for a noise and the lamppost I tied the dog tee was right in front of her with it crying an whimpering!"

All the lads burst out laughing, Stevie saying: "So she fucked you out for being a cruel Scottish twat!!"

Willy, taking a tab out of Davey's box and lighting it, confirms: "Aye, she went fucking berserk, kicked me oot!! Have you ever tried getting a taxi out of Consett at 4am in the fucking morning!!"

***

Jimmy walks down to the site and into the fabrication shop to give Davey a new walkie talkie phone. Davey is up on the ladder measuring with down below Jonny holding a small note pad book.

"Here Dave, you're doing that all wrong," Jimmy standing at the foot of the ladder shouts up at Davey.

Davey takes the pencil out of his mouth and marks a line on the wall next to the string line and tape. "How did you work that out, son?"

Jimmy puts one foot on the ladder to get up closer to Davey. "Because, you silly old bastard, the young lad should be up there and you should be down here with the pad!"

Davey climbs down the ladder and puts his mirrored safety shades back on tucking his ten pound folding prescription glasses into his shirt pocket. "Jimmy, if you wanted a shite would you send me?! No, would you fuck because you know that even though I'd try you still wouldn't be satisfied!! That's why I do the measuring because if daft arse put the wrong mark on that wall and I made the pipe up to fit and it didn't, I wouldn't be satisfied! Do you get my drift or what!?"

Jonny being more than capable of the task keeps quiet and winks at Jimmy, showing Davey the book. "So I took the measurement down and I will give it in to Marky and Calla but more importantly what do you want from the deli?" he says as he prepares to write down in the pad Davey's order.

Davey sighing out loud and turning to Jimmy says: "You were the same as this cheeky twat, thought you knew it all!!"

Jimmy laughs and gives Davey the walkie talkie. "Turn it on, channel 1 gets you anywhere on site to me, Keegs, Willy, Mally and safety then go to 2 for the chit chat! It's also a mobile phone!! Give me 20 minutes to get them handed out then round the troops up. I am knocking them off at 8 and taking them for a pint at Nelly's!!"

***

Anto was tucking into a steak roll – "A say these steak sarnies are deadly!" – showering Jacko with chewed bread.

"Anto, you eat like a fucking pig, shut your mouth when ya talking!" snaps Jack; all the lads laughing.

"You numb twat and how do you suppose I feckin do dat?" Anto replies.

"Jeez you know what I feckin meant!" Jacko replies.

Tom leans into Chuck. "Who says the Irish are stupid?" Both guys sniggering.

"Just be taking ya shot, Tom, ye big bollox!" Jack snaps, waving the cue as Tom rattles a ball against the pocket.

"Yee fucked now." Jacko sinks 3 balls off the trot lining up for the black and stroking the black the full length of the table for the black to come to a dead stop a fanny's hair from dropping in, dropping the cue and running round the table to investigate. "Ah for feck sake, Anto there's a fucking chunk of rib eye on the table of your sandwich, ye prick!" –as Tommy doubles his ball in then smashes the black in throwing his cue on the table.

"And that's how it's done, son...that's how it's done. What's that Jacko? 100 bucks? You Irish should learn when to quit!" He laughs while taking a swig of his beer as the lads leave the pool table and sit in the large booth, Tommy signalling to the barmaid for another round.

"I'm gunna get mash potatoed tonight, boys, feck graft! H&S meeting tomorrow then it will be meet n greet, fecking appearance money easy!" Jacko snides chumming back half of his unsettled Guinness, the other guys lighting their glasses in a toast.

"I'm not staying here late, guys, just here until she aint got no more Bud to pull!" Tommy winking gulps back his lager and tops it up with the pitcher.

"I bet you that 100 dollars I owe you I could drink 5 pints faster than you and I'm drinking Guinness!" Jacko flicks the dollars onto the table, Tommy snatching them up.

"Bet you a 100 bucks you couldn't get them back?" He grins at Jacko while thumbing through the dollars just as Jacko smacks him square on in the mouth sending the dollar bills fluttering down and Tommy's head crashing off the wood frame of the booth, shocked and dazed wiping his mouth.

"Oh now y'all gunna ay for that!" standing up, flipping the table of drinks over as Jacko and Anto leap backwards from their chair.

"Now hang on, big man, a bet's a fecking bet!" Jacko tries to explain with his hands out as Tommy crashes a windmill against the side of his head sending him sideways across the bar and onto the pool table that had just been re-racked by 2 Americans who were just about to break, balls scattering as Jacko lands palms down across the table to the upset "Hey" of the 2 guys. Jacko just about gathers himself, ear ringing, when Tommy grabs him by the shirt and drags him backwards ripping his shirt.

"How, that's my favourite t-shirt, ya cunt!"

Anto gets hold of Tom around the neck as the three bundle backwards into another table sending glasses and drinks crashing everywhere and involving 3 angry Americans who immediately start on Tom. All the while Chuck is rescuing what dregs there are from the pitcher and pouring himself a pint, rocking back on the chair watching the fracas. "Ahhh beer and a show ring side seats!"

Tommy, struggling with the 3 Americans punching and holding him down, leaves Jacko a chance to wriggle free of Tommy's grip. Getting to his feet he kicks one of the Americans in the ribs and grabs the second in a choke hold allowing Tom to land a punch into the side of the 3rd guy. Anto is curled up in a ball trying not to be trampled on as the door swings open and in steps Calla who steps over the blokes rolling on the floor and without an inch of concern says "2 pitchers there love" as Dave, Kris and Jonny walk in.

"Whoa, what's the story here?" pulling Tommy back who is cocked ready to splat one of the American guys, Kris getting hold of Jacko and Jonny holding the other yank back. Davey barks: "Now settle it down lads, I'm thirsty and tired and don't need this shite, so everyone say sorry, buy each other a drink and enjoy your neets!"

The 3 Americans dusting themselves off are about to wade into a speech about who started it. Calla nudges him. "Eer man, he said leave it so chill the fuck out and get a beer!"

The other lads stand the tables up and help the barmaids clean up the broken glasses. Tommy dusting himself down heads for the toilet following Jacko. Dave gestures to the Americans to get a drink; battered and bruised they decline and leave.

"What the fuck went on there?" Davey quizzes, helping Anto to his feet.

"Jacko smacked Tommy in the mouth!" Chuck barks, still sat with his feet up, beginning to laugh.

"Eee did what? What for?" Calla shouts, walking to the table with the pitchers, Jonny behind him with the glasses.

"Yup, clean shot straight in his keeshter! Fucking brilliant! Thought it would be funny – it was!" He pours himself a beer as all the lads sit down.

Calla pipes up: "Eer man, them fucking Irish, nowt but bother and thick as pig shite. You hear the one about Paddy and Murphy? Paddy is walking down the road with a bag of doughnuts. Murphy meets him and says if I can guess how many doughnuts are in the bag can I have one? Paddy says, if you can guess how many is in the bag you can have them both. Murphy says 4?"

All the lads roar with laughter as Jacko and Tommy return, Tommy heading straight to the bar to apologies for the chaos, Jacko flopping down shaking his head! "Fecking Texans are mad bastards so they are; Tommy shook my hand and thanked me for taking 2 of the yanks out!"

The lads all laughing with Chuck piping up: "Yup sure love a bit roll around, beer, babes n battling! Hey y'motherfuckers crack me up!"

Tommy sits down with another pitcher and a pint of Guinness for Jacko, putting his arm around him. Jonny quips, "Fat bird served me in McDonalds today – she said sorry about the wait, sir. I replied, don't worry fatso you'll lose it eventually!"

All the guys are crying with laughter as Jimmy and Mally walk in. "Oh here's the heavy artillery!!" Calla shouts over at the lads.

"Two pints of Guinness, flower," Mally orders as Jimmy walks to the toilet thumping Calla in the arm, sending him through the service hatch. Mally sits down and lights a cigarette and shouts the lads over. "Look lads, there's been an email sent through today; we have to have this first phase done by Christmas!"

Davey jumps down off his bar stool and looks very concerned with the month of October approaching an end.

"Which Christmas though, Mally?" Jonny spits his beer out with laughter as Willy and the lads walk into the bar.

"Here Willy, we have to have this first phase done before Santy Claus comes." Mark setting the pool balls up informs Willy, Keegs, Hags and Screamer of the new news.

Willy squeezes through past Mally to get to the bar –"excuse me, tradesman coming through!!" – Mally looking at Willy, shaking his head.

The barmaid comes through to serve Willy with some more news: "Have you guys heard of a beer from Europe called

Kronenbourg? Because we have just had it put in and the first 2 barrels are buy one get one free for a promotion with every 2 pitchers getting a polo t shirt!"

Calla's ears prick up immediately!! "Two pitchers of krazyburg, gorgeous!! Have we heard of Kronenbourg!? We have it on our cornflakes back where we come from!!"

The barmaid starts pulling the lager into the pitchers and looks up. "On your cornflakes? Really? Wow!"

Calla nudges young Mark and winks, putting on an American accent. "Andrew!! Does it really go up there! Do I really have to cook your mince and dumplings afterwards too!! Ok!!"

Mark shakes his head with disgust and goes over to break the balls with Jonny singing out the words of Mark Knoplfer into the queue looking over at the barmaid: "Girl, you look so pretty to me!! Like you did last night..."

Davey joins in with a heavy northern accent: "Like the Spanish City to me which was a load of shite!"

Jimmy, returning from the toilet, responds: "Classic tune, don't change the words, lads! That's blasphemy! Girl you look so pretty to me like you always did, like the Spanish City to me when we were kids!" Jimmy continues to sing out the melody.

"Stick to ordering people around, bigman, you sound like you swallowed a pool ball!" Willy mocking the voice of the fuhrer!

"Aye and you actually will swallow the pool ball, four eyes, any mare of your shite!" Jimmy continuing: "From Cullercoats to Whitley Bay out to rockaway, rockaway!!"

All the lads were fed and well into the Kronenburg when the American football came on. "Ah for fucks sake do we really have to watch this shit!" Andy Jones standing up puts his quarters in the pool table.

Chuck Weaver shouts to the barmaid to turn it up "This is the Cowboys against the 49ers, dude, a real good game."

Calla coming back from the toilet with his new t-shirt on and the jukebox off: " Wah hey I'll have one for each day by last orders!! Ah for fucks sake what's this shite on the telly!!"

All the lads are more interested in the pool when a big shout from

Tommy Jones –"TOUCHDOWN!!" Dallas cowboys scoring early on!

Just minutes later Jemma and Rosie the barmaids bring out four huge baskets of chicken wings for the tables. "What's these for, gorgeous!?" Keegs puts his hand on Jemma's waist.

"Well I would like to say they were for you but they aint; every time there's a score it's half price pitchers and chicken wings for 5 minutes!"

Calla nearly breaks his neck tripping over Davey's stool to get to the bar. "Two pitchers of Krazyburg, flower and a large t-shirt!!"

Screamer lifting his pint up says: "This bar's shite, man, is there nee others we can go to!"

Willy laughs at him. "Screamer, you could be sitting on a fucking yacht in the Maldives getting your boaby sucked, drinking pina bastard coladas and you still wouldn't be happy!!"

Screamer taking a drink of his pint and brushing crumbs off the Hugo boss shirt he wore for work (being a diehard poser for his age) replies: "Willy, as long as you and Calla weren't there I'd sit on a fucking dinghy on the Clyde in December quite happy!!"

Calla looking up at the bar with froth on his lips off drinking out of the jug says: "Haweh!"

Willy as well not happy adds: "Aye Screamer, you're some cunt."

"TOUCHDOWN!" Mark shouts out along with Kris AND Calla!!

Mally looking over at Jimmy says, "I can see these Monday nights being a bit messy mind."

"Right who's up for a game of Killer?" Willy walks over to the bar with a big smile on his face. Calla in full swing at this stage with the two for one pitchers replies: "Only if it's 20 bucks a man and a double if the last man misses the black!"

Davey and Jimmy are whispering in the corner. "We'll go halfs lad and try and get big Keegs out – if you are in front of him, he's a fucking shark," says Davey forming the old partnership with Jimmy.

Young Mark breaks off with Mally following him as Calla walks over to the jukebox. "Make sure you leave that cunt nowt on Mark, we know you're fucking shite at pipefitting so you'll do anything to get in Mally's good books!"

Keegs follows Mally's miss, shouting over at Calla: "Aye Calla, speaking from experience you couldn't put a pipe in a fucking snowman's mouth!!"

All the lads loving the banter, Willy stands on the stool so he's the same height as Keegs as he is bent over pretending to ride him from behind. "Aye big man, speaking of putting things in mouths your mother used to like a bit of that they reckon," as Andy swipes the stool sending Willy on his arse.

"TOUCHDOWN!" Big Tommy smashing the balls as he shouts at the top of his voice.

Jimmy shakes his head. "Ah for fucks sake, this has to be the highest bastard scoring game of the season!!"

Calla shouts over: "Aye Jimmy, the 49ers have had nearly as many touchdowns as Krissy Robinson with his welding – he's fucking yepless!!"

Kris is up next to pot his shot and remains with only 8 of the lads left. "Calla, are you surprised I have as many welding with your set ups? Fuck me Stevie Wonder could set the cunts up better."

Davey, taking the cue off Kris, replies: "You know, Kris son, Calla was my apprentice and he wouldn't set up what he couldn't weld himself."

Andy Jones gives Kris his pint. "Take no notice, brother, it's the pipe-fitters union. We'll see how clever they are tomorrow when they want summit welding and I have to go for a 2 hour shit!!"

Keegs takes the cue off Davey. "Davey, you dirty old bastard! Snookering me in a game of pool."

Davey walks back to Jimmy with a wink. "Well we got rid of the Irish, there's one Texan left and one jock against me, Jimmy and Calla and now you only have one life–" just as Keegs wobbles his shot in the pocket.

The doors open into the bar and four girls walk in and sit down in the corner. Davey nudges Screamer. "Look at the cunts, there are like meercats! The first sniff of Minge and the antennas are up!!"

Screamer nodding looks over in disgust. "You're right there, Davey. I bet that dark haired one with the nose like Gonzo has a fanny like Davy Crockett's hat!!"

Jimmy chirping in: "Aye if I get her back to my billet I will have to give it a whack with a cane and send the red setters in to clear out the pigeons!!"

Calla walking up to the table pissed and lashing the stripe into the far pocket without it even touching the sides then not breaking stride heads over to talk to the girls "How's things, girls? I'm the new entertainments manager in this pub and was wondering if you would be interested in speed dating!?"

The blonde one of the three smiles and asks where Calla is from.

"Boro. What ya doing tomorrow neet!!" Calla spurting out the sentence very quickly, laughing. "Fancy going out with me? bet that's the fucking quickest date you were ever offered!! Speed dating!! Does what it says on the tin!!!"

"Right, daft arse, it's your shot!" Jimmy passes the cue to the table where Calla is sitting. Only five left at this stage: Davey, Jimmy, Keegs, Texan Tom and Calla who stands up gyrating around the pool cue.

"How many lives have I got left father!?" Davey giving the barmaid the finger gesture signalling get the round in for the troops.

"Three, son and you will be going some to pot this!" The white ball lies behind one on the cushion of the remaining two. Calla puts the cue behind his back and lashes the ball as hard as he can!!

Screamer was never the happiest soul but to be hit on the side of the head by a pool ball at 100mph was never going to be one of his happier moments! "You fucking big useless twat!!" Rubbing his head as the lads piss themselves laughing.

Willy pulls Screamer's head down and kisses it reassuringly. "There there, Brian, it's only a little lump!"

Screamer is not at all pleased. "Aye like you, ya fat little specky fucker, get off me will ya!!"

Gordon Haggerty walks into the bar all suited and booted to the roar of approval off the half pissed squad. Keegs shouts over "Oy Oy, Hags, are you due in court or have you a hot date!"

Hags walking over to the bar nudges Jimmy. "Alright big man, let me in there for a swally! The queen is due in from Scotland in an hour and I've to look my best if I want a ride!" Jimmy moving over takes a drink of the pint and tells the lads: "Gordon's missus is due in tonight; he has his best underpants on so he can get his hole and looks like a respectable husband wanting his wife in town for a fortnight!"

Willy walks over sitting down at the table full of girls with young Mark. "Aye Hags, you fucking two faced bastard! You said to me you had no clean underpants yesterday when I needed some!!"

Davey spits his beer out laughing. "Willy, you're some operator! Why have you no clean kegs like!!"

Willy lifting his head right back so he can see Davey through the jam jar glasses, replies: "I have now, turned them inside oot!" The girls pausing to drink their drink with disgust as Willy turns to them with a smile: "Now would you ladies like a drink?"

Texan Tom rattles his ball against the pocket throwing the cue over to Davey as he walks over to Anto and places his big hand on his shoulder. "This is a Monday, Anto! What the fuck is a Friday night like?!"

Anto laughs and tells Tom the only difference is the day after. Jacko chirps in: "It's like new year's eve every night, lads working hard through the day and trying to forget what they are missing back home."

Chuck laughs and nods his head. "Guess it's the same all over the world then! I go out every night to try and forget the big motherfucker of a shit kicker that I got back home eating all my wages in Texas!!"

Anto falls backwards off his stool laughing. One of the four girls walking past the lads gives off a huge waft of perfume. Tom sniffs out loud and picks Anto off the ground. "You know what, guys: if it smells of cologne leave it alone!!"

Down to 2 on the Killer – Jimmy and Calla – Davey not happy at all about his miss and exit from the competition. "Make sure you beat him, Calla son." Davey gives Jimmy the wink being spotted by Willy. "Oh aye wink wink nudge nudge, you pair of cunts are halfing!! Calla, wop it right up the fat Geordie bastard!"

Three balls on the table, one life apiece, with Calla wobbling as he chalks the cue. "I'm fucking bladdered on this Krazyberg!"

Screamer shouts over. "Just keep the white ball on the table this time, you dozy bastard!"

Calla looks over. "Haweh," potting the 5 ball in the middle bag with a smile as he hands Jimmy the cue.

Willy, pissed with his arm around the blonde girl: "We've seen some drama tonight but this is fucking Shakespeare!!!"

All the lads are now interested as Jimmy misses the long 7 ball leaving Calla with a tough shot. Screamer getting excited at the fact his assailant is going to be beat shouts: "Go on, Jimmy, what a shot, bigman!"

Calla, downing the last of his pint, says: "You know what? I'm going to start and call you thrush, Screamer!!"

Screamer looking confused as Calla eyes the shot up with a sway. "And why's that Calla? Because my voice is like a beautiful bird?!"

Calla, bending down, firing a double across the table into the bottom bag and standing up with his chest out in satisfaction: "No, Screamer, because you're an irritating cunt!!"

The lads all laugh as Jimmy walks over with the pot of dollars. "Well won, lad, ya Jammy twat."

Calla, thumbing through the wad of dollars, replies: "Call me what you want, Jim yee fat shite, I still have your MONEY!" He tries to jam it all in his back pocket.

"Giz that here, silly bollocks – you'll lose it!" Davey takes it all from Calla and folds it neatly into his wallet. Calla gives Davey a big slobbery kiss on the cheek before staggering to the toilet. "Or ya dirty bastard, get some fucking mints, ya breath is hooching!" Davey wipes his face with his sleeve and shakes his head.

The lads begin to settle laying the cues on the table, Anto and Jacko leaning against the table, Tom, Chuck, Davey, Jimmy and Mally sitting in the booth opposite leaving Mark and Willy sitting with the girls and Hags at the bar like the monopoly guy talking to Screamer. Calla bounces past. "Here Hags, where did you get that suit? Barnado's bin?"

"Or fuck yee, Calla you wouldn't know style if it landed on ya heed!"

"Something must have landed on your head if ye think that suit's got style; did you rob a tramp?" Calla laughs back into Screamer who props him up.

"No, the tramp didn't want it!" Screamer snaps, putting his arm around Calla. Both of them guffaw laughing.

"Or here, bollocks to this, might've known you twats would be on my back. I'm off to the airport to fetch the queen!" Hags flicks a twenty onto the bar and necks his whole second pint dribbling some down his shirt.

"Whoa whoa whoa, hang on a minute, ya Scottish tramp, don't go wasting the krazybourg – it's travelled a long way and commands respect; you should be ashamed of yourself. Get back in the gutter with ya buckfast, ya Scotch twat!" Calla takes the pint glass from Hags, shaking his head laughing.

Hags taps Screamer on the shoulder. "Catch you later, lads, al see you the morn!" as Hags walks out.

"Hags, Hags, wait!" Jimmy shouts. "Have you got your welding mask?"

Hags turns, confused. "Eh it's at work – what the fuck do I want that for the neet?"

"For your lass so no-one recognizes her with you wearing that suit!" All the lads laugh again as Hags straightens his suit and walks out giving Jim the bird.

"Me and the boy are taking these two treacles out for a bit scran so we're off. There's 50 bucks for our drink!" Willy fastening his belt throws the 50 to Calla.

"Hawld on a minute, there's nee leaving the boozer with a bird before kicky out time like! It's in the rule book!" the rest of the lads agree with a roar.

"Calla lad, one asked if you fancied coming as well – straighten yourself up though and get some mints!"

"Do a fuck, I'm not leaving the troops in the trenches while I run a solo mission with a virgin and a jock! A dint care how nice the blurta is!" Calla sways to the rest of the lads' cheers, raising his glass like a key speaker and takes a large gulp. "Just out of curiosity, which one like, Willy?" Calla quizzes.

"The dark haired one with the big tits!" Willy replies walking off.

"What the!? Hawld on, Willy, am coming!" Calla necks off his pint and slams it down on the table, tucking his Kronenbourg t-shirt in. "Eeer, Davey, giz that dosh, man!"

"Or aye, so much for not leaving the troops spiel, fucking traitor!" Davey stands up and thumbs Calla half of the money. "Al give ya the other half the morn, lad."

Calla snatches the money and swaggers off, putting his arm around the dark haired one asking, "So darling, what's your name?" as the seven of them begin to troop out.

Anto knocks his pint down and stands up, throwing dollars on the table. "Well it's just not reet, lads! Seven is an odd number and anar she has a nose like gonzo but she has a fanny and that's enough for me!" He winks and jogs after Calla and the rest of them.

"Fucking stroll on! Bet that makes for an interesting dinner table: 2 smogs, 2 jocks and 4 naïve American lasses, god help them!" Jimmy laughs into his pint.

"Not like you not to be amongst that, Keegs – what's up lad?" Davey asks.

"You know me, Davey, like to fly below the radar!" Keegs smirks back, glancing across at Jemma the barmaid.

"You do fuck, Keegs! You're as subtle as a smack in the mouth, you're after a slice of that barmaid and mind you she pulls a canny pint so she gets my thumbs up!"

"Got big tits too," Tom chirps in.

"Motherfucker! What I'd do for a shot on them!" Keegs standing up finishes his pint. "Eye come on, Big Tom, that's our lass you're talking about! Same again lads!" as Keegs heads to the bar with a smile.

Meanwhile...3 taxis are on their way to Noche, a very swish restaurant on the water's edge

The restaurant looked onto a lovely backdrop of water and low lights of the candles gave the table of 8 the best view in the house. All were sat girl-boy as the waiter arrived in his pristine white jacket and black dickey bow tie. Calla sits up in his chair. "Excuse me, captain dooby, could we see the drinks menu?"

The waiter is not very impressed. "No problem, sir, but I must advise you this is a very respected establishment and I must request a small bit of decency."

All the lads look at each other very sheepishly. On his return the waiter hands out the drinks menus and advises what the specials are for the evening. Willy and the lads are already tucking into the bread on the table, Calla trying to speak with half a loaf in his mouth. "Never mind the specials, get us 4 pints of Coors and 2 bottles of wine for the lasses."

As the waiter walks away, shaking his head– "Oh hold on a minute..." As the tired waiter turns round at Calla with the look of 'what now?' on his face– "What time do we fucking set sail Captain!?"

Willy has the paper napkin rolled up and stuck either side of his top gums doing a Don Cheech expression out of the Godfather while all the girls cry laughing, the Scotsman doing a sterling impression of the Italian godfather. Calla gets up to go for his 10th piss of the night stopping at the door taking a French loaf stick out of the basket and walking back to the table swinging it full force and smashing it across the back of Willy's head. "Ha ha, you're wacked, you spick bastard!"

The girls and the table next to them are covered in crumbs as Calla walks off to the toilets eating the stump of loaf left in his hand.

Into the third round of drinks and the food is placed down in front of the party, hands on legs and arms around the girl's shoulders at this stage with all the smooth talking and bull dust in full flow. The captain places a big steak in front of Anto. "Fuck me, boy, it's still breathing this fucker, ask the chef to burn it some more!"

Calla lifts the steak up with his fork off Anto's plate. "He's fucking right, captain – a good vet would have that back on its feet!!"

More food is brought to the table with Mark's fresh cod hanging off the side of the plate, Willy and Calla in full throttle now with the patter and the drink! It's always a sign Willy is gunning for some action with the ladies when his glasses come off. Calla laughs as he takes a big drink out of Gonzo's wine glass. "How Mr Maggoo, we are over here, Willy – taking them glasses off isn't going to make you look like Brad Pitt! More like shit pit!"

Willy folds them in his pocket putting his arm around Anto. "She is lovely isn't she, my date for the neet." Everyone is pissing themselves laughing as one more plate of fish arrives at the table.

Calla stands up with his wine glass and taps it with a fork. "I'd like to propose a toast to Captain Birdseye here for bringing us the catch of the day, ooo arrrrr captain, you miserable twat!"

Calla sits down to the drunken laughter of the rest of the group as the manager walks over. "Excuse me, sir, can I ask you and your group to please control yourselves? We have received a number of complaints about your behavior – this ISNT some night club!"

Calla brushes the crumbs from his t shirt and stands up. "It isn't?" he quizzes. "Wayy that's another strike against you... so you don't serve Krazybourg–" (holding out the badge on his t-shirt)– "you don't sell chicken parmos, the wine is weak as piss, your waiter has a face like a smacked arse and now you're saying the table and chairs can't be moved so we can push out a few shapes to the Pet Shop Boys? What sort of joint is this, or – you know what – we aren't paying!" Calla signalling to the rest of them who begin to stand up.

"SIR, your bar bill alone is in excess of $200!! Should you not pay I will call the police!" the manager leans into Calla with an embarrassed whisper.

Calla signals the manager to the side of the table away from the others for a private word and the two begin to speak while the others assemble themselves at the door – except Willy who is at the bar getting 2 bottles of wine to take out.

Calla comes booming over. "Haway folks, there's away, oooooh you're in for treat the neet, love!" Grabbing hold of the dark haired bird and leading her through the door towards a taxi, the rest following, a confused Willy shouting "Call, how the fuck did you pull that one?"

"Piece of piss, told him we weren't paying, simple as it went, said I'd settle the drinks bill but that's it! Snapped my hand off; think he was glad to see the back of us!" Calla laughs while nudging his date into the car.

"Did he not say owt like?" Anto asks.

"Err ay, none of us are allowed back!" Calla says, jumping in the car as it begins pulling away and the rest of the lads follow suit, jumping in the taxis with their lass, Mark and Anto sharing their cab as the 2 girls lived together. The taxi slows up and the driver swivels round. "$43 please."

Calla jumping out shouts: "Square ya man, I'm busting for a slash." He jogs into the garden and pisses in a bush as the girl pays up and walks past and begins unlocking the door with a tut. Once inside Calla noses round. "Posh place you got here like – live on your own?"

"Yes my fiancé moved out 6 weeks ago. Can I ask you a question, Andrew? Do you remember my name?" Calla's head slowly appears from the fridge with a can of bud and he replies sheepishly while swaying. "Way aye daft arse! How do you spell it though?" He slams the fridge shut and slurps the froth from the can.

"L I S A!" she replies putting both her hands on her hips in disgust.

"See that's my favourite name in the whole world. I might not be able to spell but I know a pretty name when I hear one, Sandra!" as Calla puts his arms around her and moves in for a kiss. Lisa drops her arms and begins to laugh. "You're funny, Calla!" as the two kiss for a while.

"REEET! enough small talk and foreplay, get up them dancers and get ya kit off, you've got till I finish this can then I'll be up!" pushing Lisa away who is looking confused.

"Would you say you like beer more than me, Andrew?" "Haweh... don't be daft, get up the stairs and I'll be up when I finish this – it's only a half can so get a wriggle on!"

Apprehensive and confused, she darts up the stairs. Calla, crushing the can and leaning two handed on the bench, steadies himself and takes a deep breath. "Fuck me I'm pissed!" Looking straight at him from the couch was a Chihuahua dog. Confused, Calla opens the fridge for another can. "Fucking rats are big oot here!" Pinging the can, he takes a Kamagra out of his pocket and rips the wrapper open and squirting it into his mouth washing it down with his can thinking to himself he better have 5 minutes while it kicks in, flopping on the couch, shooing the small dog away. "Fuck off, ya rat!" he says as he sticks his hand down his strides for a bit clart on with the old lad to give himself a rut on.

Meanwhile Anto and Mark are climbing out of their taxi, Mark left with the bill as Anto walks with his arms around both of the girls towards their place.

Opening the door into the condo is a huge smell of rotten eggs or something that died. "Jesus Christ almighty, what the fuck is that smell?" Anto covers his mouth while Mark does the same on entering.

"Oh that's my cat – it has a disease in its ears," the blonde girl explains and continues: "I'll make you guys a coffee; make yourself at home."

Anto sits down while Mark rummages through the CD collection, stuffing a couple of classic Rolling Stones CDs into his jacket pocket.

"Ah, for fuck's sake, Marko, would you ever give it up!" Anto gives the young lad a bit of stick as the cat comes purring around his ankles. With one swift sweep Anto lifts the cat off the floor with his right foot, the cat flying over the top of the TV with a big cry as it hits the wall and scampers away.

"You cruel twat, it's sick." Mark sits down next to Anto, not very happy with the cat being volleyed over the telly.

"Never mind the fucking cat, I'll be sick if I have to smell anymore of it!" Anto shouts through to the girls. "Are you making coffee or what?"

The 2 girls come in with 4 coffees and baileys with a bit of cream on the top. "We thought we'd make Irish coffees for you, Anto."

Anto stands up, taking a mouthful and twisting his face. "You know, girls, that's very nice and all, but we have to leave – that cat fucking stinks!"

Jimmy is sitting in the minibus smoking a cigarette when Davey comes out with his flask and bait box. "How many casualties, Davey, or missing in action?"

Davey climbs in next to Jimmy. "Well believe it or not, Calla is ready – he'll be out in a minute."

Jimmy blows the smoke out. "Fucking hell, he was like a rubber doll when I seen him last!"

Davey laughs. "I know! He hasn't had a wink of sleep, look at the twat!"

Calla closes the door fastening up his denim shirt as he walks towards the van. "Jimmy, will you stop at the garage? I need a tin of pop – my gob is like an Indian taxi driver's flip flop!"

All aboard the bus with only one missing. Jimmy pulls away waving at Earl on the gate. The big fella stops the van. "If you guys are looking for Anto I think you want to call over the road to Tracey's house! When young Mark and him got out of the cab he peeped Tracey walking across the road out of a cab and hot footed it over!"

Andy Jones pipes up in the back: "I had feelings for that girl; I was supposed to take her out tonight!!"

Kris lifts his baseball cap and laughs. "Well we'll have to get the Scottish contingent over tonight – she seems to be fond of forming her very own Great Britain!"

Jimmy pulls up in the street outside the house Andy pointed to and blips the horn. "Y'na it's like babysitting you lot!" Jimmy's not happy at the 10 minute delay.

Calla opens the door and spews in the street. "Get oot!! I knew that fucking steak wasn't done properly!!" Calla wipes his mouth as he shuts the sliding door.

Davey looks over his shoulder. "That would have nothing to do with the 2 gallon of Kronenbourg you had, son, though, would it!?"

Calla sitting back folds his arms. "Davey, I could never blame the drink for sickness; fair enough it might not have been the steak!"

Silence in the van as Jimmy honks the horn again then Calla pipes up: "I had some cheese last neet before bed, that's it!" As he opens the door to be sick again, he says: "Jimmy when we get Anto can we go back to my digs for 5 minutes!!"

Jimmy not happy at all already turning round rages at Calla! "What the fuck for!!!"

Calla smirks at Jimmy. "I've shit my pants."

Kris, opening the door, confirms it. "Or here he fucking has Jim. Calla you fucking dirty bastard," as he squeezes out onto the street cackling! "You lot are trying my fucking patience here. Calla, you can walk back, hurry up!" Jim snaps.

"Or eer man, it's fucking squidgy, Jim, you're some cunt!" He squeezes out holding his arse and waddling back towards the digs.

Anto is peeping out of the window, wearing a pink dressing gown as Jimmy waves him out. Anto holds his hands out and mimes '5';

Davey jumps out for a smoke. "Tuesdays eh? For fuck sake!" He lights a cigarette!

Andy arrives like he got ready on the waltzer. "Areet lads!" throwing his shirt on a chair as he straightens his Kronenbourg t shirt. "Lads, she knows how to fuck by the way." Anto winks at Mark.

"I was supposed to be taking her out the night, Anto, you twat!" Andy snaps.

"Oh! erm that's not gunna happen, ha-ha! Unless you want to go fetch us a takeout while I nudge one into her!"

"Where's Calla gone?"

"Shat his pants!"

"Oh right!" Anto rocks back in his chair, fastening the seatbelt. Calla jogs towards the minivan.

"Haway, shitty arse, get in." Davey flicks away his tab and pats him into the van.

Calla stops and puts his hands on his knees. "Hawld on, Davey, av got a stitch!" The minivan roars out towards the site with Calla and Kris singing 'hi ho hi ho'. Once on site the lads crack on getting busy, Calla storming off for a shite. "My fucking back eye is like a bastard Japanese flag!"

***

(A number of weeks pass, with the Job moving well on.)

"What do you mean we have to send 4 lads to Fucking New York!!?" Jimmy fumes at Mally.

"Jimmy, there is a small job that needs specialists and to be honest we can't afford to send the yanks!" Mally explains to Jimmy the urgency of the request coming from high level.

Jimmy wanders down the site with Mally and Keegs through into phase 1 where Jacko and Anto are up on scaffolding with Screamer and Andy Jones. Jimmy and Mally look up, assessing what the lads have done in the morning with Screamer bent over the scaffold looking down. "What's you 2 daft cunts looking for? Pigeons!!"

Mally signals Screamer down off the scaffold. "We need a

supervisor, 2 fitters and a welder to go to an old facility in New York on a shutdown!"

Screamer's scratching the back of his head with a pencil. "What now?? I've got the groceries in for the week!! Spent all my keep money!!"

Jimmy shakes his head. "You mean you have bought a pound of cheese, a packet of crackers, 2 pot noodles and a tub of Bryl cream, you tight old twat!!!"

Screamer leans against the scaffold. "Do you want me to go to look after this job, bigun and pay me more expenses, are what?!"

Mally shouts Chuck Weaver over asking him what the lads are entitled to leaving one job onto another.

"Well they are entitled to expenses up front and travel money."

Jimmy shaking his head. "That's utter bollocks, I just give them expenses last month!!"

Chuck shakes his head. "Welcome to the US, brother – this is where the unions looks after tradesman not like Europe where you get pissed on from the so called unions!!"

Jimmy is not very happy but admits defeat. "Screamer, bring Podge and Rodge with Andy fucking pandy up to the office after dinner and I'll sort you out with dosh for the job!"

Jacko looks down the scaffold. "That's fecking brilliant news, I always wanted to go to the parlour up on 85th street!!" Him and Anto do a high five as Jimmy and Mally walk off through the job.

Jimmy turns to Mally as they walk up. "Make sure the accountant gets these bastards to sign for their money only when they sign for their flight ticket!!"

Mally sniggers as he offers Jimmy a tab. "Jimmy, you're one miserable twat, do you know that!!"

Jimmy lighting his tab walks through to the fresh air as Calla's hand squeezes down off the scaffold placing a tin foil shaped knob on Jimmy's hard hat!

Screamer starts to pack up his tools and slide out of his boiler suit as Calla walks in. "Eer, where you going, Scream?"

"I'm away to the Big Apple, young'n, seems they need the A-team to pull a job oot the shite, I'm away to get some expenses and pack me bags!"

"Hawld on, pack your bag? You only have one pair of jeans, 3 t-shirts 1 pair of kecks and one sock!"

"Ay true, that's why am gunna get a half shift in Nelly's to score a few Kronenbourg t-shirts for the capital. Heard the birds are a bit more classy up there!"

Throwing his welding mask on the bench and wiping his brow, Calla shakes his head. "Screamer, you couldn't pull in a brothel with a hard on wrapped in hundreds!"

"Ay? Wey a couldn't give a fuck, lad. NYC here a come with a new roll of expenses to spend on getting pissed!" Winking as he heads out of the workshop to round up the lads to tell them the good news!

Calla catches Davey outside. "Eer Dave, Screams is away to New York, mare expenses, the lot, probs a few more dollars an hour. How come you never got us on it?"

"Or, Calla man, ya never bastard happy, are ya? It's bastard freezing up there anyhow, it's about -4! It's 38 degrees here, digs are nice, near a good boozer, what mare do you want?" Davey pats a puzzled looking Calla on the back.

"Did you not hear me, they're getting another wad of expenses! Mine's just about gone!"

"Calla, you're some fucking unit when it comes to scattering cash! What the fuck you spent it on?"

"Drink!" Calla snaps.

Davey walks off cigarette quivering on his bottom lip as he laughs. "Why at least you haven't wasted it!"

Back in Jimmy's office Jimmy takes his hat off. "Al string that bastard Calla up! If he sticks owt on my hat again!"

Mally shakes his head. "Jimmy, why did you agree to him coming out here? He's a clown!"

Jimmy going to light a cigarette pats his pockets looking for his lighter. "Because he's a great hand and Davey asked me to sort it out; he was in bother back in smog land. He is a silly bollocks but a quality silly bollocks!"

Mally gathers some drawings from the table. "Jim, your big heart's gunna be the ruin of you!"

Walking out the office, Jimmy gets his hat and heads back to site to get a lighter off one of the lads. Back down at the workshop he collars Davey. “You got a spare lighter, Davey?”

“No. You can have these though.” He throws him a box of matches and Jimmy strikes one, lighting a cigarette, thanking Davey and walking out as Calla leans down and sticks an empty strawberry yoghurt pot on his hat as he trundles off up to the office bumping into Tom on the way down.

“Yo! What’s the story about the NYC, Jim?”

Stopping to chat, he says: “Tom, av manned it up now, would you’ve fancied it like?” He sniffs the air around him looking confused. “Can you smell strawberries?”

Tom picking up his bag and small transformer looks at Davey’s hat and, shaking his head, walks off. “Nope! The New Yorkers are all like Billy Idol’s brother BONE!”

Jimmy heads off sniggering.

Tom gets into the fab shop where Kris, Marky and Calla are all busy grafting away. “Hey any of you guys got a flashlight I can borrow?!” Tom on his way down site forgetting his up at the bait cabins.

Kris lifts his welding screen with a puzzled look on his face. “A fucking flashlight!!? Do you not mean a torch, Tommy!?”

Tom leans against the metal table. “Nope, I mean a flashlight!”

Mark walks over with a tape measure and a piece of pipe in his hand. “Tommy, do you mean this!?” As he turns his Maglite on: “This, Tommy, is a torch and you can borrow it nee bother!”

Tom accepts the torch and places it the top of his Carhartt shirt pocket. “Well guys, I’m from the deep south, torch’s are what them KKK folk use! this is a flashlight!!” He winks and walks out; the lads piss themselves laughing!!

Calla shouts Kris over. “Kris, hurry up and get a tack weld on this bastard!!”

Kris struts over with the welding screen on. “Keep it fucking still then daft arse!!”

Calla laughs. “Just tack the bastard! Next time it comes around, nail it!!”

Mark laughs. "Calla, you're shaking like a dog shitting razor blades!"

Calla holds his head away while the bright arc of the welding machine fires the UV light into his face, kneeling down with his Stabila level and tape,eyeing the flange up with the pipe and not even touching it. "It's like a Scotsman's purse, hammer, nail it!"

Kris crouches down to obey his command.

Mark looks puzzled. "Scotsman's purse?"

Calla winks. "Sporran son, fucking sporran!!"

Keegs walks into the fab shop with a bundle of drawing. "Alreet chaps, fuck me you've done loads! We are trying to get a turkey out of this job!"

Calla gets the highlighter pen and draws a huge green line through the drawing on the wall. "Where we going the neet, Keegs, you big handsome bastard!!"

Keegs, studying the drawing and checking the pipe on the floor with his tape, replies: "Oh, I am having a quiet one tonight, Calla – staying in is the new gan oot!"

Calla walks over to Keegs shaking his head and sticks a paper cup on the kneeling Scotsman's head. "I'm staying in n all!"

Kris lifts his welding screen up. "You're staying in, fuck off!!"

Mark joins in. "I'm not having that!"

Calla, putting another drawing on the wall, says: "I am lads, until 7 o fucking clock!! Then am oot!! It's Karaoke night in Nelly's!!"

Big knock on the door of Jimmy Hunter as Screamer, Anto and Jack with Andy Jones troop in. Jimmy is smoking and looking at the monitor of his PC. "Take a seat, lads I'll be with you in a minute; it's the Arc de Triomphe this weekend and I fancy that O'brien mount!"

Jack chirps in. "My cousin's dad's best buddy has a share in that and he says it's fuckin no chance!!"

Jimmy looks up at the Irish lad with a frown. "Your fucking cousin's dad's best pal!! Well that's going to be useful information!! How many cousins do you have? Is it the right one!!?"

Jack takes out his mobile. "Can I use the phone der, Jimmy? Oh and I've 52 cousins!"

Willy and the lads are laughing and Anto is punching Jack. "You have got 52, ya bollocks and I reckon you have rode at least 5 of them!!"

Jack's not even breaking movement as he takes the phone offered by Jimmy. "I have indeed, Antony you tramp, and I rode your sister and your ma!!"

Anto pauses for a response but decides to leave it.

Jack hangs up the phone after a 5 minute conversation that had even his pal Anto confused. "If you can get 10-1 for him back him each way!"

Jimmy leans back on his chair rubbing his chin. "Ok but what if he is as it says here online 7-1!?"

Jack looks confused. "Well back him only to win cos the odds are too short for an each way bet!"

Jimmy smiles. "You seem to have a bit of knowledge, Jacko, anyway, I'll back him!" Jimmy stands up and shouts of Michelle. "Right lads, the wages lassie will give you 1500 dollars each expenses, and a return ticket leaving tomorrow to La Guardia. Simon 'Cockmuncher' Harrison will be there to collect you all!"

Andy Jones rubs his hands together. "Fucking casino tonight, brothers and big yahoo."

Jimmy quickly turns to the lads. "You sign for this dosh and a ticket; if anyone doesn't make the flight the dosh gets deducted from this month's pay cheque!"

Michelle walks in with the 4 envelopes and an A4 piece of paper all named for the lads' signature. Jack gives her a nice wink as she smiles. "So lads you are leaving us, are you having a farewell drink – anywhere I can say bye?"

Screamer straightens his Hugo shirt and touches her hand, beating Jacko to the prey. "Aye hen, see you in Nelly's Irish bar at 7pm. I'll be buying you whatever you like, sweetheart."

The girl's eyes light up and she smiles right into Screamer. "A date! I will see you there!" She walks back into the office with all the lads leaning, watching her go.

Jacko looks at Screamer. "You dirty old bastard! You warm her up tonight and I'll put the fire out back at mine afterwards!!"

Jimmy shakes his head. "Right lads, sign the dotted line and piss off back to the digs and get sorted! Oh and I mean what I say about tomorrow's flight!!"

Andy puts the envelope in his pocket. "Are there any flights today to Vegas and we could go from there tomorrow!!"

Jimmy snaps. "I mean it, you Manc bastard!! Oot!!"

The 4 lads are on the way home to pack, with Screamer driving when he slows the van down putting the indicator on towards the digs. "Ah come on to fuck, Screamer – straight to Nelly's for a quick pint then we'll go back early and get sorted for tomorrow!"

No reaction from Screamer as the car heads to the gate where big Earl salutes the car as it passes. "I've been around the block a few more times than you 3 bastards and I am getting things sorted the now and will stroll down Nelly's at 6!"

Back on the dance floor Calla and Kris are up in the ceiling space finishing up the last of the tie in welds, Calla leaning back against the unistrut supports while Kris finishes off welding.

"Do you think you could buck any of the barmaids in Nelly's!?" Calla asks as he colours the inside of Kris's safety glass frames with black permanent marker to leave the outline hopefully on his skin when he puts them on.

Mark shuffles his way down to the lads. "Right, Calla, all the brackets are on and buttoned up. How did I get that job anyway!?"

Calla puts Kris's shades back next to his bag. "That would be, young Marko, because you are a fucking numpty!!" He kisses the welding mask shaking as he laughs.

Mark heads down onto the scaffold as Calla places a plastic cup on his hardhat. Kris lifts the welding screen up. "Right, hammer, that's us, and as the shepherd says to his sheep–" –Both lads saying at once: "Let's get the flock oota here!!"

Kris puts his safety glasses on as Calla struggles to hold his laugh in and offering his hand "Give me your bag, Krissy, you fucking legend I'll throw it in the box!"

Kris climbs down and wraps all the gear up just as Jimmy and Davey walk in through the curtain, Davey winking at Calla. "Alreet son, how did you get on up the gods!?"

Calla holds the end of a 110 cable as Mark wraps it up towards him. "The eagle has landed father!! It's time for booze!!"

Jimmy is impressed. "So will we submit that for test at the weekend? Double bubble Sunday!?"

Kris's eyes light up. "Count me in for that!! 70 buckaroos an hour!! We couldn't spend that on drink!!"

Calla looks at Kris. "Haweh!"

Kris stands corrected. "Sorry everyone, bar numb nuts couldn't spend that on drink!!"

Calla throws the keys to Mark for the toolbox. "Lock her up dipshit, we are oota here."

Jimmy laughing looks at his watch. "You're a bit pig's tail, Andrew, but haway we'll get a tab and fire the happy bus up!"

Mark looks confused. "What's pig's tail!?"

Davey's deep accent answers the young lad's question as they all head for the exit. "A bit twirrly son, a bit too early!!"

All aboard the happy bus, Jimmy pulls out of the job, seeing Mally and Willy walking up the bank from the site, and lowers the window. "Lads, we are headed straight for Nelly's, will you get the ball rolling for permits to test at the weekend?"

Mally walks up to the window. "So I have to get all the permits organized while you take this shower on the piss!?"

Calla dives through 2 seats. "Aye you smoggy bastard, cos this shower has just piped up 200 metres of pipe while you walk around pulling the squad to bits when in reality they are paying your fucking wages!!"

Davey turns around. "Righto, Calla, sit the fuck back down." Mally walks off shaking his head.

Calla sits down. "He does my fucking Swede in, that bastard. I'd love to crack the fucker!"

Mark winds Calla up some more. "Calla man, Mally would kick the shit out of you anyway!"

Calla biting straight away responds with: "I'll hit him with you in a minute, you scrawny little twat, if you go on much more!"

Mark and Kris: "OOOOOoooooooo."

Willy walks away from the minivan giving Calla the finger "Is that four eyed bastard winding me up?"

Jimmy pulls onto the interstate, lighting up a cigarette. "Aye Calla, so sit down and relax. I'll buy you a Kronenbourg for hoying all that cherry ripe in today, son."

Calla looks out of the window at Willy walking away then turning back. "Now you're talking my lingo, Hunter, good darts!"

Kris opens the door into the bar, the smell of stale ale and smoke wafting up his nose. "Ah Bisto!!" he quips as he walks in.

Calla has his arm around Davey. "I need to go to the ATM, Father, can you sub me a couple of hundred until tomorrow?"

Davey opens his wallet giving Calla 3 folded 100 dollar bills. "I want it back mind!" giving Calla the money, Calla with an astonished look on his face.

"Your some micey bastard Davey having a spare 300 sheets in your wallet! Fucking happy days!! Pitchers are on me!" He bursts pass Kris to the bar. "Rosie, 4 pitchers of Krazyberg, 3 plates of wings and your telephone number please!"

Rosie gets up off the chair smiling. "No way would you get my number, Andrew, you are a dooshbag!"

All the lads laughing with Davey shaking his head, wallet still in his hand. Jimmy pulls his stool into the same position as where it goes every night against the bar. "You're not wrong there, pet, he's a dirty rat!!"

Calla puts his money on the bar, looking at Jimmy with hurt. "Haweh!"

Jimmy laughs. "He's a fucking good grafter though, I'll say that for him."

Calla smiles as the sound of 'Gimme Shelter' by the Rolling Stones comes on the jukebox as Mark sets the pool balls up and Kris selects the music. Davey coming back from the toilet, his hair all wet and face washed. "Some tune this, Krissy son, I seen these live at St James' Park! What a show!!"

The 2 Irish lads and Andy walk in the door all cobbled (washed and shaved). Calla is straight onto them. "Fucking smell off you 3, is that the famous aftershave Oh my god by Hugo boss!?"

Andy laughing over at Calla: "I tell you what, our kid, I won't miss your bullshit when we get up to New York!"

Calla walks over to Kris and Mark with the pitchers. "Aye, the chosen ones get a jolly to New York and we get to slum it down here. Where's Screamer anyway!?"

Anto pulls a chair up next to the lads. "He has a hot date tonight, the old bollix, with the wages girl so he'll be a while getting ready!"

Kris turns from the jukebox. "Michelle?! Fuck off! I thought Jacko was tailing that!"

Jimmy takes a huge mouthful out of his pint. "Jacko is too busy giving me tips for horses – it runs in 30 minutes! Rosie, put the racing channel on will ya!!"

Just then the door opens and Screamer walks in, dark blue Hugo boss jeans and polished leather cowboy boots underneath with an Armani leather jacket covering a white crisp shirt underneath. Calla shouts over. "Fuck me, it's Butch Cassidy. Howdee partner, where did you tie the hoss!?"

Screamer shaking his head as he finds a nice space at the corner of the bar, strategic for when his date arrives. "Calla, you wouldn't know style if it cracked you in the face!"

Screamer sits up on his bar chair, young Mark breaking the pool balls. "Screamer you have some Plymouth Argyle, hammer, I'll give you that!!"

Most of the boys are in the bar, some reading the song books while the Karaoke is being set up, the others cracking on when the door opens. Michelle and her 3 friends walk in and a fine quartet they are, Keegs being the first to break the ice, moving his stool to one side and standing up with a smile, "Hiya, Michelle, ladies! What can I get you to drink!?"

Calla is on his way back from the toilet into the bar when Mark leaves his leg out and Calla trips into the Karaoke speaker, knocking it sideways for the small American lad setting it up to just catch it.

"We heyyyyy!" Kris and Mark shout as the 4 girls turn around, the lads were expecting the usual swearing and outburst from Calla who got to his feet next to the tall brunette. "Hello beautiful, my name is Andrew and as you can see I have fallen for you!!"

All the girls burst into laughter and Calla moves right into the middle of them winking at Keegs. "Trevor, would you be so kind and get me a beer please?"

The brunette is all over Calla and his posh English accent like a cheap Matalan suit.

Screamer taps Michelle on the shoulder and smiles. "You wanting to sit somewhere a little mare comfy, hen?" Michelle smiles back and walks towards Screamer.

Davey nudges Jimmy., "She can sit on my face if she likes, marra!"

Jimmy chuckles. "There's more chance of you winning the lottery, Davey; them days are long gone for you!"

Davey takes a huge mouthful of his pint, emptying half into his gullet and burping out loud. "Aye son, I'd rather have a good shite anyway."

Mally and Willy are last into the bar, work gear still on, Mally wearing the company t-shirt and Willy wearing a t-shirt with a print that has the girls laughing 'Rehab is for quitters'!!

"Alreet, lads and lasses." Willy is looking directly at all of the girls' tits through his jam jar glasses, one of the girls commenting on the little fella's stare. "If you want I can get them out for you!! How rude!"

Willy lifts his head back with a disgusted look on his face. "Get them out, you'll be wanting me to sign my name on them once you hear me on this fucking karaoke, pet!!"

Michelle asks Calla if he sings.

"Me, sing? Are you joking! Good looks, personality, great in the sheets and sing!? Well of course!"

Jimmy shakes his head. "Can you fuck sing – you couldn't sing for your supper, you skinny twat!!"

Anto shouts over. "I'll bet you 20 bucks you can't sing. Go on, get up there first, ya gobshite!"

Calla drinking back his beer and rubbing his hands together shouts over at the karaoke man. "How, fanny chops, when's this gunna be ready!?"

The small guy getting the microphone and switching it on, holds it to his mouth. "Ladies and gentlemen, welcome to star karaoke nights! My name is Dean, I'm the man on the scene, looking for the next karoake king or queen," as Calla grabs the mike off him ."Giz the mike, you fucking balloon!!!"

The small guy is trying to wrestle with Calla as Calla's strength pulls the mike out of his hands. "Right boys, and girls–" winking at the ladies– "This is a dedication for Anto, all the way from Ireland! Soon heading to pastures new!!" Calla clears his throat

"Put your 20 where I can see it, buddy!" Anto shouts over as all the bar turn towards Calla.

Calla puts his hand in his pocket and reaches over to Davey. "Here, father, get Anto's and keep them safe." Calla pulls a chair up and lifting one leg, points to a song on the book to the disc jockey. As it starts Calla looks over to Anto. "Happy birthday to you, happy birthday to you!! HAPPY BIRTHDAY dear Anto, Happy birthday to you!!"

To the rasping applause of Davey, Jimmy and Mally – "That's not a song, you bollix!!!" – Anto stands up as Calla takes the 40 dollars off the bar and drinks his pint.

"Calla, you are one lousy bastard, you know that!" Anto is at the bar ordering the drinks.

"Haweh!" Calla's smile makes Anto smile, patting him on the back.

"Gunna miss the crack, bigun."

The blonde girl turns round to Anto. "Hey, you're Irish, I'm quarter Irish!!"

Anto puts his hand on her waist as she smiles. "You are, and where else are you from?"

The girl puts her drink on the bar. "Well, quarter American, Italian, English and Swedish."

Calla looks puzzled. "Well I was never the brightest at maths, pet, but 5 quarters!?"

The girl looks confused as Calla whispers in Anto's ear: "You're in there, pal, loves the Irish and thick as a whale omelette!! Perfect"

The music starts, Folsom Prison Blues as Willy starts it all off! "I hear the train a coming, it's coming round the bend!"

Davey claps his hands together joining in then stopping. "Oh fuck, it's our lass's birthday, Jimmy lend me your phone!!!" And he waddles out of the bar quickly.

Outside Davey listens as the phone rings and the voice of his missus says hello. "Happy birthday to you, happy birthday to you hap–" Stopped mid chorus she tells him she's the pan on and the grand bairns are running riot! "Aye alreet, pet, I miss you and get yourself out for a bit scran, never mind cooking! Take 30 pound out the bank, my treat."

She thanks him and says for him to go out and get a beer too. "Ah, I'm knackered, pet, I'll just watch a bit of telly and have an early neet. Tell the bairns I love them, and you."

He walks back in the bar and gives Jimmy his phone back. Jimmy puts it in his pocket. "Everything tickety boo at the ranch, pal?"

Davey waving over the barmaid– "2 pints of Guinness, darling, aye Jimmy, nee bother!! Thank fuck daft arse got up and sang that song or I'd've forgotten and been reet in the dog house!!"

Jimmy chuckles and flicks Calla on the ear. "Mally has them permits sorted, fanny chops – in at the weekend, nee failing!"

Calla turns round from his conversation with the lovely American girl. "Jimmy, look! I'm busy! Will you fuck off annoying me and make sure you look after that housekeeping like I told you earlier."

Calla turns back to the girl explaining that he was Jimmy's boss and Jimmy was very lazy. She peers around Calla's shoulder to see Jimmy shaking his head, turning to Davey. "Y'naa Dave, if brains were dominoes Calla would be knocking!!"

Davey laughs. "Aye he's about as bright as night the clown but he'll dee a bit for ya, marra and it will be done reet!"

Keegs is getting on famously with one of the quartet while Michelle is deep in conversation with Screamer, commenting on how nice the lads are. "Aye just be careful there hen, they seem nice but they'll get up on you as quick as a flash at half the chance."

Michelle laughs. "And you wouldn't, Brian?"

Screamer is not even fazed "Half a chance, no; a solid chance, definitely."

Michelle goes all red. "Well I like the men in my life to be refined and classy so we'll see."

Calla getting wind of the conversation joins in: "Well that'll fucking rule yay reet oot the equation, Screamer!!!"

Keegs, Krissy and Mark piss themselves laughing. Michelle taps him on the knee telling him not to mind them. Willy is in full flow on the Karaoke with a nice little crowd gathering in Nelly's. "You were always on my miiiind" finishing off an Elvis classic and handing the mike back, all the lads clapping.

Calla shouts over: "Tonight Matthew, I'm going to be a Scottish, five foot five, four eyed, cockend Elvis!!"

Willy, taking his voddie and coke off Anto, replies, "Aye Calla, let's hear you up there, full of wind and pish!!"

Calla bouncing a peanut off Willy's head says: "I've been up once, you berk, I would hate to show you up!!"

Just then a black girl gets up on the stage and nails Whitney Houston – "Didn't we almost have it all". All the lads are watching as she is tasty too, Davey turning round to Jimmy. "Is this song about Newcastle in 1996!"

Jimmy puts his pint down. "Haweh! Dave, we normally don't talk about me being a Geordie and you being a sad mackem bastard!!"

Calla is now dancing with the brunette slowly as Willy bounds over and separates them. Calla fumes. "What you fucking doing, Elvis! Fuck off back to Specsavers!!"

Willy pulls at Calla. "Andrew, you know the rules! It's not even fucking 9 o'clock and you are smooching in a public place!! Now get the shots in and go and stand in the corner for 5 minutes on one leg!!"

Calla looks at Willy. "Haweh!"

The girl smiles assuming they are joking as Calla turns to the bar. "10 slippery nipples, and I'll see you in 5, darling," as he trudges into the corner necking his pint and the girl watches in disbelief as Willy puts his arm around her. "Alrighty, ever been to Scotland, hen? I'm William."

The girl is already fascinated by Willy and his ability to charm her as Calla marches back over. "Right goggle box, fucking shots in and go in the other room for 1 full hour with your socks and shoes off with matches in between your toes!!!"

Willy sighs. "Ah for fucks sake man, is there any need!?"

Jimmy and Davey both piss themselves laughing, Davey wiping

his eyes. "Aye Willy, that's what you get for cutting someone else's grass!!!"

Willy shouts the shots in and trudge through the lounge door with his vodka as the girl's mouth hangs open in disbelief.

Jimmy shouts the round of drinks in with one circular movement with his fat sausage finger and a wink with his bloodshot beng eye to the barmaid who receives the message loud and clear without one word between them. Jimmy leans into Davey. "Didn't even have to say a word, hammer, and the barmaid knew exactly what I meant."

Davey looking over at her winking, arrows his thumb at Calla and making a wanking sign to which she nodded. "Same here, no words and she knows Calla is a complete wanker!!!"

Both of the lads are laughing as Calla walks over from his time in the corner. "What's the two Ronnies laughing at?"

Davey puts his arm around Calla. "Which one of them lot are you riding tonight, son?"

Calla looks over as he takes a massive mouthful of his pint. "They are off to that wine bar after this. I'm opting out – it always costs me a bastard fortune, the lasses have Champagne taste with rola cola pockets." Davey easing back laughing.

Keegs walks into the conversation, taking his pint off the bar and clinking glasses with Jimmy. "Are you going to say a few words, big man about the 4 lads leaving us?"

Jimmy gets up. "Aye, suppose I will." He takes the mike off the DJ who just finishes the Van Halen classic 'Jump', tapping it to get everyone's attention. "Right girls and boys, a round on the bar to wish Screamer and the 3 boys best of luck on the job up the road! We would've loved to keep them here but we are carrying enough arseholes as it is!!"

All the lads let out an 'oooooooooo'!! "So I thought I'd sing you all a little song to get you on the way!"

Willy walks in from the lounge with his socks still off, smoking a tab and holding a glass of voddie. "Go on, bigun, what you gunna sing?"

As the music starts, Jimmy starts the chant of Sweet Caroline, all the boys joining in loud, standing on chairs as the door opens and

Warwick Mathews walks in with a Thai girl. Calla looking nudges Mark. "Fuck me, Marko, there's that wretch of a safety officer with Suzie Wong!!"

Warwick walks over, clapping his hands, singing along to Jimmy's chant and squeezing past Mally into the bar. "Hi guys, bit late for a school night isn't it!!?"

Davey looks down his nose and decides not to say anything as Calla walks over. "Warwick Hunt, the site safety officer!! Areet, here let me buy you and Suzie a drink!"

Warwick doesn't get time to say anything as Calla jumps in again. "2 Tom Collins for my 2 friends, Rosie and make them a double."

"I am driving so I can only have one," Warwick tells the lads, "and I'm only buying you one, y' balloon."

Calla hands the glasses to the couple as Mark hands the girl the song book. "There you go flower, get a song picked out!"

As she walks off looking at the book, Kris chips in: "Endless love, 5 dollar more, by Lionel 'I can dance on the ceiling' Ritchie is in there!"

Davey gives Kris the finger on lip signal as Jimmy finishes off his song to a massive applause.

Michelle gets off her stool and tells the lads they are heading on if the boys want to join the 4 of them and Gordon. Calla spits his drink in his pint. "Gordon!! Ha-ha, fucking hell, Hags you sound someone important!"

Hags straightens his collar cool as a breeze. "Aye son, I bet your father wished he was impotent having a bairn as ugly as ye!"

Calla leans against the bar smiling. "Your daughter never thought that when I–" Davey kicking him in the shin just in time.

The brunette had her jacket on and said to the girls she was going to put some make up on. "Howld on there, gorgeous, lipstick I presume? Give me it here!" Calla's big stride over to the girl taking the lipstick off her with a smile. "You see, I do this so well. Pucker your lips, darling."

The girl leans forward and everyone goes quiet. Jimmy leans past Mally. "What's the daft bastard up to now!?"

Mally, miserable as ever, replies: "Fuck knows but it'll end up in tears!!"

Calls steadies his arm, holds the back of her head firmly and from the left draws a huge moustache over her lip and stands back as she struggles. "Now!!" he says. "You look as fucking ridiculous as Hags does with them bastard cowboy boots on!!!" as the bar erupts and a half a lager covers Calla...

Calla, without even wiping the lager off his face, takes a drink of his pint and smiles. "I suppose a kiss goodnight is out the question now, raggy tash," as the remainder of the lager is thrown at him again. The bar is in fits of laughter as the girls and Hags walk out the door. Davey brings a beer towel over to Calla laughing. "Here you are son, wipe your bracket!"

Calla wipes his face and turns back to the bar where all the lads are grouped at this stage. "What a fucking waste, she could've threw pop over me! Fucking nee respect for beer, some lasses!"

Keegs finishes his pint. "Right fannies, are we going to this wine bar or what? That blondey one is mad for a bit of Glasgow sausage!" Kris and Mark both neck their pints, Kris checking his hair in the mirror. "Someone's getting it the neet, I'm sick of wanking!!"

Calla ordering another pitcher of lager tells the lads, "I was brushing my teeth this morning and when I looked down I had a wankerchief stuck to one of me plates!!!"

Jimmy retches. "Calla, you're a fucking dirty bastard!"

Calla laughs. "Oh aye, Jimmy, I forgot you don't wank these days."

Davey jumps in straight away. "Aye Calla, Mally does all that for him!"

All the lads are laughing as Mally stands up. "Well I can see the waves of abuse are just going to get higher so I am out of here chaps!!"

Willy puts his arm around Mally. "Ah c'mon, bigun, just one mare!! Little voddie for the road and we'll get a pizza on the way yem."

Mally is not taking the bait. "No way, Willy, one more!! I have fell for that before and the morning is not pretty!!"

The DJ gives the last request shout over the mic as Warwick and his Thai bird get up to sing 'Up where we belong' duet, Calla shouting, “ Tonight, Matthew, Warwick Hunt is going to be Joe Cocker and Suzi Wong is going to be Jennifer what’s her fyace!!”

Just the 4 lads left as the bar quietens down, Jimmy thumping Calla in the arm as he comes back from the toilet. “Jimmy, one of these days I’m gunna knock you right out!” Jimmy chuckles. Calla heads to the toilet as the music stops and a few claps echo for the absolute shit version of a classic. Davey shakes his head. “That was fucking shite. I could’ve sang it better with a dust mask on!!”

Jimmy giving the ‘one more, knife across the throat order’ signalling one more then that’s it as Rosie walks over. “Last orders, mind, guys, some of us have been working all day!”

Willy takes his glasses off and moves towards the bar with a smile as Jimmy laughs. “Willy is going in for the kill, Davey with Rosie, the specs are off, hammer!!”

Davey looks back. “Aye she will need the cunts on if she does owt with the ugly little twat!!”

Calla gets back to the bar seeing more beverage on the counter. “More drink!? Get in!! I thought we were done! I’ll definitely piss the settee the neet!”

Davey gets a fit of giggles. “Jimmy, do you remember that time you locked your car keys in the bedroom and Calla went out on the slate roof in the pissing rain to get them?” Jimmy and Calla listen on, laughing, as Davey continues: “The houses were old pit houses and the slate roofs joined two houses. Daft arse there got on the roof to shimmy along from his window to yours, Jimmy!”

Jimmy chirping in: “Aye I went down in the backyard and held a clothes prop up so he could get a howld of it, he had nowt to hold onto! He was gripping the wall with his nails and the slates were damp!! I could see the stupid bastard’s feet slipping so I ran back upstairs!!”

Calla was pissing himself laughing! “Aye me feet slipped, there was nothing I could do, man! I went into the skier position and screamed as I went off the roof heed first into the coal house!!”

Davey taking a big gob full of lager wiping his tears. “Aye Jimmy comes into the bedroom asking if he fell. I had watched the silly twat ski off the edge!! Aye, a said to Jimmy, he’ll be ok, the ground broke his fall!!!"

Jimmy chuckles away. "You know the frightening thing? The stupid twat came up the stairs face all black off the coal bunker he went into and said I'm gunna tak off me trainers, Jimmy. Honestly, I'll not slip in me socks!!!"

All the lads are laughing as Warwick's missus comes over. "Hello guys, Warwick has stormed off. We had an argument so I was wondering if you have any cab numbers?"

Calla's eyes light up. "The best bet for you, darling, is to get one from mine."

Jimmy and Davey leave Calla trying to sweet talk Warwick's bird and head off. "Mak sure you lock the door when you get in, Calla," Davey shouts back to Calla who is sitting at the bar trying to persuade the pretty little thing back to his place.

"Look you can get one from here but they always take ages. I'll even make you a nice cup of coffee."

The girl smiles. "You don't even know my name!?"

Calla sits back on his stool. "Fair point, hinny, I would need it for the taxi driver! Go on then, what is it!? Oh and I'm Andrew by the way."

Willy is chatting away to Rosie behind Calla, seeming to be getting on making her laugh. "Ah haway Rosie, I've got a nice bottle of red there and Hags is away clubbing, we'll have the place to ourselves with a nice DVD!"

Rosie leans over the bar. "Calla, do you want one last order before I lock up?!"

Calla turns round. "Aye get me four Belfast bombers and 2 packets of salty crisps!"

Willy shakes his head. "Last of the fucking romantics, you, Calla."

Calla chirps back to Willy: "Howld on Elvis, at least I am only taking Suzie back to get a taxi, you're trying to woo our Rosie in the sack with a bottle of QC sherry and 4 episodes of Cracker – you're some smooth talking cunt!"

"My name is Lai me," says Callas prey, Calla trying to keep a straight face. "Well that's a lovely name that pet."

Willy laughing his cock off behind Calla. "Aye big man you can re phrase that straight away for yasel! Me Lai!" Both of the lads burst out laughing and chink their glasses together.

After half an hour of persuasion Calla, Willy and the 2 girls head back to Willy's for a nightcap, Calla linking arms with Lai me as Willy links Rosie. Calla eating his second packet of crisps held to his mouth pouring the crumbs down his neck, says, "Willy, is there any bait in yours, I'm fucking clamming! I could eat a fried tramp!"

Willy walking behind Calla replies: "Your lass will be deein that in an hour's time once them Belfast bombers kick in!"

Calla turns round. "She'll be deein what?"

Willy and Rosie laugh. "Eating a tramp! You, ya scruffy cunt!!"

Calla turning back to keep the march on. "Haweh, nee need, I got a wash last night."

Back at Willy's the iPod goes on straight away, Willy bopping around. "Bit of Del Amitri, lasses, I'll enrich your musical taste buds with some quality tunes!"

Calla is in the kitchen trying to open the wine with the bottle between his feet and a spoon.

"Give me the bottle, Calla," Rosie shaking her head as she takes it off him. "You need minding like a kid." She carries on as a bottle opener is produced from the drawer.

"Aye, you're right there, Rosie, how do you fancy tucking me up with a bed time story tonight!!?" Calla chances his arm.

"I would love to but Willy's feelings would be hurt. Let's try and leave together and we'll see."

Calla rubs his hands together. "Happy days, I reckon I'll be tired in 20 minutes."

Rosie laughs as she pats his arse on the way into the sitting room. Willy and Lai me are getting on famously as Rosie winks at Calla and he hands them both a half glass of red wine. "There you go, Jesus, Willy, you and Me lai make a smashing couple!! Don't they, Rosie!"

Lai me holding Willy's hand says: "I like William, he makes me laugh."

Calla sits down on the sofa.

3 bottles of wine later Willy is out for the count with Lai me fast asleep next to him on the sofa. Calla and Rosie are upstairs

and already in the heat of passion, the lamp falling off the cabinet smashing on the floor! Willy wakes up to the commotion and gets his glasses out of his pocket and lights a tab. "Fuck me, ma heed!" He looks down and sees Lai me asleep on the sofa, he puts his hand on her breast and gives it a feel. "Aye, nice tits, hen," and jumps up to walk upstairs to see what the commotion is. On opening the door his tab drops out of his mouth in disbelief as Rosie is riding Calla. Calla leans round to see Willy standing there with a surprised look on his face "Oy oy, pal," as Rosie tries to pull the cover over her and Willy slams the door and storms downstairs as Calla carries on regardless.

Alarm clock at 6 wakes Calla, and Rosie is asleep still so he gets up and walks down the corridor scratching his balls and sniffing his fingers, smiling just as Lai me walks out of the bathroom naked, covering herself and pinning herself against the wall as Calla walks past without breaking stride. "Morning Lai me, canny neet eh?"

Willy opens the bathroom door and Calla is sitting on the pot having a shite. "Alrighty Willy, gan get me some bog roll, hammer, or I'll have to use the bath mat."

Willy puts his hand over his mouth then vomits in the sink next to Calla. "Calla, you smelly bastard!!!" Willy wipes his mouth and retches again as he runs out of the door, coming back 2 minutes later with a tin of lynx and a toilet roll. "Here are, Calla."

Calla takes the roll off Willy. "How did you get on last neet, Willy?"

Willy stripping off and jumping in the shower. "Aye Calla, I ripped the liver oot of Warwick Hunt's lass cos you were poking fuck oot of Rosie in ma bed!!"

Calla laughs as he uses any razor and foam to get a shave. "Ah Willy, I didn't mean to. I fell asleep in your bed, pal and woke up and Rosie was riding me. Is that rape!?"

Willy pulls the shower curtain back with soap all over his face. "Rape? I will show you what rape is when we get to work and I do fuck all for 12 hours and get 300 quid!!"

Calla walks into the bedroom to find Rosie already gone, Willy walking into the room, whipping Calla's arse with a towel. "Where did she go, Cal?"

Calla rubs his arse cheek. "Fucking bitch just used me. I was after a cuddle and a kiss this morning. Here, Willy, give me a clean pair of scuddies and socks will ya, oh and a t-shirt."

Willy getting ready replies: "Calla, you only live 2 doors up, you fucking goon."

Calla laughs. "Aye Willy, but I like it when I get a bonk on wearing your tight underpants at work!!" =

Willy opens the drawer, pointing and walks downstairs lighting another tab.

The kitchen smells of toast and coffee, cereal poured with milk and juice on the table with Lai me in Willy's t shirt smiling. "Morning William, you eat breakfast please."

Willy gives her a kiss on the cheek and a slap on the arse. "Fucking fair fucks to ya, hen."

Calla walks into the kitchen all smelling nice, shaved and rubbing his hands together. "Well fuck me, breakfast! I chose the bastard wrong one!! Mine gets up and leaves skid marks on the sheets and Willy's gets up and makes the breakfast!!" Calla sits down taking Willy's buttered toast off his plate and stuffs it in his mouth, washing it down with coffee.

The bus pulls up outside the house and blasts the horn, Davey letting the window down. "Jimmy, these two cunts will never be up!" Just then the door opens and 2 fresh faced, new t-shirted lads walk out, Willy getting a kiss on the cheek off his new lass while the rest of the lads look in disbelief.

Calla opens the sliding door, kissing Davey on the head and slapping Jimmy on his. "Morning, fanny chops!!"

Willy gets in behind Calla, taking Jimmy's tab out of his mouth and sits down. "Drive on, Jeeves." Both Davey and Jimmy say nothing but shake their heads in sync.

Halfway to work before the first word is spoken Kris, Jonny and Mark all dying off a heavy night at the club with Hags missing in action. Davey leans back. "Where's Hags, is Keegs bringing him in the Keeg mobeel!?"

Calla, with his head against the window, says: "Well he never came back to the ranchelero last neet."

Jimmy flicks his tab out of the window. "Well what happened in the night club? Are the 3 stooges awake or what!?"

Mark lifts his baseball cap up. "Last thing I remember was drinking them tequila slammers! That Michelle was all ower Hags, the lucky old bastard!!"

Willy smiles and nudges Calla. "A hatrick of gold for team GB."

Calla sits up straight and leans against Davey's seat. "If you two old cunts started enjoying life away from them 2 stools at the bar you wouldn't need to ask what we got up to!"

Jonny shouts down the bus. "Why like Calla, what did you and Glasgow Elvis get up to, bit of bum fun!?"

Willy laughs. "Aye ma ring piece is fucking in pieces!! Don't know if that's off Calla's boaby or the 2 bottles of red wine and 2 dozen chicken wings!!"

Jimmy chuckles in the front. "Right lads, plan of attack today! Sunderland are playing Newcastle tomorrow at 6 am our time so if we get those VNBs hooked up and a test late tonight we'll have tomorrow off – I'd say it'll be a 16 hour shift!"

Calla shakes his head. "I'm as rough as a badger's arse and I've to work 16 fucking hours!!"

Kris pipes in. "Jimmy, fuck the match, I couldn't give a fuck!!"

Davey steadies the ship. "Look lads, we are going to get 24 hours' pay for 16 hours' graft."

Willy wangling his old bartering tactics comes in with: "So Jimmy, if we finish at say 6 o'clock the neet the deal is the same?"

Jimmy turning into the site replies: "Aye, there's nee chance you'd get all that done but aye."

The van pulls up and the lads jump out to see Hags talking to Michelle outside the office, Calla shouting over, "Morning, how's beauty and the beast!!?"

Willy laughs. "Aye, Hags, I had a lonely night last night – I had to ask Andrew around to keep me company!"

Hags smiles over, not fazed at all. "You better not've touched my red wine ya pair of bastards! Good stuff that!!"

Calla galloping on his way into the office joins in. "It's now touching the cloth of Willy's underpants. Hope nee one's in the trap!!"

Hags grabs Willy's arm. "Did you let him drink my red wine!"

Willy shrugging Hags off, replies, "I had nee choice pal, it was either that or let him try your Armani suit on!"

Hags is going red with rage. "He wasn't in the wardrobe, was he, Willy!"

Willy laughs. "No, that's why I got him pissed, man!"

Hags says his goodbyes to Michelle and walks down to the site with Willy.

"Hags, we have job and knock today, I have all the spools pre fabbed and big Jimmy doesn't know!! We'll be in the bar for 2!! Where's Keegs!?"

Hags points over at the car. "Engine's running, he's fast asleep – was riding the blonde one last night!!"

Willy laughs. "Four golds for team GB in one night, I'm going to phone fucking Roy Castle!" Hags half understanding what Willy meant.

Davey, Kris and Jonny are already lashing pipe in as Calla walks into the hive of activity. "Davey I think my arse is honestly burst!"

Davey with a pencil in his mouth replies: "Your fucking nose will be if you don't hurry the fuck up and get that pipe into the sky."

Calla throws his hi vis vest on the bench and puts a harness on.

***

Jimmy puts the phone down and shouts Mally in. "Mally, get the paperwork over to the client with Keegs for a test later on."

Mally is smoking his tab. "Jimmy, there is no chance at all we will do all that before tomorrow. It's dinnertime and the lads will be up soon for an hour."

Jimmy shakes his head. "Mally, that's 5 hours they have been down there now; let's take a wander."

Keegs sticks his head in the door. "Alreety lads, client happy with paperwork. Willy reckons test will be on by 3, 2 hour hold and signed off by 5 all going well."

Mally looks at Keegs. "Are you on drugs, Trevor, or what?"

Keegs laughs. "I must be to come in to graft and listen to your pish!!"

Jimmy and Mally walk down to the site. All the lads are busy. Willy marches over. "Just got to hook it in upstairs and that's us!"

Jimmy smiles as Mally looks above his head, Mark shouting over: "You lost them pigeons again, Mal."

Jonny walks through the door. "All done up top."

Mally shaking his head walks off with the phone to get the client, Calla shouting down through the roof, "Get me pie and chips with brown sauce, you horrible cunt!"

Davey laughs. "Jimmy, we'll be in the battle cruiser by 7, marra."

Jimmy walks off down site smiling, with 2 paper cups stuck on his white hat.

Tommy and Chuck walk down from the live building into the new site to find all the lads standing around. "You guys on strike or what?" Chuck walks over shaking hands with Davey.

"We are officially off for 24 hours now, cut a deal with the big fella and everything has gone tickety boo, so we are done."

Kris is wrapping everything up and locking the box. "Where you two off? Nelly's?"

Tom shakes his head with a smile. "Nope, we are heading to South Carolina, boys!! Myrtle beach!! Paradise for golfers!!"

Davey shakes his head. "You jammy bastards, when did you book that!? What's the script?"

Chuck opens his bag pulling the brochure out. "Only booked the mother fucker yesterday, boys, special anniversary deal!!" All the lads are admiring the beautiful scenery and facilities. Tom continues, "Yep, flights, 2 nights and one round of golf for 250 bucks!"

Mark looks through the brochure. "Which hotel? Bet it's like Fawlty Towers!!"

Chuck shakes his head. "Marky, there are no shit hotels in Myrtle. We are staying at the Hilton."

Calla is sitting on the metal bench. "Davey, when do the wages gan in?"

Davey looks at his watch. "They will be in now, son – why?"

Calla jumps off the bench. "Fuck Nelly's, am gan to Myrtle

beach!!"

Davey smiles. "Are you fucking joking me or what?"

Calla opens his wallet and takes out his cash card. "Meet Mr Plastic Fantastic!! He's taking me away for the weekend!!"

Kris and Mark laugh. "Calla, you're mental. Jimmy will never let you go!!"

Calla looks down his nose. "Unless he has a fucking machine gun he's got nee chance!! We get there the neet, we are off the morra anyway and we fly back Sunday! I only miss half a day!!"

Willy walks through the door from upstairs with Hags. "Right, baw bags! It's a wrap. We are off to see the wizard!!"

Kris shows Willy the brochure. "Chuck and Tommy are heading here tonight for 2 days. Calla is going there as well."

Willy squinting through his jam jars. "How much like?"

Chuck shows him the special deal receipt. "250 Bucks, buddy, flights and hotel!"

Willy looks at his watch. "Davey, when does the wages gan in?"

Davey smiles. "They're in."

Willy rubs his hands together. "FOUR!!!!! I'm going too!!" Calla and Willy hug like long lost brothers.

Calla puts his arm around Davey. "Haweh, father, come to Turtle beach!"

Davey shakes his head. "It's Myrtle, you balloon! If Hunter says it's sound I will but I don't want to rock the boat, we have to keep some lads here!"

Just then Keegs and Jimmy walk through the door into the group of want-away lads. "What's going on here, boys, you went very quiet!? It's got nowt to do with the wages – they are in plus overtime shortfall!!"

Davey moves Mark out of the way. "That's exactly it, Jimmy, the wages are in!! Tommy and Chuck are away to Myrtle beach, 2 nights for 250 bucks!"

Jimmy sighs out loud and leans his arse against the bench. "Who wants to go like?" Then shaking his head: "Calla!"

Davey nods with Calla walking over. "Howld on I just said we are back Sunday dinnertime – it's only half a day away"

Jimmy laughs. "Back to Orlando Sunday dinnertime, Calla, back to graft Monday!!"

Calla smiles. "I'll come straight from the airport, Jimmy."

Jimmy smirks. "Aye right!!"

Davey shows Jimmy the brochure. He flicks through it. "So who doesn't want to go?"

Hags, Mark and Keegs all shake their head, Keegs taking the brochure. "No interest, Jimmy, I've to graft tomorrow anyway to get the paperwork done."

Mark and Hags agree. "Aye, fucking hate golf! It's for faggots."

Jimmy stands up. "So that's Hags, Mark, Keegs and the 10 local lads on site, with Mally."

Davey puts his denim shirt on. "So you don't mind me and Calla going with the 2 Texans, marra?"

Jimmy takes his phone out of his pocket laughing. "Me, you and Calla, hamma!"

Willy smiles in Jimmy's face. "Aye and me bigman!!!" As Jimmy hears the voice down the other end, "Michelle, ring this number and book me 4 flights tonight to Myrtle returning Sunday mid-morning, with the hotel and golf deal! Aye, me, Calla, Willy and Davey." Jimmy winks at Dave.

Just then Jonny walks in the door, Calla seeing him first. "Here he is, the new man! No drink, working with the client on design but still an ugly cunt!"

Jonny laughs. "Aye, saving a fortune and making a fortune, Calla, something that you will never do!!"

Calla laughs. "Av changed, man Jonny, after I get back from Turtle beach this weekend am stopping in."

All the lads laugh. Jonny picks the brochure up. "How much?"

Calla tells him the full deal. "Davey, when does the wages gan in?"

Davey laughs. "Am I the only cunt that knows that on this job?"

Mark chirping in: "Dave, we all know that you check the on line banking as soon as it gets to 12 o'clock in England!!"

Davey scratches his arse. "Aye and I'll be gan to the ATM before wor lass n all or I'll never get a pint in Myrtle beach!!"

Calla pushes Jonny. "Are you coming, Jonny or what, you miserable bastard!?"

Jonny is thinking about it. "Aye alreet count me in!!!"

Jimmy thanks all the lads for grafting to get the job done and wanders off shouting back, "Davey, the happy bus leaves in 20 minutes!!"

Calla shouts over to Chuck, "What time do we fly, Chuck? Have we got time for a pint in Nelly's?"

Davey looks at him, "Calla, it's an 8 o'clock flight – we just have enough time to pack our bags, ya muppet!"

Calla, Willy, Davey and Jonny are sitting in the bus as Jimmy climbs in. "I've got all the itinerary, lads. Jonny, I booked you as well. I want the 250 a man when we get to the airport or you don't get the ticket! Right, Calla!"

Calla leans against the window with his legs against the chair in front. "Jimmy, you will have the cash for fucks sake, it's not as if I haven't got it!! I owe nee one nowt and I have been paid!!"

Jonny reminds Calla: "Apart from the bairn like, a mortgage and a bit of maintenance!!"

Calla winces. "Oh aye, direct debits look after that!! Set them up before I came out!"

Davey is in shock. "Fucking hell, Andrew, getting all grown up, eh!!"

Calla smiles. "Aye, wey a cancelled them last month cos I needed all the money for owa here."

Willy laughs his head off blowing smoke everywhere!! "Ah, Calla you're some man for one man!"

Jimmy gets towards the digs, Calla and Willy sound asleep, heads against the window. As the indicator goes on Calla wakes up. "Drop me at Nelly's, Jimmy, I need a mender!!"

Davey pointing forward says, "Fuck him, Jimmy, tak us yem first, he'll have me packing his bag if we don't."

Calla laughs. "Aye, father, that's what I was going to say, anyway. I have to pack my golfing clothes, so drive on."

Jonny laughs. "Golfing clothes?! Like what golfing clothes do you have like!?"

Calla takes offence. “Whoa whoa whoa, fanny chops, av a Lacoste t shirt!! It’s a real one n all”

Willy laughs. “Looks a bit stupid, though, Calla, with jeans and Adidas sambas!!”

“Right lads, 6 bells outside.” Jimmy pulls away leaving the instruction.

Calla walks in the digs behind Davey, straight into the kitchen and opens the fridge. “Mm come to daddy, my little fizzy friend!!” He opens the cold can of Budweiser and gulps it down, burping out loud as he pulls a block of cheese out with the other hand, putting it on the bench and cutting a block off with a knife out of the sink! Still taking another massive swallee of lager he burps again putting the house brick of cheese on a slice of bread and opening the cupboard. “Fatha, where’s the high pressure!!!!”

Davey from upstairs shouts back. “It’s in the cupboard under the microwave.” Calla gets the HP sauce out and lashes it on his sarnie as he walks upstairs.

***

Jimmy pulls up to find Willy, Jonny and Davey all polished with their bags. “Where’s daft arse?” Jimmy shouts out the window.

Davey climbs in the usual co driver’s seat “He’s at Nelly’s, I said we’d pick him up! I’ve his bag and passport so nothing to worry about.”

Jimmy pulls the happy bus up outside and blasts the horn, 10 seconds later the door opens and Calla bursts out with 2 slabs of Kronenbourg and a sleeve of nuts!! “Oy Oy, golfing pals!!”

Jonny opens the sliding door and Calla drops all the booze on the seat! “Howld on, I’ve 7 baseball caps and 7 polo t-shirts.”

As he runs back in –”Nee wonder he’s got nowt” –Davey lights up a tab as Calla runs out with his t-shirt gone, a new ‘Nellys’ one on and a cap!! – the rest in a plastic bag for the lads. “Right ee o, Jimmy ya fat bastard, chocks away!! We are gan on owa hols!!”

Willy gets his t shirt and cap out of the bag laughing. “Calla, 48 cans, the fucking airport’s only 20 minutes away!!!”

Calla opens a tin, smiling. “Don’t you just love that sound, Willy!!”

Davey leans back. “Well, 48 cans, are you alreet!”

Calla lets out a massive burp. “Aye, 24 for the trip to the airport and 24 coming back Sunday!!”

Jimmy shakes his head. “So much for gan straight to work.”

The bus rocks on towards the airport with Willy and Calla singing Sir Cliff’s classic, altering the words. “We’re all going on a golfing holiday, we’re all hoping for a hole in two; drink and bucking on our golfing holiday, no more welding for me and you, for a day or two!” The bus is awash with cheers and beers, Willy squeezing Jimmy’s neck and passing him a can –”I’m driving, silly bollocks!”– pushing the can down –”can yee not multi task big man?”

Willy sits back down and throws Jonny the can. The bus pulls across two spaces in the short stay car park, all the lads jumping off, empty cans rattling off the floor, as the seven stand there stretching out.

“Fucking hell, Jim, looks like the van fell out the sky!” Jonny points at the pissed parked bus.

Jimmy slams the rear door shut and throwing his bag over his shoulder blips the central locking. “Who the fuck are you like, the Stig?”

Once inside the terminal the lads queue up at the self-check in console. “Eer how the fuck do ya gan on with this?” Calla rubs his hand across the full touch screen as an assistant steps up.

“Please sir, you need to scan your passport in the slot first!” She takes Calla’s passport and leans in to scan it as Calla turns to the rest of the lads winking while pretending to ride her.

“Mr Callaghan?” the stewardess cuts in.

“Or ayy, that’s me, hinny,” getting by as he continues to follow the screen prompts. “Here, gorgeous, what’s this asking for now?” putting his arm round her pulling her in.

“Well sir, it’s asking if you want to travel in a better class? $75?”

Calla drops the bird like a hot snot and jabs the yes button to the wiry sound of his boarding card being printed. “Woo hoo, lads – first class! $75,” while punching the air.

"Calla man, that's on my card and we're only in the sky an hour and a half?" Jimmy barks back barging forward.

"Or Jim, we'll give you it back, man, it's only 50 bob!" Willy replies thumbing his glasses onto his face as he picks up his bag, Davey with his deciding words: "Ay, Jim, haweh there's away in style."

Big Tom sniggers – "Y'all don't think this is anything like 1st class, do ya?" – laughing hard –"these seats recline a bit further and you get an extra couple of inches and a glass of wine thrown in but that's it!"

"Free wine, wayy it's worth $75 for that!" Calla barks, marching for the terminal.

The lads are all stood waiting at the gate as the stewardess begins taking passports, a real darling tall blonde with huge tits as Calla barges a few Americans out of the way. "Excuse me, lads, 1st class coming through," winking at the unimpressed woman –"off to Turtle beach, love, improve me handicap, you know how it is!" an exhausted Calla spits, handing her his ticket.

"Yes sir, I know where you're going – I'm assisting on the flight!"

"Quality, well you can assist me and my good golfing associates back there to get rather merry with the free wine in first class!" He snatches back his stub and with the biggest, cheesiest smile marches down the tunnel like a kid heading towards a toy shop.

"Don't mind him, love, we have him for the weekend – the special school has put him in my care for the weekend, I'm part of a help the disabled group!" Willy snides taking his stub back from her who is fighting back a laugh.

On board the usual pushing and tugging has gone with the extra space given in the better class. "It's nee Darras Hall but it's better than riding back there with the goats n hens." Davey flops onto his seat and stretches out beside Jim. "Ay suppose for 50 sheets it's worth it, keeps daft lad owa there happy!" nodding across to Calla who is trying every combination the chair has to a disgruntled steward who in the campest voice says: "Sir sir SIR, can you please bring your chair forward at once!"

Slamming his hands on his hips, Calla not impressed, squares the chair off and in his best camp voice replies: "I'm so sorry, handsome, I haven't been on an airplane before, I normally calm down with a nice chardonnay!" winking towards the steward who is blushing.

"Well sir, let me get everyone and I will see what I can do!" He disappears beyond the curtain, the lads all together go "Oooooooo, Andrew, stop it!"

Calla laughs. "Eer, ad tickle anyone's arse if it meant getting an early scoop!"

Once in the air the steward pops through the curtain to the sound of Willy shouting, "And tonight, Matthew I'm going to be!" –Jonny barking back, "Boy George!"

The lads all sniggering as the steward hands Calla the small bottle of wine and glass then walks back up the plane. Calla pouring the wine and reclining his chair takes a long slurp with his little finger out. "Ahhhh that's a fantastic Australian chardonnay, would go well with a nice Ritz n cheddar," the rest of the lads all reaching for the service button.

The round of drinks is quickly followed by the in-flight meal: spicy chicken rice and a trifle for desert –"fucking sound this like"– as Calla rives open the foil lid of the meal and immediately throwing the full chilli garnish down his neck which he nearly chokes on because it's that hot. Grabbing his trifle he spoons the full dessert down his neck, Willy turning slightly with a piece of chicken on his fork and a puzzled look on his face tilts his head, looks down over his glasses and quizzes: "Ynee want ya chicken, big man?"

Calla fans his tongue like it's on fire. "Noooo! Fuckin hell, that chilli! It's bastard scadza!"

Davey chinks glasses with Jimmy. "Fair play, Hunter, you turned out ok! Kind of expected you to let the lads away, being we are ahead of the game."

Jimmy smiles. "Aye, you can never forget your roots, Davey. The amount of times I turned in late pissed or missing due to shagging and you never once gave me a hard time! Maybe grinding pipes for a few hours like but that's all."

Davey laughs. "No more of the motorbike for the both of us with a bit of luck! Been round the block on that Yamaha more times than Barry fucking Sheen!!"

Calla and Willy laugh their heads off. "Here, Jimmy," Calla

shouts, "I went back to Willy's the other neet and put a mouse trap in Hags' inside pocket on his Armani jacket!! Willy's gan mad."

Willy is beetroot from laughing. "I hope the bastard has it on this weekend if he takes Michelle oot!!"

Jonny spits his wine out, laughing. "There's nee chance of his fingers getting caught, Willy, you have to put your fucking hand in your pocket for that!!!"

One more round of vino for the lads as the captain announces descent as Chuck comes back from the toilet. "So Jimbo, we get picked up off the hotel limo, straight in and then out?"

Jimmy gulps his wine down. "Why change the habit of a lifetime, hamma?"

Calla presses the call bell and Boy George comes out from the curtain. "Yes sir?"

Willy interrupting: "Would there be any chance of one mare drink, handsome?"

The camp steward toddles off for more drink which the blonde hostess brings back. "Fuck me, what a change, Matthew – he went in Boy George and he came out Madonna!!" Calla shouts as the girl puts the drink down asking where the lads are staying.

"We are staying at the Hilton, darling. Any good bars there we can buy you and your pals a drink later?" Willy is straight in, no messing. "Well, there is Sinatra's and I am heading there with my sister later so we'll maybe see you there." As Jimmy and Davey look on smiling, Willy and Calla return the smile back, lifting their glasses. "Hole in one" they both chant at the same time.

In through passport control and to the baggage counter. "Where the fuck's Calla?" Jimmy nudges Davey.

"He's got his bag, man Jimmy, while you were at the toilet and away through."

All the lads come out the door to the Hilton minibus, Calla in the back with a bag of cans. "Calla, we can wait a bit for a drink y'na." Jonny shakes his head as he gets in.

"Shut the fuck up misery guts!! I've never been on a golfing holiday before – I'm excited!!"

Jimmy climbs in the front with Davey. "Who's are them golf clubs in the back? Titleist pro – must be the hotel's!"

Calla laughs as he drinks his can. "Are they fuck, they are mine! Some cunt left them next to that sports equipment check in desk!!!"

Davey looks over the back towards Calla as Call stops the can from his mouth. "What?"

"Calla, get ya arse back with them sticks, lad! I mean it, that's bad craic!"

Calla puts his can down. "Davey, you're too honest, you like, bet the cunts fucking loaded anaal!" as he climbs out with the clubs and trudges across the road towards the airport. "Where's silly bollocks going?" Jimmy quizzes.

"Needs a piss!" Davey's a bit embarrassed.

"Could he not've nipped it like, for fuck sake!" Jimmy slides in.

All the lads are waiting as a spritely Calla jogs across the road with a smile on his face, jumping in beside Davey. "Home James," he barks to the driver.

"What you smiling about cunty, what ya done? Where's the sticks?" quizzing Calla.

"I gave them back. The bloke was inside having kittens – apparently he has had them modified or summit coz he is a stumpy cunt!"

"And?" again Davey quizzes.

"Well he saw me with them and ran over barking at me! told him to settle down, I was a hero, told him a lad I work with took them, the one with the specs, I told him it was wrong and returned them!" Calla is smirking like a Cheshire cat.

"And why are you smiling?" Davey asks, knowing there was more to the tale.

"Well, right," Calla shuffles round excitedly. "Apparently this old codger has a huge estate 10 minutes from our hotel and he is throwing his daughter a birthday party! BBQ, free lepp, everything! Invited us along! Well, except Willy!"

Davey sighs and sits back. "Way lad, give owa! We're not up for going to some fancy pants do!"

Calla is bursting with excitement. "Davey man, he said we must go, it's his daughter's 30th! I said we would pop in there after the golf.

Haweh man, bit of cooked meat, few cold scoops F O C, then out on the tiles, mebbies with his daughter if she's not a hound, hahaha!"

Jonny butts in: "What's this!"

Davey replies: "Well dopey hole has been invited to some rich bloke's party the morn after golf!"

Calla chirps in: "No, all of us! Told him we were away on a company all-paid trip for our hard efforts, told him our gaffa Jimmy was paying for it!"

Jimmy spins round. "I will fuck, lad, you're lucky to be getting away for a day, dint tak the piss!"

Calla shakes his head. "I'm winding, man ya fucking minjbag, ay and it's his daughter's 30th, free scran, free booze!" Calla snaps getting even more excited.

"Fucking quality, Calla sonny, count me in!" Willy rubs his hands.

"Ay speck face! You can't come, said he didn't like the look of you, said you looked like a tea leaf!" Calla laughs. "Eyy come on! I'm nay thief, I'll talk him round, hen!"

Calla looks at Davey and sniggers as the limo rolls on.

As they pull up to the front of the hotel Calla jumps out smacking the legs of the guy about to open the door – "Soz helmet!" – as he looks in awe at the hotel as Jonny gives him a dig in the ribs. "Oooowf, ya cunt!" as the lads troop into the hotel, Jimmy bringing up the rear sticking a $10 in the driver's hand.

Inside the lobby Jimmy goes to the front desk to check the lads in as Davey lets out a laugh.

"What's up with you, Dave?" Willy asks.

"Just had a text from Stevie the spark – apparently the yanks have pulled the plug on the work they were getting, got a cheaper price locally; says he is foaming, asks how we were getting on? Told him that he isn't missing much – we have been dragged away for the weekend to play golf at Myrtle beach! He will be screwing; he plays off 5, the sad twat!" All the lads laughing.

Jimmy trudges back reading the itinerary. "Right lads, rooms: we have 3: me and Jim, Chuck and Tom, and Calla, Jonny and Willy have a 3 bed clubhouse. They double booked the rooms and are fully booked so they upgraded you, ya jammy twats! Was thinking of making you top and tail and me and Davey having it but I didn't want you crying!"

Willy snatches the key and with a whoop heads for the lift. "Howw, shit for brains, it's a clubhouse! It's down past the pool near the course apparently they are normally $1500 a neet!"

Calla punches the air. "Fucking double whoop, that's what I'm talking about!" picking his case up and doing a shimmy and a shake towards a confused and flustered old lady before spinning and doing the moonwalk towards Willy who is also doing a jig, Jimmy looking round at the hotel guests looking on in horror

"Well lad, what we let ourselves in for here mate!" Jimmy shakes his head and puts his arm round Davey and heads to the lift with Tom and Chuck laughing behind, still looking over their shoulders at the lads, Jonny doing the hula, Willy riding a bucking bronco and Calla now on his back spinning round. "Hey Chuck, those dudes are the craziest motherfuckers I think I have ever seen!"

Opening the door into the suite Willy has no reaction as the 12 ft sliding glass windows look onto a private pool with the golf course and bar are in the distance. "Me ma's council house is better than this."

As Jonny walks in singing "hey diddly dee, a welder's life for me," Calla is straight over to the open plan kitchen opening the fridge: "Unlocked and loaded!!" He pulls three cold tins out of the fridge and walks over to the patio where Willy is having a tab. "Here are, sad sack, I suppose your mam's back garden has a better view n all, does it!"

Willy laughs. "Here Calla, there's that Sinatra's bar the hostess was on about," pointing over the pool to a swanky looking bar, lights just on with the sun setting. Jonny wiser than the 2, already picking the best bedroom, looks over the top of the lads. "How tweedle dum and tweedle dee, are ya getting polished or what!!"

Calla looks up. "I've only got to wash me balls and brush me gnashers and I'm ready!"

Willy flicks his tab over the fence. "Let's say 10:30pm ready, we'll invite the lads down."

Calla agrees. "Aye I'll ring them now while you go and iron your face, you horrible twat."

Willy walks off to get his bag. "Nee need, bigman, nee need."

Davey answers the phone. "Aye, ok son, see you in half an hour."

Calla sits on the high stool in the kitchen looking in the fridge. "Bring some tins down; there's only 10 left in here – oh and tell the 2 Texan pie eaters n all."

All polished the lads are sitting round the pool when the doorbell rings. The four lads, all shirts and jeans, walk in smelling like the perfume department in the airport. Jonny gives them all a can.

"Swanky pad, lad." Davey takes all the surroundings in. "Has it got all the mod cons?"

Calla sits outside looking very smart, white shirt, pants and a tie on loose. "Aye, father, it has the lot, although you will have to wait for the iron – Willy is upstairs using it on his face!"

Chuck and Tommy take a seat. "Nice. You guys ever seen anything like this in London?"

Jimmy looks at them. "Howld on, how come you yanks think England is London? We are from the North, man!! It's like calling you a yonker!!"

Tom is offended! "God damn, that's as well as calling us a nigger!!"

All the lads laugh. "Bit strang like, Tom, but we understand!!"

Willy comes onto the patio with a bright yellow shirt and white linen pants as Calla spits his drink out and nearly falls off his chair!! "Fucking hell, tonight, Matthew, I'm going to be that fat cunt Christopher Biggins!!!!"

All the lads are pissing themselves laughing as Willy walks back in. "I'm going to get changed! Cunts."

Willy wanders back out with a pair of jeans on and a pink shirt in his hand, just about to pull it on when Calla pipes up: "Eer ya not wearing that bastard, are ya, it's worse than your first effort!"

He pulls it on. "It's fucking Armani, you smoggy cunt!"

Calla gesturing his hand replies: "Dint believe ya, giz a look!"

Willy slipping out of the shirt hands it to Calla who throws it straight in the pool. Willy lunges forward to try and stop him as he looks at the shirt floating on the pool, the lads all laughing as Calla rocks back on his chair with a smirk. "Try again, fanny chops and this time put your gleggs on!"

Willy is foaming but keeping it cool he spins round, swiping Calla off his chair, and marches back to his room leaving Calla lying flat on his back can held straight up shouting at the top of his voice "DIDN'T SPILL IT!"

Willy finally comes out with a small checked blue Lacoste shirt on, tucked into his jeans. "Reet, a couldn't give two fucks what yee think of this – it's all I've got!"

Calla steps forward and straightens Willy's collar then pecks him on the cheek. "That's better, petal, don't be late home."

Willy shakes his head and heads to the door. "Cal lad, you were an abused child!"

The lads all trooping out after him.

Straight out to the front Tom signalling to the concierge for 2 cars as the guy waves in 2 white BMW 7 series across. The lads pile in, Calla, Willy and Jonny in one car, the rest getting in the second car. Calla jumps in the front, shouting over to Jimmy wandering over fastening his belt. "You're in the fat lads car, Jim," then hopping in and slamming the door. "Eer driver, nice skate this like, any chance of a shot?"

The puzzled driver eases the car out and asks where they are going. "Take us to Sinatra's, marra, do you know it?"

Before he could reply Calla ramps up the radio to the sound of the Goo Goo Dolls' 'Here is gone', clicking his fingers and mimicking playing the guitar. "Sounds a fucking good tune!"

The driver adjusts the sound so he can be heard. "Yeh Goo Goo Dolls, great band!"

"The Goo Goo fucking Dolls? They poofs?" Willy asks leaning forward.

"No, sir, well not that I know of!"

The two cars pull up and Calla flicks the man a $10. "Sir, it's $30 flat fare for hotel guests!" the driver holding his hand out.

"Fucking hell, we only went round the corner as Willy steps forward and thumbs off his wad 2 tens. "Ta raa driver," then slamming the door and turning toward the entrance to the club which was pumping the tunes out.

Two huge doormen are sizing up the group of lads. "Evening, gentlemen!" then looking at Tom. "We aren't looking for trouble tonight, are we?"

Tom stepping past the 2 patting the bigger one on the shoulder says: "Only if the beer runs out!" The rest of the lads follow suit.

Inside it's a bustle of groups of young and old girls dressed to the nines, middle aged men dripping with money who are mostly wearing a variation of Ralph Lauren attires, pinks and green stripey shirts with knits slung over their shoulders and tied at the front. "Fucking time warp, lads, check out the set of Happy Days!" Willy chirps as Calla does an impression of the Fonz.

"Seven beers, my dear," Jimmy shouts across to the barmaid who leans forward to hear the request again because of the loud music.

"Sorry what?"

Calla leaning in: "He said 'can he have a shot on ya tits'?"

Jimmy elbows back as the barmaid replies with an even more puzzled look, "What was that?"

Jimmy leaning in shouts, "8 BEERS AND 8 SHOTS!"

The barmaid smiles as she spins out two pint glasses and sets 2 taps away running as she racks up 7 tequilas. "This ok for you?" the barmaid waving the tequila bottle.

"FINE THAT, LOVE!" as he turns and gives Calla a stern look.

Out on the terrace the lads find a perfect spot, plenty of elbow room, not too noisy and a perfect view of the talent in and talent out. "Some spot, bit better than the Shoulder of Mutton on a Friday neet" Davey smiles and winks at Jimmy.

Tommy moves over to talk to a group of girls pointing over at the lads. One of the girls in her 40s walks over to Jimmy asking him which part of London he was from, Tom laughing in the background. "I'm not from London, pet, I'm from Newcastle! God's country!!"

She asks him for a light and tells him his accent is lovely, Calla shaking his head. "Jimmy have you got any of them clap tablets left I can borrow, my bell end is like a fireman's helmet!"

Jonny bursts out laughing as Jimmy is unfazed. "See him there, pet, he was dropped on his head at birth! A bit retarded so any adverse reactions or fits just ignore him please."

As Jimmy carries on the small talk, Davey comes back from the bar with his shirt all wet. "Why some bastard spilt his pint all owa me shirt." Davey picks his pint up.

"Who did? Did he say sorry?" Calla puts his pint down.

Davey points over at a group of young lads laughing and carrying on. "No son, leave it; they are only young'uns."

Calla walks over to the group and Willy puts his glasses in his pocket. The blonde haired lad turns around as Calla taps him on the shoulder. "Sorry to bother you, pal, but see where I come from if you spill someone's pint you firstly apologise and secondly kindly offer to buy them a new one."

The group turn around looking at Calla as the young lad looks over to see Chuck and Tommy standing behind Calla, both over 6 ft and 15 stone. "Oh sorry man, I didn't think."

He was just putting his drink down when from the side a fist cracks him right on the jaw sending him into the table and out for the count. The lads step back in total shock as Willy points down at him. "Aye and see where I come from you just get a fucking kicking, ya cunt!!"

Calla bends down and grabs the lad, hauling him to his feet and squeezing his arm and leaning in. "See what happens when you're not careful! What are you going to do in future?"

The lad stuttering out in fear, dazed and confused: "B-b-b-b-bb-be more careful!"

Calla leaves go of his arm and slaps the lad on the back just as one of the doormen comes over. "Problem here?" half turning to Calla.

"Yes mate, actually there is," getting hold of this random girl's arm. "You see this lovely lady she hasn't got your number?" nudging the blushing girl towards the doorman who with a smile leans on the table. "Hey well we best fix that right now; my name, by the way, is David!" paying no further attention to the lads or the lad walking off rubbing his face.

As the two begin to chat Calla nudges Willy back outside. "Wayy am like a big handsome Cilla Black!"

Willy laughs. "Ayy sweet move, sailor, thought the big lad was ganna spoil our neet before it starts!"

Davey shakes his head. "You 2 bastards are not reet, you could've got yasel kicked out!"

Both lads swigging the rest of their beers reply, "Wey narr."

Willy digging Davey shouts up '7 more beers' to the waitress as she passes; the lads carry on scoping out lasses and cracking on as the waitress comes back with the drink.

"Here love, keep these glasses full and you might get my number!" Willy winks at the barmaid!"

"Hey Willy, remember that barmaid in Burnley you were banging! Remember that neet ya took her yem, rattled her on the couch and fell asleep and when you woke up her Jack Russell was licking your balls!"

The lads all roar with laughter, Tom spraying Davey with a mouthful of beer.

"Haweh, Tom!" Davey jumping back.

"Don't listen, lads, that's bullshit! Calla, you're always making things up!" Willy barks, all the lads momentarily taking notice. "It was a Shitsu!" Willy says with a straight face, all the lads again rocking with laughter as the barmaid comes back with the tray of beers.

Calla pipes up. "Eer man, I was in Stockton one night and pulled this darling right up for she were, ended up back to hers, fucking filthy! Dirtier than a pigs chin, had her in every room in every position, ends up crashed out in bed with her when I'm woken up by someone smacking me in the face!"

The lads all look shocked. "You what?" Tom sniggers.

"Yeh right, so I looks up and there's this reet hard faced bastard mullering my face so I throws him off half pissed and jumps up! The bird's in the corner of the room crying saying she is so sorry! I'm stood there in just me undies bleeding like fuck! It's only her fucking husband!"

The lads roar with laughter, all eager to find out what happened. "Well what happened?"

"I fucking Usain Bolted! Shoved past him called her a fucking slut and jogged on!"

"What about your clothes?" Tom shakes his head.

"Fucking left them! Didn't have time to get hold of them, brand new phone anaal! Cunt chased me through the fucking streets! Right fit cunt. Imagine that: 7am Sunday morn, mid-November, running like a cunt with nowt on, this daft prick screaming he is going to smash my face in! Finally managed to lose him, was fucked, had a reet stitch! Banged on the first door I saw, only a fucking bungalow! Some old dear answers, thought she was going to have a heart attack. She was sound though, managed to convince her to let me use the telephone so I could ring one of the lads to pick me up! While I was on the blower she made me a brew and gave me some of her deed husband's clothes! Fucking canny old stick. Tinny, one of the lads, rocked up in his tranny van and he was nearly sick with laughter when I walked out of the bungalow, little old dear waving me off, me wearing some plum cords and a green knitted jumper!"

The lads are roaring at this point, Davey wiping away the tears from his eyes. "Ha ahh lad, what did ya mate say?"

"Or aye, that cunt wiped the tears from his eyes and said I'm not surprised I'm pulling birds that old wearing gear like that, the cunt!"

"Did you return the clothes? Or you swan around toon in them?" Willy squeezes Calla's arm.

"A did fuck, never went back to that fucking area, still haven't! I swear that loonapath would've killed me if he got hawld of me! Not many blokes wobble me but that cunt was massive! Ha-ha, bought a pay as you go SIM and texted my phone saying I knew where he lived and I was going to burn his house down with him and his tramp wife in it!"

"Calla, you fucking lunatic, did you get a reply!" Jonny asks.

"Ha, ay the psycho bastard texted me back saying he was going to burn it down with her in it anyway! Thanks for the iPhone and all the pictures and videos of my cock that was on it were going straight on You Tube and if he ever saw me again he was going to set my head on fire! I texted back nee need for the setting my heed on fire like, ha-ha!" The lads all shaking their heads laughing.

"Fuck me you're a belter lad!" Jonny swinging his arm round Calla. "Now let's knock some JD's down our necks and get some blurt! We're staying at Beverly Hills, ha!"

Calla gets his eye on the blonde air hostess walking in. "In bound at 3 o'clock!!" Calla nudging Willy who was just about to put his glasses on and quickly pushes them back in his pocket.

Jonny laughs. "Willy, why don't you get that laser treatment?"

Willy knocking his tequila back winces. "Aye I should dee summit, the amount of birds I've bucked because I couldn't really see how ugly they were is off the scale."

Calla comes back with 2 blonde girls. "Remember this one, William? The lovely girl who looked after us today!?"

Willy takes her hand and kisses it. "Hello again, is this your sister? I'm William, this is Jonny and those 4 over there are too old for you."

The girls receive drinks from Calla straight away.

"So where you two headed?" Jonny moves into the circle.

"Oh we are staying here for a while then maybe to the casino for a late drink."

Calla's eyes nearly jump out of his head. "Casino? Where?"

Davey with a bit of a wobble on wanders over to the lads. "Calla son, where we going next?"

Calla shuffles his hands together. "Casino, father, for a bet and a pint."

Davey shakes his head. "Cash card."

Calla looking at Davey sighing out, "I've only got about 300 bucks though."

Davey's hand is still outstretched. "Cash card."

Calla slaps his wallet in Davey's big mitt. "Fucking hell, I knew it was a bad idea bringing yay!"

Chuck and Tom wave the white flag. "We tee off at 8am, dude, home time for us."

All the lads walk 5 minutes through the grounds to the casino, flashing the room keys to get in. The 2 girls are laughing at Willys crack. "Aye back to mine after this and I'll give you both the best cocktail you ever had."

Jimmy smiles as the 2 girls are getting roped in to Willy's net. In the casino is a very posh layout and waitress service as Calla looks for a drink off the tray. "How much are these, love, look a bit fancy!?"

As he tastes– "They are free, sir, all cocktails are free to hotel residents"– just as Calla swallows the drink and starts taking them off and passing them off to Jonny. "Here are, hamma, pass them along."

The tray empties and Calla gives the waitress 10 dollars. "Every 15 minutes come back and find me, pet."

She smiles and winks as Calla and Jimmy walk over to cash some money for chips. Jimmy cashes 200 dollars and Calla does the same. Davey, Jonny, Willy and the girls are sitting at a table playing black jack having a good laugh, only 5 dollars a game.

Davey taking 100 dollars from the dealer, stands up. "Wey that's me, I'm gan for a piss," as he gets up and walks off scratching his arse.

The dealer asks Willy, "Is your buddy coming back?"

Willy laughs. "Nee chance, he's 95 dollars up!!"

On the roulette table Calla is down to his last 20 dollar chip while Jimmy is a small bit up. Calla takes his shoes and socks off. "What ya deein, daft arse!?" Jimmy shouts at Calla as the waitress walks back over to give Calla another cocktail. He accepts and takes it off the tray with a sock on his hand. She shakes her head and walks off confused. "Jimmy, give me 50 bucks chips, pal, I'll go cash this 100 in a minute."

Jimmy slides the chip to Calla as the 4 other players and the lovely lady look at Calla with a sock on his hand. He places 20 dollars on red and 50 on 23. "The sock of glory has chosen."

Jimmy laughs. "You're not fucking piss wise!!!" Jimmy puts 20 on the 23 as well. No more bets as the ball bumps and lands in red 23! "Fucking get in, you little white twat!!"

Calla shovels the chips into his lap with the sock. Jimmy laughing his head off as Calla flicks his 50 chip back. "There you go, flanja."

Jimmy stacks his chips up neatly as Davey walks owa and Calla gets another 50 chip and sticks it on red 23 again. "Why Calla, you have a few bob there, son."

Calla laughing –"aye and a few more in a minute"– as the ball bounces around and lands in red 23 again!! The crowd can't believe it and neither can Calla. "Fucking milky bars are on me tomorrow"

Davey gets him by the arm. "Right son, haweh, cash these in!"

Calla gets up and tips all the chips into his shoes.

With over 3000 dollars in his hand, Calla goes over to Willy at the black jack table "How ya getting on, wee man?"

Willy sighs. "Me and Jonny are 200 doon, big man and the 2 blondes have fucked off."

Calla drops 300 on the table. "Fuck that and fuck them. Haweh we are going to the champagne bar on the 55th floor."

Davey and Jimmy are calling it a night as Jimmy reminds them they tee off at 9am and not to be late. The 3 lads get into the lift, Willy nudging Calla. "You didn't even give the girl a tip and you won over 3 grand, you're one tight cunt."

Calla looks surprised. "Howld on, Braveheart, firstly I stopped Evander Holyfield from pulling your head off! Then I pay your gambling bill! Take you for champers and you're telling me to tip the roulette girl? Would she give me my taxi fare if I lost the lot??"

Jonny leans against the wall. "Probably not, Calla, but you only fucking live owa the grass."

Calla shakes his head. "You're missing my point, fuck giving tips to some cunt who doesn't know you from Adam!"

Flashing of the suite card and a table on the terrace, the waitress comes over, absolutely stunning. "Yes, sirs."

Calla smiling slides her 300 dollars. "Bottle of your finest, I see its $275! Keep the change."

She wiggles away Calla watching her. "You are one 2 faced fucker, Calla!! You just said not to tip anyone." Willy wondering.

Calla, leaning back eating an olive off a cocktail stick, replies: "A meant anyone you don't want to buck."

The waitress returns with the champagne and begins to pour the 3 glasses as Willy necks his pulling a face. "That's horrible that hen, tastes like pish! Get me 3 bottles of lager please love!"

She places the champagne in the ice bucket. "We only do Peroni, sir, is that ok?"

"Ay get me 3 glasses as well," flicking her a fifty as she walks off again, the 3 lads staring at her arse. "That is a smashing arse that cunt!"

Jonny sips his champers in disgust. "Haweh lads, that's owa lass, man!” then laughs.

The lads are all in good flow with the crack when she returns placing the 3 glasses on the table and the bottles, handing Willy his change, giving Calla a smile. Calla pours half the Peroni in the glass and half the champagne taking a large swig. “Ahhhhhhh that’s better, fucking sound that bastard get a shot of that!”

The 3 lads do the same and quickly the Peronis disappear. Calla jumps up. “Am gannen for a piss, shout the fizzy booble ades in, Jonny, ya fucking minjbag!” stotting a peanut off Jonny’s head and walking off throwing a handful down his neck.

Calling the waitress over Jonny orders 3 more beers, the lads sitting ages waiting for Calla, nearly finishing the Peroni. “Here, where’s daft arse gone?” the lads are puzzling just as Calla comes walking through the bar with not a stitch on, the place silencing in shock as Calla sits down and sips his beer and begins to tell a story. The lads are roaring laughing, Willy taking his glasses off and wiping his eyes. “Oh you crazy bastard, what the fuck ya deein?”

Calla pauses, taking a sip of his beer. “What?”

The manager comes storming over. “SIR, WE HAVE CALLED THE COPS!”

Calla swinging round and crossing his legs, sips his beer. “Well good! I would like to report a crime!”

Confused the manager stutters, “Cr Cr Crime? The theft of your clothes I presume?” folding his arms.

“No, the price of the Peroni in here!” The lads all laugh as the manager steps back.

“I suggest you leave right now, sir, the law enforcement don’t mess around!”

Willy standing up nudges Calla. “Ey let’s go, big man, dinny want to be banged up and miss ya flight.”

The 3 lads stand up and walk out, Calla grabbing the half empty bottle of champagne and taking a big swig then burping in the manager’s face. All the lads waving bye to everyone, Calla doing the helicopter with his knob next to a table to a horrified lady!

Back at the club house Jonny flops on the sofa still chuckling about Calla’s antics, Willy kicks off his shoes and unfastening his belt then heads onto the patio for a smoke. Calla throws the remains of the 3 grand all crumpled up onto the bench and gets 3 cans out of the fridge, heads outside throwing Jonny a can – who has started to nod off – hitting him on the belly.

"Ooogh, ya tit!" waking up and pinging the can.

"Here you go, Mongol features!" slamming the can down on the table for Willy. "Ynaa what, Willy, I wouldn't tak your face ratting!" Calla starts to laugh.

Willy turning round and stabbing out the tab opens the can. "That reet, big man!" Willy laughs. "All I can say is Ding fucking dong!"

Calla frowns. "Eh?"

"Ding, dong man, ding fucking dong, what dya think about them apples!"

Calla is about to ask Willy what the fuck he is on about when the doorbell goes! Willy, smiling and walking past Calla, patting him on the arm, heads for the door, opening it to find a tall dark haired bombshell stood there. Willy invites her in and gives her a kiss on the cheek. "Welcome, let me give you the guided tour, hen!" as he puts his arm round her and kicks the door shut. "This is the living room! That's the balcony with a scarecrow! And this here is my bedroom! Tour over!' nudging the bird into the room and looking over his shoulder towards a gob-smacked Calla, winking 'night boys' and slamming the door shut!

Jonny squinting through one eye looks across to Calla. "Hawld on! Where's the Scotch cunt pulled that one from?"

Calla marching in replies: "Fuck narz!" putting his ear on Willy's door! "He's some cunt!"

Inside the bedroom Willy doesn't get a chance to perform his usual ritual of brushing his teeth and giving himself a rub and semi on as the girl pounces and starts riving Willy's shirt and pants off, Willy trying to resist and hit the brakes as they hit the bed, Willy on top kicking the remainder of his jeans off. Kissing her neck he slowly tears the wrapper on one of Calla's Kamagra jellies, quickly squeezing the contents down his throat and stashing the wrapper inside the pillow case! Coming round he starts to kiss the girl who stops eases him back licking her lips and asks, "Have you had Jam?" Willy smiling begins kissing her and moving things on!

An hour later Jonny is asleep on the couch, can in hand and Calla's headed to bed when there is a knock at his door. Calla opens his door with a WHAT? to see the bird Willy had invited round stood with nothing on, all her clothes in her hands.

Calla is standing there in his Calvins with half a bar on; the leggy brunette is standing in the doorway trying to cover up her bits. "Could you phone me a cab please?"

Calla pokes his head out of his bedroom door looking left then right and ushering her in. "Why like, where's Willy at?"

She pulls on her pants and bra as she explains: "Well I went to the loo and when I got back all my clothes and bag were outside his door and it was locked."

Calla smirks. "What a horrible man! To do such a thing to a lovely girl like you!!" He picks the phone up and dials reception. "Cab for suite 1017 please, at the door in 20? Thanks."

Calla stands up, dropping his Calvins. "Could you do owt with this?"

Calla's showered sitting in his pink lacoste t shirt and chino shorts, socks and sambas on, eating toast as Jonny comes down all showered and ready with a t shirt and shorts on also. "Morning Calla, up with the larks, ham? I've a heed like a box of fucking frogs."

Calla blows out loud. "Aye that champoni is nee good for your heed or your wallet!!"

Jonny puts the kettle on. "Did you spend much like?"

Calla pulls a wad out of his side pocket. "2800 dollars and I turned out with 300 so I'm 2500 to the good!!"

Jonny laughs. "Do you want a cuppa or what?"

Calla stands up, stretching. "Aye gan on, may as well, there's nee cans."

Willy walks out in a dressing gown smoking a tab.

"Morning, ye pair of fannies."

Calla, taking a bite of his 8th slice of toast, says, "Right Willy, I've a serious bone to pick with yay like, 4 o'clock in the morning and your bike of a wench chaps my door asking for a cab as some little bastard locked her out."

Willy sits down as Jonny plonks a cup of tea and a slice of toast in front of him. "Cheers Jonny, y'see Calla, I'd blown me cocoa all owa her face and she said when she comes back from the bog she is dying for a cuddle."

Jonny laughs. "So?"

Willy flicking his tab out the patio – "So fuck that!! Cuddles!!? Who am I? Mr Sensitive, fuck that!!"

Calla jumping in with "So I had to phone her a cab and let her out!"

Willy nurses his cuppa. "Did you buck her?"

Calla standing up shakes his head in disgust! "What, stir your fucking horrible Quaker Scottish mingin porridge?"

Willy looks up. "Aye."

Calla, walking to the patio door with his cup of tea. "No I did not! She sucked my cock for a bit though!"

Jonny spits his tea all over the sink as Calla turns round. "She didn't do that either!!" winking at Jonny.

All the lads head down to the clubhouse where Davey and Jim are loading up their golf cart with the hired clubs. "Haweh dopey holes, get a wriggle on!" barks Davey jumping into the cart.

Calla points at Jimmy's golf shoes. "Ha-ha, fucking nice Scooby's, hamma! I wouldn't wear them to kick shit about!"

Jimmy shaking his head gets on the cart. "You will have to hire a pair – you can't wander round in sambas, dipshit! Haweh Davey, there's away, see ya at the first hole!"

All checked through, golf clubs and golf shoes on, the lads jump on the carts, Calla having the second one all to himself, waving his putter out of the cart. He flies past Jonny and Willy shouting "CHAAARGE!" flying towards Jim who is about to tee off and skidding the cart sideways nearly tipping it over. Davey snaps: "Calla, behave ya bastard sell, will ya!"

Jumping off the cart and getting a ball out of the bag and his driver ready for the off. "You 3 are 15 minutes behind us so hold your horses," Jimmy snaps.

"Areet flanjeroo, settle down, just getting prepared," as Jimmy drives his ball straight down the fairway about 125 yards!

"That'll dee me," he says bending down to pick his tee up, letting Davey set his shot up. Calla is chanting "Miss miss donkey's piss!" as he hooks the ball straight into the lake on the right. "Bollocks!" Davey turns to Calla who is laughing.

"Straight in the bastard drink, Davey! Fucking quality!" as Davey sets another ball up, this one going straight but falling short of Jimmy's. "Better, see you ladies later and nee piss farting around!" Davey slides his driver into the bag and heads down the fairway.

Calla puts his ball on the tee. "Bet you a $20 a could hit their cart!"

"Gan on then, Calla, hope you do. Could fancy seeing Davey punch your teeth out!" replies Jonny who is leaning on his club.

Calla drives the ball towards the cart, hitting it on the roof, the cart slamming its brakes on, Davey waving his fist back towards Calla who is on all fours laughing.

Jimmy and Davey finish the hole and Calla steps up, smashing the ball as hard as he could sending it straight into the bunker. "Orr, for fuck sake straight on the bastard beach." He smashes his driver off the grass.

"Look out, ya useless cunt!" as Jonny drives his straight down the middle leaving it just on the fairway. Looking smug with himself he jumps back on the cart as Willy stands getting ready for the shot with his tab in his mouth. He smashes the ball just like Calla but tops the ball and it bobbles about 3 ft! Calla and Jonny are pissing themselves.

Willy straightening up and putting his hand over his brow looks out towards the flag. "Where did that gan?" squinting with the sun.

"About 4 yards, ya idiot – there, look! It hasn't even made the women's tees!" Jonny shakes his head. "Ay ya mong you have to walk to ya ball with ya pants down if you don't get past the women's tees! FACT, it's in the rules!"

Willy shrugs his shoulders and drops his shorts and shuffles along like a penguin towards his ball as Calla quickly throws down a ball and drives it straight off Willy's white arse.

"Oowwwwwwww, you CUNT!" Willy jumps up and down rubbing his arse. "That fucking stings, Calla, you prick!" He pulls his shorts back on and throws his golf club at Calla, just missing him, bouncing right in front of an old couple who are waiting for their tee off, both looking very unimpressed.

Jonny nearly falls out of the cart laughing. "Should make the greens white and spotty like Willy's arse, Cal, might improve your game!" as Willy knocks his ball onto the fairway, still rubbing his arse, fag quivering on his lip.

Jonny and Willy put their second shots onto the green, leaving Calla to get his shot out of the bunker as he takes a swing at it sending a load of sand all over the fairway, ball not even moving! "That doesn't count! The ball didn't move, the ball didn't move!" Calla claims.

"Just hurry up, piglet features."

Calla finally gets his ball out of the sand and into the light rough just off the side of the green, 8 yards to the pin.

"Not bad, lad!" Jonny pats Calla on the back as they head towards the ball. "Colin Montgomery once told me to use a 7 iron when ya ball's like that, Calla. Just use the 7 like a putter and it lifts the ball just enough to get out of the rough!" as Calla uses his pitch and dinks it straight onto the green 4 yards from the hole.

"Ay and Tiger Woods told me that was an old wives' tale!" Calla smiles and heads towards his ball with his putter, all smug.

Jonny finishes on par, Calla getting a bogey, leaving Willy to finish four over! "I can't put for bastard shite! Two shots to the pin and 6 to get it in the hole!" Willy shakes his head.

"Ay and that 4 yard putt missed and ended up a 12 yard put for me next shot, the twat!"

"Calla you're just a heavy handed cunt! You could break an iron ball! A bet if they put hair round the hole you two would have nee bother! Getting it in!" quips Jonny laughing.

Onto the second hole and the drinks cart pulls up. "Can I interest you guys in a tasty beverage?" the guy shouts over pointing to the cans of Pepsi.

Calla rubs his hands. "You certainly can, me old china, but none of that shite – what lagers do you have?"

The guy frowns. "Beers, sir, but it's only 9 20 am?"

"Way aye man, giz 9 tins please!" Calla gets his wallet out, the man laughing and handing across the 9 cold tins of Budweiser, Calla handing him a fifty asking how much are they.

"Five bucks sir!"

"Eh, they're only mini tins, you jipping twat? Areet then Just giz another one then and keep the change!" The guy throws Calla the 10th can and drives off.

At the 5th, a par 3, Jimmy and Davey are waiting as the 3 lads pull up. "What's the delay, lads?"

Willy hops out of the cart. "First par 3 and first par 5, nearest the pin 50 bucks and longest drive on the 9th for 50 bucks!"

Calla swigs his tin of bud. "Sounds good to me, fat chops!! Let's see you go first."

Jimmy pops the ball about 12 yards from the pin on the green as Davey tees his up.

"Nee pressure, father," Calla shouts over. "Pressure is for tyres!!"

Davey chips his ball high into the sun; it lands 3 yards from the pin. "Fuck off!! You micey bastard!!"

Davey gets back in the cart looking over at Calla and winking. "Ka ching!"

All 3 lads land just off the green and Davey is sitting on the cart rubbing his hands together as the lads weigh over the bet.

Jimmy gets in the cart and tells the lads they will wait on the 9th for the longest drive, Calla shaking his head. "Davey's one fly cunt, always pulls a stroke to win the dosh!"

Willy laughs. "It's only 50 bucks, Calla you can get that back on the longest drive, man."

Calla gets in the cart. "Miss me out this hole, I'll take one over, I'm bursting for a shite!" As the cart speeds off in the other direction.

The lads are just teeing up for the 7th hole as Calla arrives back in the drink cart. "Oy oy, lads, would you care for a snack?"

Willy hits the ball a good way for a change. "Fucking hell, Calla, you robbed the drink cart, are you mental!"

Jonny laughs. "I can answer that one!!" Calla jumps out of the cart. "I was coming out of the bog and the driver was going in so I hid mine behind the toilets and stole his – he'll not know where the fuck it's gone!!"

The lads get another can and Willy takes a packet of nuts off the side of the cart. "Whoa whoa whoa, four eyes!!! 8 bucks for the nuts!!" Calla demanding the dosh.

"Fuck off, Calla, you retard, I'm not paying for he haw off that cart!!"

Calla gets off the cart. "Is that right? Ok hold on a minute I have something special for you!!" Calla reaches down into one of the compartments of the drinks cart.

Willy walks over peering over Calla's shoulder. "Well give me it then!"

Calla turns round and rams a Cornetto straight on Willy's nose, pushing him back at the same time so he lands on the grass. "There you go, Worzel" As Calla bends down and takes the packet of nuts out of Willy's hand, Jonny gives Willy his hand picking him up, Willy's face covered in ice cream.

"I swear, Calla, you're as bad as the bairns!!" Calla takes a club out of Willy's bag and tees off down the fairway.

Davey and Jimmy are at the 9th waiting on the lads with the course marshal. "No we haven't taken any drinks cart from the 3rd hole amenities!" Jimmy is arguing with the man.

Davey is sitting in the cart. "And how could we, none of us used the toilets and we left the lads on the 5th hole?"

The marshal is still on the walky talky as Calla comes through the bushes on the drinks cart blasting the horn. "Drinks ladies!!!"

Davey, putting his hand over his eyes, says, "Ah for fucks sake!!!"

Calla skids the drinks cart sideways as the marshal storms over. "What on earth do you think you are doing?" screams the marshal pointing in Calla's face.

Calla totally flips out, jumps out of the cart and towers over the short fat bloke. "I TELL YOU FUCKING WHAT YOU LITTLE PIECE OF SHIT IF YOU DONT LOWER YOUR VOICE STOP POINTING I'M GONNA SMASH YOUR UGLY FAT FUCKING FACE ALL OVER WITH MY 9 IRON!"

The bloke wilting down responds: "B B But sir, the drinks cart, you can't take it!" taking small steps back from the red faced Calla.

"IT WAS ONLY A FUCKING LAUGH, WE HAVE HAD 3 CANS AND THE CUNT WITH THE GEGGS ON HAD A CORNETTO! THE MONEY IS IN THE CART – SO FUCK OFF!" Calla steps forward.

"Ok ok ok calm down, sir, I'm sorry for losing my temper. So long as you have paid for the drinks we will overlook the theft of the vehicle and let you on your way!" The man gestures his hand out towards Calla who turns his back and lifts his golf bag off the roof of the cart "fuck off fatty!" a slightly embarrassed marshal radioing to the drinks man to bring a cart to hole 9 and his had been found.

Davey and Jimmy shake their heads and play their shots again. The man tries again to apologize to Calla who returns it with a simple "Shut your mouth, fat chops!"

Willy sniggers as he drives his ball straight into the trees. "Or man, fuck right off, will ya! This game is fucking shite!"

"That's coz you're hopeless!" replies Jonny as he out drives Davey and Jim, leaving it on the fairway.

Calla steps up and smashes his ball which flies straight over all the balls and lands a good 50 yards past them in the bunker. "GET THE FUCK IN!" as he punches the air.

"Not a good lie for your next shot though, numb nuts!" Jimmy snides.

"A couldn't give a monkey's fart helmet, fairways for show, distance for dough! Now get ya spuds out!"

All the lads hand over the money as Calla's cart arrives. "Here fatty, before you piss off get me 10 cans!"

The marshal is busying himself getting the cans as Jonny whispers to Willy, "He's fucking shitting himself! Bet he has never seen an angry smog before!"

Calla snatches the cans and distributes them to the lads, Calla handing over the 50 dollars as the marshal and the drinks guy hop in their carts and drive off!

Jonny leaning against his club watching the cart drive away looks at Calla "I like how he was the one apologizing Cal, you stole his cart you fucking arsehole!" laughing with Calla stotting a golf ball towards him.

"Calla lad, that was some speech – have you ever considered becoming a peace negotiator? You're some charmer." Jimmy laughing nudges Davey who is shaking his head.

"Fuck off," Calla swiftly replies.

Further down the course the lads are waiting for Willy who has disappeared into the bushes. All the lads are laughing at the sound of a ball hitting trees, bushes being chopped and trampled and the reoccurring "FUCK! PISS OFF! CUNT! and TWAT!" as the ball finally pops out of the trees, a flustered Willy wandering out lighting a tab! "This fucking game's enough to drive a man to drink!" Taking a large swig of his can!

The lads are all pissing themselves. "Haway, ya scotch cunt, your gan still!" Jonny signals.

"For fuck sake I've just took 12!" as he swings for his ball and totally misses. "CUNT!" then steadying himself he swings and misses again spinning all the way round, the lads all killing themselves now, Willy spitting his fag out and concentrating again totally missing the ball. Calla is now on his knees, all the lads laughing, as a red faced Willy bends the whole club over his knee and hammer throws it straight in the bushes. "Fuck this! Fuck you lot. I'm going straight to the 19th hole!" He throws Jonny's bag of clubs on the floor, jumps on the cart and fucks off! To Jonny's moans of "Haweh, Willy, ya cunt!" Willy drives off with his middle finger up.

The lads finish the hole with Calla getting his first par and demonstrating his delight by pretending to ride the flag like a horse! The lads stop for a piss at the 10th as the lads all empty their bladders in the woods. "Any mare cans?" Jimmy asks.

"Aye, al get you one!" Calla walks over to his cart and opens a Bud then squeezes a Kamagra Gel in on the sly, handing it to Jimmy.

"This can tastes like Jam, Calla you cunt. Have you put one of those Viagra jellies in!!?"

Calla takes the can off Jimmy and swaps it with his. "There you go, fat chops, I 'll have this one."

Jonny's laughing over. "Calla, you're a sick man, there's nee fanny out here!?"

"There's 5 fannies out here, Jonny! You been the biggest! I took it to make me play better! I'm going to start putting with my cock on the next hole!!" All the lads laugh as they move onto the next tee.

Waiting on the terrace, shades on and 2 cold beers, Tom and Chuck watch as the lads appear on the hill of the 18th hole. Calla is preparing his shot and Davey's looking on as his and Jimmy's ball are already on the green. "Calla, I'm not coaching you son but that's going to go miles away from the green."

Calla takes the club back to swing. "Aye but it'll not be far off hitting them two fat Texan bastards and that little specky cunt on the terrace." He swipes through the ball and all the lads watch it in slow motion head towards the decking.

"Calla, you stupid twat." Jimmy's tab is lying on his bottom lip as the ball bounces about 6 ft next to the lads, they both jump up and watch it hit the sun canopy and bounce like a pinball all over the decking, people standing up to get out of the way as Calla holds his hand to his eyebrow. "FORE!"

In for a quick wash, Calla and Davey are using the free toiletries. "Nice shave, deodorant and aftershave, fair play." Calla nudges Dave as he sprays his sack with Hugo boss deodorant jumping. "Oww ya cunt, it stings like fuck."

Davey laughs. "That's cos they're sweaty, you mad bollix, wash them first!!"

Calla straightens himself up. "Wash them? They smell fucking champion!"

Walking out of the toilet area out onto the terrace where the lads are sitting– "There's a cold beer for you, Calla!" Chuck slides a pint over to Calla who lifts the glass with a 'cheers', sways and downs half a pint letting out a massive burp!

Jimmy turns round. "Calla, what do you say, man!!"

Calla catches his breath. "Cheers, Chuck, that's fucking lush!!!"

Davey joins the lads for a beer. "So what's the plan for tonight? We are on the 10 am flight! It's 1.30 now, back for a snooze?"

Willy moves to make himself comfortable. "I'm going to have to go rip the bald champ's heed off before I do owt!!" All the lads laugh.

"We'll go to the shop, get some supplies and have a few tins round our pool, get showered and head to that posh BBQ, eh?" Calla making a solid plan.

Jimmy nods. "I'm game for that, we are spending no money which makes a change!"

Calla remembers about his winnings. "Fuckin hell a forgot about these bastards."

Davey puts his pint down. "Pass it owa, son, so I can count it." Calla is like a kid handing it over as Davey thumbs through the wad. "There you gan, son, 1000 dollars, 700 more than you had last night and a few bob in my pocket for when you're skint!"

Calla pushes the money in his pocket.

"You mind him like he's one of your own, Davey." Jonny salutes Davey. "Some cunt's got to, he's fucking yepless."

Calla pausing with his drink – "Haweh!"

The lads begin to flick out score cards. "Hey Tom, what you get?" Jonny leans in.

"Two over, a little disappointed – 4th hole and 5th hole a bogeyed them, hit a 72!"

"Out of 70. Fucking hell, you still should be playing in the Masters!" the lads all working out with handicaps where people finished.

"Willy, what's your score?" Calla quips.

"Fuck off!" Willy snaps.

The scores finish pretty much as predicted: Tom, Jimmy, Chuck, Jonny, Davey, Calla then dead last Willy.

"That game is fucking shite, costs a fortune as well. Had to pay $100 penalty for losing an iron, they said it will be reimbursed if it's found and handed in!"

"I'm doubting that son, it's a bit George Michael!" Davey swigs his pint off and stands up smacking his belly. "Time to get the big man a piss and a bit bite to eat!"

Calla snaps. "Haweh man, dinnit waste stomach space on non-essentials like food!"

Davey shakes his head and heads for the toilet. "Eat to live, Calla son. Eat to live!"

Calla rocks back. "Fuck that, you don't put solids in a hydraulic system!"

The lads chuckle. "Jimmy, you going to stick this trip on your company expenses, man? Bet you could wangle it somehow?" Calla pushes Davey's arm spilling his pint.

"Calla man, a cant and then what would the lads say back at the ranch who never came – they would want the money instead – and then there is Tom and Chuck who have paid for themselves!"

Tom pipes up. "Hey Calla, I will pay for your trip if you're short!"

A shameless Calla replies: "Fucking champion, sound that Tom. I tell you what! I don't believe all them stories about you being a cunt!"

The lads laugh as Jimmy chirps in. "You won 3 grand, Calla, you should be paying for the rest of the lads!"

Calla squares himself off. "Hawld on, I earned that money! It took my mathematical brain to work the odds down to red 23 – why should my hard earned education and talent benefit you thick twats!"

Jimmy bites back. "And you're prepared to let Tom pay for golfing trip?"

Calla rocks back. "He offered and my mam always said don't look a gift horse in the mouth!"

The lads are laughing as Davey comes back. "Reet, menu, what we got?"

He picks the menu up and squints at it, deciding straight away. "Philly cheese steak sub with potato wedges! Sounds tremendous!"

Tom, Chuck and Jimmy all agree as Jimmy waves the waiter over and points at the menu. "Three of them please!"

"Anything else, sir?"

Jonny says, "Ay get me a bowl of fries and some onion rings!"

Calla snides, "Eating is cheating!" as the waiter takes the menu and wanders off. "Eer Willy, ya miserable twat! What was Warwick's bird like in the sack?"

Willy chirping up and swinging back on his chair: "Cal, all I will say is wasp's lug!"

The lads laugh. "Way, that's handy, Willy, as all I'll say is wasp's whisker!" Jonny laughs as Jimmy chirps in: "Willy, if Warwick finds out, mind, he'll get you off the job! So try and keep it quiet!"

The food arrives and Davey rubs his hands and tucks the napkin into his shirt. "Pass that sarce, youngen!" he shouts to Jonny who is drenching his onion rings with ketchup and Calla's hand who's pinching them, wiping his hand on Willy's back as he goes to the toilet. "Get the beers in while you're on, ya dancers!"

Willy turns round. "You just sit there, your majesty! Ya some lazy cunt, Andrew!"

Standing up–"Whoa whoa whoa, what's with the Sunday names, William!' Calla winks and smiles at a bird sat opposite who is with a bloke.

Davey hits Calla with a chip. "Behave!"

Calla picking the chip up and sticking it in his mouth hold his hands out. "What?"

Willy walks back from the bar laughing. "And anyways, get me off the job, Davey? For what? Bucking his lass!! He wants removing from existence for having a name like Warwick!!" Willy looks over at Calla. "Aye but what about..." As Calla jumps in: "Whoa whoa whoa, remember the code of silence, Willy, Haweh."

Jimmy asks, "What code of silence? What have you done now, Calla?"

Willy bounces a nut off Calla's head. "It's not what he's done it's who he's done."

Calla gets up, pinching a wedge off Davey's plate. "Lend us your phone, Jimmy. I'm going to phone yem and speak to the bairn."

Jimmy hands Calla the phone as he walks off, pressing the digits. "0044 is it?"

Davey with half a cow hanging out of his gob mumbling: "Aye then the area code."

Calla comes back after 10 minutes. "Cheers, Jimmy, y'na that lazy fucking ex of mine!! I hope her fanny heals up and she drowns in her own piss!!!"

Jonny says to Calla: "Bit harsh, like, why? What's she done now, flanja?"

Calla sits down, taking another wedge off Davey's plate as Davey tries to speak with the other half of the cow in his mouth. "Haweh Calla, you greedy cunt."

Calla laughs. "Fucking go go gadget gob!! You fat bastard, father!"

Jonny nudges Calla. "So?"

Calla swigs his pint. "Ah she comes on the phone – can you ring back in half an hour? Coronation Street is on!!!"

All the lads burst out laughing. "Fuck off, honest Calla!!"

Calla nods. "A tell ya, man, if she could get that fucking settee into Morrison's she'd lie on the cunt with the bairn pushing her down the tins aisle!"

Davey wipes his mouth still with a massive dollop of ketchup on his eyebrow. "Aye son, you're well shot. Lang as the bairn's areet."

Willy wipes Davey's forehead. "Fucking hell, Davey, it looks like you were fed that piece with a fucking catapult!!"

Calla gets his money out. "I'll get this, lads, my treat!!"

Tom is amazed. "Jesus Christ, Calla, you have a wad there that would choke a donkey!!"

Calla whips his cock out and hits Willy on the head. "I've a fucking cock like one anarl!!"

"Behave, Calla, you dirty bastard." Willy waves Calla away as Calla does the helicopter. "Willy, it's clean, I just sprayed it with Hugo Boss, and the hoower you called last night cleaned it this morning!!"

Willy stands up. "So much for the code of silence!!"

Jimmy pokes Willy. "How much did she set you back, Willy!"

Willy brushes the crumbs off his shorts. "Money and fair words, Jimbo, money and fair words!"

The lads walk towards the lobby, Chuck putting a lump of tobacco in his mouth. "What time we meeting at your place, boys?"

Jonny grimaces. "Fucking hell, Chuck, have you put a bit of dog shit in your mouth?"

Calla peering round adds, "That's what the lass-who-sucked-me-off's breath was like this morning!!"

Willy laughs. "Aye and you had to kiss her goodbye for me!!"

Calla shakes his head. "Did I fuck, I shook her hand and pointed her in the direction of the door!! I know where my cocks been, hamma!!"

Into the bus and away back to the Hilton, the lads take a minute to chill. A few of them are just nodding off when Calla's voice wakes them. "Pull owa, driver, pull owa!! Off licence at 2 o'clock!! Wa hooo!!!!"

All the lads pile off the bus as Calla hot-foots across the car park, riving out a trolley. Jimmy nudges Davey. "Fucking trolley? That lad's not plum; we head back the morn and we are oot the neet!"

Davey doesn't answer, just shakes his head as the lads arrive in the supermarket looking at Calla who is dashing round. "Howw, soppy bollocks, you on supermarket sweep or summit? Just get a few cans and some Doritos – we're oot the neet!" Davey barks owa to Calla who is loading a box of Miller, a box of Skyy vodka lemonade and a box of Crabbie's alcoholic ginger beer.

"Davey man, look, it's all pop only got one box of beer!" as Davey pulls out a bottle of Crabbie's. "It's bastard 5.5%, ya dopey twat! Wadn't give the bairns that!"

Calla snatches it and heads to the crisp aisle, Davey shouting, "You won't drink all that, man! You're wasting your money!"

Calla barks back: "Rather look at it than for it," then disappears round the corner.

At the till the lads are stood with their own 6 packs and mixed array of crisps, nuts and beef jerky, Calla ramming his trolley against the back of Jonny's legs.

"Oooof, ya cunt!" He turns round looking in Calla's trolley. "What the fuck! Cal man, you can't take it back with ya!" looking at all the alcohol, the 5 big bags of Doritos, 3 mega gulp Gatorades, bottles of water, nuts, cheese slices, ham, pastrami, bread, bacon and eggs!

Calla glancing up says, "Ya reet!" lifting the four pack of water out and cramming them on the sweet shelf to the left!

"That money burning a hole in ya pocket, Cal!" Jimmy asks!

Calla with a straight face replies: "Tell you what, Jimmy, when you come round mine after your homo snooze, and me, Willy and Jonny are drinking icy cold Crabbies on the balcony while eating a platter of sarnis you can keep your little fat trotters off! You can just stand in the corner and contemplate your actions trying to tell another man how to spend his winnings! Especially when the man has spent most of it on keeping others in alcohol!"

Jimmy laughs as he puts his items down and asks for some cigarettes, sniggering. "Tell you what Cal, you're all heart, I apologize!"

Calla shakes his head. "Nope, too late fatty! You've spoiled yourself – none for you!"

Jimmy lifting his bag of swag off the counter and change walks to the car still sniggering. "Nee bother, Cal, ya sacked!"

The lads all laugh at Cal who standing up straight shouts: "Hey Jimmy it was a joke – you can come round mine for crabs!" holding the bottles of Cider up.

Everyone in the store pauses and looks at Calla laughing and loading his items on the counter.

Back in the van the lads again are waiting for Calla who looks like he is carrying the groceries, struggling to get in and throwing the bags down. Tom asks: "Where you been?"

Calla lifts a porn mag out of the bag and flicks it open. "Went back for this!" as he rummages round for a Crabbie's, twisting the lid off and taking a swig, then carefully sliding the pages across and studying them! Tom and the rest of the lads shake their heads!

Back to the hotel and the doors are opened. The lads get out with their groceries to the horror of the hotel staff! The lads troop back to their rooms. "Call, ya ganna have to lend me a shirt – you fucked mine in the pool last neet!"

Calla doesn't look up, still thumbing through the mag, twisting it long ways to get a good look at the centre spread. "Ay ginger bollocks, nee bother!"

In the room Calla removes the shelves from the fridge and lifts the bags straight in leaving no space for Jonny's drink who turns. "You've some class, you Calla! Where do I put these!" holding out his beers.

Calla still engrossed in the filthy book walks back to the freezer, lifts out the ice trays with one hand, empties them in the sink, puts the plug in and turns on the cold tap! He walks off to the balcony leaving Jonny mouth open rolling his eyes.

Willy's leaning on the balcony looking at the pool. "Hey, lads, I'm away in there for a dip!"

Calla puts the book down and looking out at the pool then onto the next club house where there is a lylo stood up against the wall –

"Ay me anaal!"

He heads back to his bedroom. He is changed and heading out towards next door with a pair of England shorts on, checking no-one is in he grabs the lylo and a pair of Ray-Bans that were left on the table. Putting the sunnys on, getting his mag and drink, he carefully lies on the pool kicking himself off the side towards the middle. He adjusts as Jonny cranks up the volume of MTV and heads out with his shorts on.

"This is the life, lads!" Calla drifts on the lilo as Willy sits on the side drinking his Alco pop.

"Aye not a bad weekend. You have to live a little on these horrible long trips away or you'll go mad."

Jonny sits down putting his feet in the water. "Take the rough with the smooth, we'll be 12 hours hoying pipe in Monday!!"

Willy swims over towards Calla and puts his arm on the lilo. "Some tits on her, Calla."

Calla looks down his nose through the Ray-Bans. "Aye nearly as big as your man tits."

Willy flips the lilo over as Calla kicks his legs holding the magazine and the bottle above his head as he gets to the side, out of breath. "You little twat! I could've ruined me mag."

Jonny laughs. "That'll be ruined with jizz shortly."

Tunes belt out and the lads are sunning it up when a girl pops her head over the wall. "Do you mind giving me my lilo back."

Willy is stretched out on it with his hand down his shorts. "Why don't you come and get it, hen, there's plenty of cold drink too." As a big massive black head towers above her – the doorman off last night.

"How about you just pass it to her and I'll take my sunglasses too!!" his voice booming over to the lads.

Calla stands up "get round big man, we have plenty drink!" then goes into the kitchen returning with 2 cold bottles. "There you go, big un, and here's your shades!" The big guy accepting as Willy climbs out the pool handing the lilo over. "Bet you wish you had a body like this, big fella!"

The giant looks down at Willy's whiter than white scrawny torso laughing.

"We are expecting others!" the big man booms "having ourselves a small party before the concert!" Willy flicking his tab over the fence "get on the blower and tell thin its this side of the fence instead! Haway big man makes sense ouja ouja!" the big man raising an eyebrow at the last part of the sentence before shrugging and flicking open his phone.

Big knock at the door, 3 of the lads standing there as Jonny opens up. "Alrighty, where's Jimmy?"

Davey mooches past. "He's on to Mally about something at work; think it's about some company do next month if we get the first phase done."

Tom pats Jonny on the arm. "Yep then it's off home for Santa Claus!!"

The lads walk outside chinking bottles with Calla and Willy, "Who's the neighbours, Willy?"

Chuck peers over the fence. "The doorman and the waitress from Sinatra's, sound skin."

Willy replies. "Aye aye, Willy saved you from another kicking!"

Calla puts Willy right.

Tom pulls out a bag and has 2 huge metal horse shoes. "You all fancy a game of ringers!?"

Chuck puts his bottle down and walks over past the pool sticking a metal rod in the grass, marching out about 20 metres and pushing the other one in.

"What's this like, some hillbilly game of pitch and toss!" Jonny asks the lads.

"Well Jonny, seems like you are the wise ass – why don't you come and see if you can beat me."

Jonny puts his beer down. "Nee bother, pitch and toss champion 88, 89 and 90 at school."

Calla laughs. "Fuck off now, I know you're lying, Jonny – you didn't go to school, you thick cunt!"

Jonny walks over. "So what are the rules?"

Tommy shouts over, "We'll keep it simple: you gotta throw the shoe, nearest pin gets a point, if you get a ringer it's 3! Where it hoops around the pin, and if your opponent gets a ringer, it cancels yours out! First to 11."

Chuck goes first, the shoe swinging round the pin in a circle down to the grass off an elegant 20 metre throw. "There she is boy!!" Punching Jonny lightly in the arm.

Jonny throws his – it lands 6 ft short as Calla and Willy both start laughing. "Ah, you're about as strong as a half sucked mint, ya cunt!" Davey ribbing him.

"Haway then, Davey, let's see how good you are!" Jonny shouts over as Chuck and him walk to the pin to collect the shoes.

"No son, I retired when I won the world horse shoe championship!!"

Calla finishes his bottle. "Fucking horse eating championship, ya mean!!"

The door swings open and the big guy from next door troops in with a large cooler box under his arm and the girl on the other, putting the box down. The introductions begin, Davey the first. "Areet bonny lad, Davey."

The guy nearly shakes Davey's arm off. "Real pleasure, Davey. Boys call me Rama!"

"Nice to meet you, Rama. So you're a friend of silly bollocks owa there?" pointing to Calla jumping out the pool soaking.

"Yup, your boy got me hooked up with this beautiful thing!" Giving the girl a kiss on the head and squeezing her hand.

"Ay way pleased he is good at something – meet the rest of the lads!" as Rama walks round them all shaking hands.

Jonny cranks the music up louder as the door opens and 8 stunners walk in and one fat lass bringing up the rear, all with floaty dresses on and bikinis underneath. Calla and Willy look at each other and bash fists. "Way mind howw!" Calla says doing a back flip into the pool as the girls all hug and kiss everyone, one of the girls saying the rest are outside just as another 7 girls walk in with 3 other guys carrying cooler boxes. Calla just catches sight of this as he climbs out of the pool, half pausing in amazement as Willy kicks him straight back in.

Time and drink passes as most of the girls are in the pool and others drinking and dancing as Jimmy walks in with his six pack.

Getting to the decking area he checks over his shoulder and looking puzzled he scratches his head. Davey walking past with a cold beer says, "Ay lad, yav come to the right place!"

Jimmy shrugs. "Sound, Davey, where will I put these?"

Davey walking off says, "Or over there in one of the cooler boxes, Jim, can't believe how much drink is there and how much talent is kicking around," spotting Willy pulling the string from a bird's bikini as he gets out of the pool. "Ayy Jim this is one of the best weekends I've had!" He skins a beer and puts his wet arm round him.

"You've changed ya tune – golf was for shite and for gays this afternoon!" replies Jimmy sipping his beer.

"Hang on, big man! Where is the golf sticks? Nay golf here – just fanny and drink!"– necking off his beer then throwing a bird in the pool and jumping in after her.

Jim and Chuck still well into the game as they both head over for a piss and a beer, Calla using the horse shoe to knock both iron rods flush with the grass then tossing the horse shoes over the fence onto the course. "Fuck that for a game of soldiers!" – then hopping back into the pool.

Hours pass and the sun starts to cool off as most of the girls all have seats now and are talking away in their individual groups. Few groups are talking away and dancing, Willy sat with one on his knee and two others on their hunkers in front of them listening to his lies. Tom and Chuck are also in good flow with two of the birds leaving Jonny, Davey and Jimmy in the biggest of the group talking to the rest as Calla appears at the door showered and changed wearing a Lyle and Scott polo shirt and a pair of blue jeans with flip flops, hair all combed back, stinking of Armani aftershave.

Davey shouts over. "Howw, bollocks, where you gannen?"

Calla points to his watch. "It's 6 bells, heading owa to that bloke's BBQ!"

Davey walks over. "But Calla lad, there's a party here? What about all these?"

"Bollocks to these, they will still be here when we get back. Told Rama to keep an eye on things – he reckons a load more are coming later! Let's go to this BBQ, then head back after a couple of bars!

Haweh man, we're headn back the morn!"

Davey walks off towards the rest of the lads. "You're not piss wise!"

Calla shoves Jonny into the house. "Get changed, cunt. Haweh man! This will be like this when we get back!"

Jonny trying to push back –"Calla man, am pissed, let's stop here!"

Giving up, he heads for a shower. Tom and Chuck head past. "Davey says we are headn to that BBQ?"

Calla agreeing bangs him on the back. "Ay see you out front in 20 mins. I've the address in my sky!" as they walk out, Jimmy and Davey following, Davey still shaking his head.

Calla rubs his hands thinking just Willy to sort as he walks over. "Excuse me ladies, need a word with chock heed! Haweh, Willy man."

"Fuck off!" Willy is still sat with a harem of ladies.

"Willy man, we will be back later."

"Calla, fuck right off!" Willy is not budging.

"Or fuck it, you stay here! You weren't invited anyway. He said he would rather let leather face in with his chainsaw, ya scotch cunt!"

As Calla storms off Willy gives him the loose wanker sign as one of the girls takes a draw of his cigarette and blows the smoke down Willy's neck. Willy flicks the top off a bottle with his thumb. "Ahhh this is the life! Gals, did I tell yee about the time I was working in Norway?"

Calla, Jonny, Tom and Chuck are all stood out the front of the hotel, Calla stood hands on hips frustrated as Jimmy and Davey come tappy lappying along. "Haweh, you two old twats!"

Davey stops. "Hold on! Get me a camera, Jim, this is a first! Calla the horrible waiting for us?"

"AY Haweh, this party started at 4 – bet it's fucking class! Hurry up!" Jimmy and Davey shake their heads as they pile into the hotel's 6 seater Tahoe.

As the gates open the jeep drives down the gravel drive to a line of fancy motors lined up around the circular driveway. "Looks a decent kip." Jimmy flicks his tab out of the window as the jeep pulls up and the lads bail out.

Davey stretches himself and lets out a massive fart as Jonny hold his hand over his mouth. "Jesus Davey, you might have to get that stitched!!

"Aye son, that dinner was a bit heavy, I'm bursting for a shit! I've got 5 bar on the clacker valve already." As all the windows drop and the lads poke their heads out fro fresh air Davey laughing "Haweh lads man its only a bit shite!"

The lads mosey on up the stairs and Calla presses the bell, Chuck peering through the windows. "Are you sure we are invited here, buddy"

The door opens and a small man in a tuxedo opens the door. Jonny nudging Calla – "It's him out of Fantasy Island, hamma."

Calla laughs. "Hello Tattoo, is the main cheese around? We are guests of honour from afar."

The small waiter gets a fancy book and looks down through it. "Name, sir?"

Calla turning to the lads says, "I never gave him my name – I just took the address and the invite like?"

Jimmy pushes Calla past. "Ger by, daftarse. If you could do us a favour, my friend, and go and get the owner of the house? Just say we found his golf clubs at the airport."

The man closes the book, asks the lads to wait and walks down the hallway. "Fuck this shit, haweh, we'll go round the back." Jonny ushering the lads as they walk to the side gate and through down some fancy steps into a huge garden. A view of the bay and a huge marquee with chefs at various BBQ stands and ice sculptures on the patio.

"Who is this fucker anyway, Calla – he has some pad." Tom walks down behind Jonny over the grass.

"There he is, talking to Tattoo!!" Calla points and shouts over as the lads follow. "How Stroker!! Remember me!?"

The small waiter is pointing as the old fella stands up and walks over from a table full of people. "Well you made it, sorry about the confusion. I thought when you left I didn't catch your good name, my name's Bob."

Jimmy and Davey offer their hands as they all get acquainted. Everyone is in smart casual clothes as the guy tells the waiter to fetch a fresh bottle of champagne for his new friends, Jonny winking at calla. “Look at them two over there, hambone, they are fucking gorgeous.”

Calla looks over as Bob asks them all to join him so he can introduce everyone.

“Where’s the toilets, Bobby?” Davey nipping his fat arse cheeks together as a small wet one creeps into his underpants.

“Just up the stairs and left,” as Bob sniffs a horrible smell. “I must get Lawrence to stop using that fertilizer – it’s rancid.”

Jimmy smiles looking at Davey walking like a penguin up the stairs.

The 2 girls are all about the lads, both 30 and 32 with lots of very nice friends. “Nice pad your dad has, Jennifer, what does he do?” Jonny chats to the birthday girl.

“He owns a golf course, but he isn’t too stuck up.”

Calla nudges Chuck. “I will be stuck up later the neet, Chucky, right up her cackbox!!” Tom and Chuck both burst out laughing.

“So, men, what do you all do in here?” Bob chinks glasses with them all.

“We are economic refugees,” Calla points out.

“You are what?”

Jimmy shakes his head. “Dont mind Andrew, he’s crackers, Bob.”

Bob is smiling. “Crackers? Economic refugees?”

Calla taking the bottle off the table tops everyone up. “Crackers is for cheese, Bobby. We are economic refugees because the economy is fucking shit in England so we come here to make a good wedge!”

Bob looks confused. “Times are tough in the UK so we come here is what he is saying, pardon his French,” says Jimmy putting Bob and the girls wise.

“So Bob, any chance one of your lovely daughters can show me around the place??” Calla nudgs Chuck.

Bob waves an older woman over. “Liz dear, have you a moment! Better still my wife can give you the guided tour – she practically designed it!”

The woman arrives in a long low cut red dress, quite attractive, ears and neck sparkling with the largest diamonds the lads have seen. "Oh she...Damn I forgot to ring my mam, it's her birthday. Jimmy, giz your phone, I'll not be a minute, I'll catch you up!" Calla takes the phone from a smirking Jimmy then heads towards the hedge maze shouting: "Ay mam, happy birthday!" winking at Tom who with Chuck, Jonny and Jimmy are trudging behind the lady towards the house, Calla pretending to be on the phone some time asking questions, laughing and smiling towards one of the girls next to Bob. "Well mam, enjoy your day wish I could've been there, love you so much!" then blowing a kiss. He pretends to hang up walking back towards Bob he asks, "So Jennifer what is you do? By the way that dress is stunning – is it new?"

He rubs his hand across the shoulder smiling, the blushing girl giving a little curtsy. "Thank you, yes, I bought it last weekend in Paris!"

"Paris? The city of luuurrve, always fancied there. My mate went for his anniversary with their lass, went on a romantic cruise up the Seine, ended up throwing his wedding ring into the water arguing!"

Jennifer is laughing."Should we get a drink and you can tell me some more about your trip there!" Calla puts his arm around her heading for the drinks stand.

"Andrew, I thought you wanted a tour?" Bob is shouting after him.

"I will later, Bob!" not even turning round.

Davey uses the full toilet roll to wipe his arse knowing there's still some shite to wipe. "Orr fuck, I bastard hate the scatters!" Shimmying across the floor he flushes the chain and decides to wash his arse in the bidet. Sat swilling his hoop he notices the toilet back up. "I can't believe this! How quickly things can gan rang!" Still with his pants down he tries clearing it with the toilet brush. "By Christ that fucking stinks and it's me own shite! Things can only get better!" Shoving hard with the brush, "get doon man!" as the door swings open to the sound of Liz "–and this is the spare bathroom."

All the lads burst out laughing as Davey swings round, cock out. "Cold, Dave?" Jonny quips to Davey trying to hide his knob with the toilet brush caked in shitty tissue – the smell could choke a donkey! Slightly embarrassed and with a gag – "Oh I'm sorry, didn't realize anyone was in!" – slamming the door with her hand over her mouth.

Davey finally getting the turd to flush squares himself off and opens the window and heads downstairs embarrassed. All the lads are back outside after the tour as Davey heads straight towards Liz to apologize, all the lads still sniggering as she reluctantly shakes his hand and smiles saying, "It's ok, happens all the time."

Slightly red– "Get me a bastard beer!"

Jonny gives him a playful push. "What you been eating?"

Davey necking his bottle in one, burping into his hand. "Wasn't that bad! Just a bit chowed bread!" as they all start laughing.

Looking round Jonny asks, "Where the fucks Calla?"

All the lads scan the grounds which are busy now with nice shirts and nicer dresses, place dripping with quality as Bob wanders over. "Gentlemen, have you seen Andrew and my daughter?"

The lads look at each other. "No, sorry Bob!" all shaking their heads.

"Oh, well sure they are around – please enjoy the drinks and help yourselves to food–" as Tom comes back with the largest plate of meat ever stacked, 3 tiers high. A rib wobbles, Tom grabbing it and sitting cross legged, plate on lap, sets about it, tearing meat off the bone like a dog at hot chips. Bob frowns. "Ha, well it seems you know where everything is. I will speak with you later on!"

He walks off back towards the handshakes and fake laughs, Jimmy grabs Davey's arm. "Giz ya phone! Calla's still got mine; let's bell him before he gets us all slung out!"

"Is it ringing?" Davey asks Jimmy as he chomps on a rib.

"Aye but nee answer, he'll be up to no good, Davey! Send Jonny on a scouting mission!"

Davey wanders over to the lads. "Nice BBQ, eh lads. Jonny, gan find Call man, son."

Chuck with a corn on the cob in one hand and a rib in the other: "Too right, Davey, free food and drink always tastes sweeter."

Tom sniffing his nose up as a woman walks by – "If it smells of cologne, leave it alone."

Davey laughs. "I'm going over to try some of that seafood!"

Jonny walks down past the drinks bar, down the path with the steps leading onto a wooden boardwalk through some lovely garden. "Calla!?" Jonny shouts as he hears laughing and giggling through the hedge. Looking through the hedge he can't see a thing and jumps at a tap on the shoulder. "Jesus, Bob, you frightened the shite out of me!!"

Bob apologizes. "We need to find Jennifer, it's the cutting of the cake."

Jonny reassures Bob. "You go back and prepare everything. I'll look for her and Andrew, I'm sure they won't be too far, Bobby."

Bob shakes his head and marches on. "I know these gardens like the back of my ha–" Stopping in his tracks as he walks round the corner to see Calla sitting on a crate of beer, drinking a can with his daughter on her knees giving him a blowjob.

"Areet, Bob." Calla lifts his bottle as Bob lunges forward, Jennifer jumping up.

"Dad I was just, erm..." Bob pushes past her as Calla stands up moving to run, tripping with his trousers around his ankles, his bare arse in the air!

Jonny grabs Bob with his arms around him. "Whoa there, Bobby, it takes two to tango man!!"

The old fella is raging. "Get out my house you dirty bastard!"

Calla stands up, pulling his strides up. "Calm down, Bob, it was just a bit of fun." Calla is trying to rescue the situation.

"I'll let you go if you are calm." Jonny releasing Bob slowly as Jennifer is nowhere to be seen.

"Ok I'm calm, just leave immediately!"

Calla reaches down into the crate, getting 2 cans out, offering one to Bob. "No hard feelings."

Bob lunges forward, Jonny grabbing him. "I said out!!"

Calla pinging the can open walks by him. "Please ya sell."

Jimmy, Davey and the 2 lads are sitting with Liz and a few of her pals getting on famously as the lights come on and the band start playing Careless Whisper when Calla comes marching over. "Haweh, there's away!!"

Davey stands up. "Sit down, son, we are only just here! Why do you want to leave!?"

Calla taking a massive garlic prawn off Davey's plate and biting it, replies: "Firstly the prawns are fucking freezing and secondly George Michael songs are wank!"

The lads laugh. "Sit down man, Calla."

As Bob comes charging out of the bushes with Jonny trotting behind him, Calla throws the prawn on the table and takes a swig of his can. "And thirdly I just got caught getting my helmet sucked off the birthday girl's dad!" All the lads jump up!!

Bob is trying to shrug off Jonny – "Get out of here now, you animal!" – pointing at Calla. Tom and Chuck sit back down and pick their plates back up, starting to eat as Bob slaps Tom's plate of food out of his hand sending food everywhere. "I want all of you OUT!"

Tom slowly stands up and with a calm voice looks down on Bob still ranting and waving his finger. "Hey now, just hold on a minute, little man! Nobody speaks to me that way and no man has ever slapped my plate out my hand!" He steps forward towards Bob who has begun to shit his pants, Tom grabbing him by the lapels and lifting him onto his tip toes. Everybody stops what they are doing and stares. "Now I'm unsure what's gone on but you see me, I've been good as gold sat here enjoying the fine food and the company of your fine wife!" Liz blushes and smiles towards Tom. "Now I suggest you lower your tone, apologize and let us leave quietly!"

Bob agrees, shaking his head as Tom lets him down, straightening himself up. "I'm sorry, I'm so sorry, please leave!" giving Calla a look of hate as the guys put their plates down and begin to wander off outside.

Calla grabs the arm of a girl walking past. "Tell Jen I'm at the Hilton private party at clubhouse 8, tell her to head for the noise! You're welcome too, gorgeous – in fact tell anyone!"

Outside the front of the house the lads stand as the small butler is waiting with a SUV's door wide open. "Please gentlemen, this way, the car will take you anywhere you like so long as it's outside these gates. Calla stooping down behind the man shouts "Boo!" in his ear, the butler nearly jumping out of his skin, his hat falling off his head as the lads pile in and the car heads out down the drive.

Davey spins round. "Well that's the shortest time I've spent at a party, Calla, you couldn't help yourself! Always acting the bastard goat, what a waste of time!"

Calla leans forward. "Had on, Davey, what would you do, we got chatting about Paris and I asked if she spoke French. She said Oui Je bien com sava Jennifer com en tapple blah blah whatever – I just laughed and asked her what's French, has two thumbs and likes blow jobs! Holding two thumbs up I said Moi! She nearly rove me bollocks off and dragged me into the bushes!"

Jonny butts in. "Dragged! Bollocks, Cal, you had a case of Miller under your arse?"

Calla laughs. "Straights she did! I popped back for the box of beer the grass was damp!"

Tom chuckles then swivels round. "Say man, you got a BJ, Davey flashed his cock and some dude smacked my meat! Say we did ok in a half hour!"

All the lads are crying with laughter as the driver pipes up: "Where can I drop you?"

"Hilton, cock features and step on it!" Calla snaps.

The lads arrive back into the kitchen to find the place deserted, the place surprisingly tidy and Willy lying on the settee, porn movie on and his shorts round his ankles!

"Fucking hell where's everyone at!" Calla goes outside looking round as the place is immaculate yet empty.

Davey kicks Willy's feet. "How wanker, wake up." Willy not even stirring as Calla comes back from the fridge with a bottle of Alco ginger beer. "Why am fucking sick, where's all the talent gone? Two hours tops we've been gone!!"

Jonny and the lads raid the fridge and move onto the patio as Calla pours half his bottle of ginger beer over Willy's balls.

"Whoa, what the fuck!" Willy jumps up covering his balls.

"Wake up, sad sack! Where's every cunt at?"

Willy pulling his shorts up and getting his tabs from the table – "Where's my glasses?"

Calla sits down on the settee. "On your heed, you balloon!"

Willy puts the glasses on, lighting up. "Aye they all fucked off to

Sinatra's for some band, Rama had to work but he says they'll be back in a few hours!"

Calla getting up turns the telly over to MTV and cranks it up. "You can pay for the porn film n all, Willy, you dirty little bastard."

Willy stands up stretching. "Nae bother, big un, although I only watched 2 minutes of it and fell asleep!"

Jimmy shouting in from outside – "20 dollar wank!" As calla walks outside and volleys a can over the fence. "Who tidied up, Willy?"

Willy walks outside. "Fuck knows, the maid probably."

Jonny laughs. "She probably laughed her way through the task looking at your little cock."

Chuck nods. "Yep, scarred for life."

Just then the doorbell goes and Calla spins on his heels. "I hope to fuck this has a pair of tits," as he marches towards the door.

Jennifer and her pals are there with boxes of drink and balloons. "Hiya Andy, my dad is in bad form so we decided to take you up on your offer."

Calla waves them in taking a bottle of champagne out of the box and stopping Jennifer in her tracks as her pals walk out to the patio. "Where yay gannin?" Jennifer looks surprised as Calla takes her by the hand towards the stairs. "You have some unfinished business to attend to." He pops the cork with his teeth and spits it through the railing s on the stairs.

Drinks are flowing by the pool as the 2 Texans show 3 of Jennifer's pals how to play shoes, Tom putting the pin back in. "We had these all set up but Calla threw everything over the fence earlier."

Jonny is chatting to a blonde girl while Willy, Jimmy and Davey are painfully listening to some girl bleat on about the American way of life as Willy shakes his head and stands up. "Ma heed is mince listening to her pish." He walks into the kitchen taking 2 eggs out of the fridge and putting them in Calla's shoes. "Calla, are you finished yet? I might head up t Sinatra's for a pint and a fight!"

Calla shouts down stairs: "Willy, get me a pint in, I'll be finished in 10 minutes," as a laugh and a headboard against the wall echoes from the room. Jonny closing the fridge looking up at the ceiling shouting "Calla, you bucking in my bed?" no reply as he wanders back out shaking his head.

Willy walks back out with a fresh shirt on and a bit of aftershave. "Right bawbags, am away for a pint in Frankie's."

Jonny looks over. "Is that my fucking shirt, Willy!?"

The girl standing up from Davey and Jimmy says, "I thought it was called Sinatra's."

Willy looks over at Jimmy. "Is she wired to the fucking moon, bigun or what?"

Jimmy stands up. "Aye. Haweh Davey, we'll gan for a couple, we have to be up for 6."

Davey gets up. "Lads we are heading up the bar if you fancy it."

Tom and Chuck not even thinking twice as they walk over, Chuck tucking his shirt in. "Yep count us in, although I have seen enough beer to last me a lifetime this weekend."

Jonny turns round. "I've spilt more down my green tie! Aye a think I'll have a wander up too."

The head wrecker shouts over to Willy: "Are you going to wait for me, Willy, I just have to get my picture taken here next to the pool." Her and the pals line up handing Jonny the camera, Willy walks over. "Watch you don't get too close, fall in and fucking drown."

The girls are posing as a massive "Wahey" as Calla flies through the air with a pair of knickers on his head, bra on and a sheet wrapped round him for a cloak, lands in the pool from the balcony, soaking all 3 of them. He comes to the side and pulls one of them in by the dress as the other 2 move away raging: "You god damn fucking idiot!"

Calla climbs out as the head wrecker does the same, her black dress right up her arse showing all her bits with her hair stuck to her head and make up all over the place. All the lads are pissing themselves laughing

"On that note I'm off for a pint." Willy heads for the door with the lads following. "Hurry up, daft arse, we have to be in bed early for that flight the morn." Jimmy and Davey head to the door.

Jonny gets another 2 cans out the fridge and passes one to calla "Calla you would fuck a good fanny up, you, by the way"

Calla sits down. "Cheers pal, ah haway girls, just a bit of fun!! There's hairdryers upstairs!"

The girls storm past Calla as Jennifer comes down. "Andy, did you see my underwear?" Calla leaning back on his chair points to the pool.

Jonny and the girls are at the door waiting for Calla, "Hurry up, shandy, it's last orders in an hour," Jonny shouts up the stairs.

Calla comes down all polished. "Ta daaa, all clean, darling," as he sits on the settee putting his shoe on, the crack of the egg making him pull his foot back out. "Bastard!!"

All the gang laugh. "What goes around comes around, Calla" Jonny laughs.

"Aye and wait till I get that specky little fucker back!!" Calla throws his sock in the pool putting his other foot in as the egg cracks. "The little cunt" All the gang laugh again as the second sock hits the pool and Calla walks out wearing flip flops.

Jimmy and the lads are in the regular spot looking onto the garden where a huge crowd are listening to a band play out from a bandstand, very personal and very acoustic setting with what looks to be a good few hundred in the crowd, all ticketed but no hustle or bustle, Sinatra's bar being in an ideal spot to overlook the event.

Big Rama walks over to the lads as Davey asks, "Some chanter him chor."

Rama bends over. "Whaathe?" Not having a clue.

Jimmy puts him wise. "The lad's singing is very good!"

Rama nods. "Yeah that's why it's a guest list only in the bar tonight, or manager's discretion."

Davey drinks his pint down. "So the crowd down there can't get in?"

Rama shakes his head. "Nope that's why there is 2 can bars in the area below – they paid minimum 100 bucks for tickets too! You guys are just lucky you are in the hotel."

Calla, Jonny, Jennifer and her pal walk up to the door, the other 2 girls leaving because they were soaked through off Calla's diving antics earlier. "Sorry guys, it's full, and you have to be a hotel resident," the doorman tells the lads.

"We are hotel guests, you balloon, go and get big Rama!"

The girls start singing out loud hugging each other at the excitement of the band. "I heard this song before somewhere." Calla tells the girls. "This song is by the Goo Goo Dolls!" Calla looking to see where Rama is says: "I'll Goo Goo this bastard bouncer all over if he doesn't hurry the fuck up!!"

Jonny nudges Calla. "Here you go, Calla, here's Rama."

Rama lets all the gang in, Calla having a word. "Rama, if you weren't built like a brick shithouse I'd be having serious words with you!"

Calla gets the drinks in. "Put me loads of tabasco and chilli around the rim of that vodka glass, pet, gan on, lash it on!!" Calla laughs and carrying them over to the lads. "Oy oy!" Calla pushing through the crowd to get to the lads.

"Alrite, fanny balls!" Willy winks at Calla. "This band is egg-sellent, Calla!"

All the lads laughing, looking at Calla's feet. "Where's the shoes, son?" Davey laughing looking down as well.

Jimmy punches Calla in the arm. "Are you saving them to wear with ya shell suit!!" Everyone laughing.

"Aye aye, very funny, here Willy, they gave a spare vodka – you're in luck!"

Willy takes the drink. "Fair play, big man, wasn't egg-sactly looking for another but cheers anyway." As he takes a big mouthful, swallows half and spits it out. "Fucking hell, me mouth's on fire!! Bastard, pass that pint, Jonny! My lips, aaah ya fucker."

Calla leans back. "What goes around comes around, four eyes. Bit chilli in here, isn't it lads!!"

All the lads piss themselves laughing. "You're some can of piss, Calla." Jimmy chinking glasses with him.

The band carry on with some more great tunes as the lads crack on with the drink. The band finish off to their signature song Iris with the girls all screaming. "Fucking stroll on, bet them bastards get some clunge!" Jonny nudges Willy who is still chewing ice cubes.

"Ay hen, some bucking that front man will dee! Cunt!" fingering another ice cube out of his glass as the place falls silent then a steady cheer as the band head off. The bar is jam packed and all the lads are in one corner, Calla sitting on the stool with 2 bottles of Bacardi Breezer in his hand with straws in. "Right Willy, I can drink these faster than you can drink your pint."

Willy with a pint in his hand replies: "Nay bother, pal, say when."

Jimmy gives the 'go' as Calla sucks one down in a second and another by 3, Willy still with quarter of a pint to go as Calla digs him in the gut, drink spraying everywhere. "Fuckin hell, Willy, the bairn can drink faster than that."

All the lads are getting pissed and starting to sing songs, Jimmy standing up on the chair. "For all the fish that live in the sea!! The heron is the one for me!!"

All the lads join in with the old especially Davey. "And what d'ya dee with the herrin's heed, we'll mak it into loaves of breed!! And all a matter of things!!"

Jimmy going on – "Hawa ye the day, hawa ye the day, me hinny!! And what dya dee with the herrin's eyes, we'll mak it inte puddins and pies!"

Davey is up on the chair now with his arms around Jimmy. "Herrin's heed loaves of breed, herrin's eyes puddin and pies and all such things!"

Chuck and Tom look confused. "What the fuck is that all about, Jonny?"

"Ah it's an old fishing song from Newcastle, lads, you won't've heard it!"

Chuck laughs. "Well it sure sounds like a drinker's song too!!"

Calla laughs. "Better than that goo schmoo dolls or whatever they are called."

Jennifer pulling Calla. "They are the best, Calla. So are you going to take me back to yours or what?!"

Calla shakes his head. "I'd like to but we have an early start. Give me your number and I'll ring you so you can come to Orlando next week for a visit."

Willy looking confused, whispers, "Calla, are you losing it like?"

Calla leans over. "Nah Willy, she's fucking shite in the sack! I mean show some fucking interest like! I made her look good in the end!"

Willy laughs. "Fuck off, you balloon!! So we going to the casino, bigman, or what?"

Calla nods. "I'm game, fuck everyone, back to ours! Let's see if there is any loose snatchage in the casino, and I've got 800 bucks to blow!!"

Jimmy and Davey are cuddling each other at this stage. "We are heading yem lads, back to graft the morn." Jimmy and Davey are necking the dregs of their pints.

"Aye we'll be down yem after these." Calla gives Davey a kiss on the head.

"Dinnit sleep in mind, we are away at 6!" Davey giving Calla the orders.

"Are you coming to the casino?" Willy asks Jonny and the Texans.

"Aye am game." Jonny swigs his beer.

"Not for us, we are wounded! Need some beauty sleep." Chuck finishes his pint.

"You should've went to bed when you were 7 if you wanted that." Calla slurps his pint down and tells Jennifer again he'll definitely see her next week as she moves in for a cuddle. "Whoa whoa, Haweh! Can't be cuddling, pet, bad for the creases in the shirt!"

Jonny and Willy drag him away. "What about the party?" Jonny asks Calla.

"Aye it's all bullshit, Jonny, sounds good at the time but fuck all them Yankee bastards coming back for the big I am. We are bollocks! let's gan to the casino and win some dosh!!!"

Half an hour in, 4 Tom Collins later, Calla is over to the black jack table. "Willy, lend me 200."

Willy turns round. "Are you fuckin joking? Thought you had 800 dollars."

Calla smiles with his hand out. "I did – now that cunt has it!" Pointing to the dealer on the roulette table. "Nah bigman, Jonny has a few hundred and I have 150, let's cash em in and go to the champagne bar." Calla takes a chip out of Willy's pile and a cocktail off the tray of the waitress walking off. "One more spin, Willy, you know it makes sense." Calla walks over with big strides to the table. "Right, fanny chops, 25 on zero and 25 on 32."

He sits down next to 2 women. "You sound like James Bond." The dark haired girl smiles and moves closer to Calla. "What's your name?" she asks.

"Andrew," Calla replies looking down her gorgeous tanned leg to the tattoo of a dolphin with the name Cindy written next to it. He lifts his drink and chinks hers. "That's a nice name." She looks into Calla's bloodshot eyes. "Thank you, Cindy."

She looks surprised and smiles looking down at her foot. "What sharp eyes you have, Andrew."

Calla winks at her. "Wait till you see my fucking teeth!!" As the ball bounces into the zero, Calla jumps up to see the ball pop out but into 32! "Get in there, always cover your bets!!" Calla puts his arm around Cindy and gives her a kiss on the cheek. "You fancy a night cap in the champagne bar?"

"Yes I do, are we going there now?"

"Yup!" as Calla is raking in his chips.

"Halfs, Calla!" Willy holds his hand out!

"Hawld on! Would you've only asked for half it back if I had lost? Would you fuck so get your minging scotch trotter out my face before I snap it off!" Calla pushes Willy aside!

"Way give me my $50 back, hen!"

Calla signals to the table. "It's still on 0 & 32!" as Willy spins to see the ball being spun round.

"Bastard!" as Willy watches in disbelief as the ball pops into the 32 again. "Fucking get in!" as Willy pushes people out of the way to collect his chips, all smug. Willy walks over to the counter to cash them in. "Squits, big man!" as Willy dumps the chips on the counter, the pretty young girl asking how Willy wanted it. "In my hand, sweetheart! In my fucking hand." He holds out his palm, the confused girl beginning to count out the amount in hundreds.

"Reet, cheesy bollocks, there's away, let's get Cindy pissed!" putting his arm round the girl and heading for the elevator.

Jonny nudges Willy. "We halfys like, Willy?"

Willy smiles at Jonny. "Are we fuck!"

Two steps into the champagne bar and the manager races over to

Calla. "Sir sir!"

Calla walks on. "Bottle of your best helmet, oh and some Peronis!"

The man eventually getts near Calla and puts his hand on his shoulder. "Sorry sir, but you aren't welcome here! We don't permit taking your clothes off in the bar! You will have to leave!"

The girl steps back from Calla. "Ok then, mate. Haweh Cindy, there's away back to mine!"

The girl looks across at Willy then back to the manager. "Do we all have to leave?"

"Not at all, just this gentleman!"

Willy steps forward and sits down. "Bottle of your finest please," kicking a chair out for the girl. "Sit yourself down, pet! See ya the morn, Andrew!" Willy laughs as Calla is walked out.

"You're some cunt William! Here Mindy, he has herpes!" Calla barks.

"It's Cindy!" the girl replies, Calla shrugging his shoulders. "Whatever dog features!"

Jonny walks with Calla. "I'm away, Willy, not leaving Calla like!" Willy shrugs and lights his cigarettes then pours his charm on the girl.

Outside Calla has half a storm on. "Some cunt, Willy like! Wouldn't trust him at times like!"

"You would've done the same, Cal, you're as bad as each other! He'll probs fetch her back and you both will go through her!"

Calla stops. "Don't be fucking stupid, I wouldn't follow that cunt into a lake if I was on fire! Let's get back to the scratch, see wees shown up!"

Back at the clubhouse it is jumping. You can't get moved with people inside and outside. Big Rama is stood in the garden with a huge cigar then holding his arms up "HEYYYYYYY! What's up brother!"

Calla points up to the sky. "That's fucking up Ram and we aren't brothers!" getting a bottle then storming into the house.

Jonny walks over to a confused Rama. "Tak nee notice. Willy's in the Champers bar cutting his grass!"

"Huh?" Rama is confused.

"Calla pulled this bird but wasn't allowed in the Champagne bar so she stayed in there with Willy!"

Rocking back with a laugh – "Haahahahaa! You guys crack me up!"

Three girls are sat on the patio from the bar as Calla walks out. "Evening lasses, I thought Cinderella only had 2 ugly sisters?"

The girls look over with disgust. "Haweh, just joking like! Anyone fancy a swim?"

Willy pushes Calla from behind as he sneaks up behind him. "Aye, you, captain sad sack!!"

Calla's arms are flapping as he hits the pool. "Me money, you stupid jock bastard!!"

Willy sits down with his tab hanging out of his lip. "Put it on the table – it'll dry, you balloon."

Calla gets out, soaked, taking all his clothes off, the girls staring at him bollick naked. "What? You never seen a bit Hartlepool rock before?"

Willy laughs. "Aye, but not with a pair of hairy plums as big and chaffed as them two cunts!"

Calla rubbing his sack and putting his hand over to one of the girls – "Smells like dry roasted peanuts, smell!"

The girls are all pulling away as Jonny passes Calla a towel. "Anyway, Willy, what happened to your date you robbed off Calla?"

Willy shakes his head. "She wasn't the one for me, too vain! Up her own arse!"

Rama comes out giving his girlfriend the nod. "Well guys, there's my cell number. I'm calling it a night. It was great fun meeting you."

Calla letting the towel drop gives Rama a hug. "Aye, I'll miss you, bigun."

Rama pushes Calla away. "You're fucking crazy!"

Calla sits down drinking his bottle, naked again. "Haweh."

The 2 girls stand up with Rama and his girlfriend. "Where you 2 going?" Willy asks.

"We are going home, we are tired."

Willy stands up, putting his arm around the blonde. "We are

going to have a bit dance now and a few cocktails – you should stay"

The girls look at each other. "Ok maybe for an hour."

Calla stands up. "I'll just go and slip into something comfortable." Calla grabs Willy and spins him into the pool. "Get in there, you grass cutting little specky fucker!!!" – as Willy lands in the pool.

He climbs out with his glasses all lop sided. "Calla, I've got nothing to travel in tomorrow, you goon."

Calla laughs as he comes out with one of the girl's coats on, a cap and a pair of sambas. "I'm travelling like this."

The girls laugh as Jonny grabs one and launches her in the pool, jumping in after her!

"Now the party is starting." Calla takes off the coat and trainers, chasing the blonde girl, picking her up and dangling her over the pool – "blowjob, I save you."

The girl screams: "No way!" as Calla drops her in the pool, jumping in after her. Just then 2 policeman walk to the back fence with torches. "Ok guys, we have had 3 complaints and hotel security refuse to come down, time to call it a night."

Calla getting out the pool with nothing on says: "Complaints about what?"

The policeman shining his torch on Call's balls: "that!"

Calla looks down at his self and helps one of the girls out of the pool. "Ok officer dibble, we'll turn the music off and go inside."

Calla wakes up and looks over at the dark hair. "Fuck me, who's that? I thought mine was blonde," he says under his breath. Climbing out of the bed and sneaking into Willy's room he opens the bathroom door. Willy is half shaved and his bag all packed on his made bed. "Areet Willy, who is that in my bed?"

Willy turns from the mirror. "How the fuck am I supposed to know that, you simple bastard!? Eyeballs not crystal balls."

Calla turns on his heels. "Miserable Scottish twat."

The 2 lads shout upstairs, "Leaving in 15, mind," as Calla walks in the room and Cindy is sitting on the edge of the bed. "Ah it's you!"

Her face is one of disgust. "Who were you expecting?" She stands up reaching for her dress.

"Whoa whoa, sugartits, just rewind one minute, I have all day with you," as Calla pushes her back on the bed. Within 7 minutes he is up getting a piss.

"Oh Andrew, we have all day of that, and that was my starter."

Calla pulls up his jeans and rolls the remainder of his clothes into his hold all. "That was your starter, pet – you will have to wait until next weekend for the main course. I have a plane to catch."

Cindy jumps out of the bed. "You lying dirty bastard!!"

Calla putting his shirt on and winking says: "I resemble that remark."

All the boys are waiting in the foyer as Calla and Willy bring up the rear, Jimmy shaking his head. "Apart from the golf, drink and fanny, did you have a good weekend?"

Willy shoving his hand in Calla's jeans and pulling out a ball of 100 dollar notes says: "Aye and we have enough to jog on somewhere else for a few, bigman, if you fancy it!!"

Davey lifting his bag up as the minibus pulls up outside – "No chance, a deal's a deal: we promised to be back to graft today so get the fuck into the bus."

The bus is a shade quieter than the inbound one with Jimmy and Davey in the front, the 2 Texans, Calla, Jonny and Willy in the back. "Can we gan to graft straight away, Jimmy?" Calla winks at Willy.

"Well I was on to Mally this morning and there is a hookup today so it might be a good idea."

Jimmy nudges Davey as Willy chirps up: "Do we get paid all day?" Willy winking at Calla as Jimmy grimaces looking at Davey trying to hold his laugh in.

"Will we go to Nelly's to discuss matters further?" Calla trying to talk posh as the full bus bursts into laughter.

All checked in and at the bar in the airport duty free the lads are tucking into burgers and chips, Jonny's burger dropping out of the bun onto the floor. "Bastard!" shaking his head as Calla picks it up, wipes with a napkin and puts it back on his plate. "Eat the cunt and stop whingeing."

Johnny putting a handful of chips in his mouth and pushing the plate away replies: "No chance, you animal, covered in shite."

Calla sticking his fork into the half chewed burger and biting it

– "There are starving bairns back in Teesside and you waste a burger because it's got a few bits of chewy and fag butts on it."

Davey drops the beers on the table. "I am off the beer after today lads."

Texan Tom bursts out laughing. "Yep, so are we."

Calla and Willy nodding – "Aye us as well, father."

Jonny and Jimmy both agree. "Aye we'll stay off it as well."

Davey taking a massive 'chum' out of his glass leaving a quarter left says, "I'm serious lads!" as they all laugh and Calla stands up. "Reeto if the world is going to end tomorrow as we know it I may as well get them in again."

Davey finishing the dregs off and handing him the glass – "Good lad."

Calla wanders back with the beers with a small dark haired waitress in toe with the other 4 pints, taking them one by one from the tray to Jimmy, Davey, Tom and Chuck, Davey chirping up "thanks, bonny lass", Calla slamming the other 3 down in front of Jonny and Willy splashing beer onto both of them.

Jonny jumps back from his seat. "You ham-fisted cunt!" as Calla shrugs, turns on his heels and marches off towards the duty free shop.

Minutes pass and Calla skips back to the lads and slams his full pint down his neck, burping out loud to the disgust of a man and woman walking past, digging Willy in the arm. "William please have some manners!"

Willy with a sigh gets up. "One more for the plane, boys?"

"Weyaye lad, you think we are a bunch of hermurs, or summit?" Davey cracks a smile then pointing at Calla's back, says, "What's in there, Calla?"

"Well, captain sad sack, I'm pleased you have asked, it is a bottle of Hennessy XO Cognac Jerry bomb!"

Davey lurches forward to take the bottle from Calla. "It's Jeroboam and it's about 500 quid a bottle!" Spinning the bottle round and squinting at the label through his glasses!

Calla snatches it back and slides it in the bag. "That's right and I'm going to enjoy a glass with coke when I get back!"

Davey looking over the top of his glasses – "Of all the years of stupidity I've taken from you, son, I won't allow you to pour one drop of coke into that! I'm telling you!"

Calla laughs. "Nar Davey, I know you like a drop of Brandy so I've bought it for you!" passing a speechless Davey the bottle. "You've kept me straight and a narr you wouldn't take halfs of me winnings."

Davey touched by the gesture leans back and takes a sip from his pint then coughing out, says: "Wayy, you're only happy when you've got nowt, ya know that!"

Calla jumping forward and hugging him replies: "I've got you!"

Davey trying to swerve his slobbering kiss – "You're spilling my pint sackless, ger by man! Calla!" pushing him back to his seat, Calla laughing.

"Ay and by the way I wouldn't spoil the taste of a good coke with that piss!"

Willy hands Calla his pint. "Here sosaj broth, what you two English arsewipes laughing about!"

"The toilet paper on your foot, you berk!" Calla quips.

Willy twisting his foot round and looking down. "Nor man, the other one!" Davey barks Willy swivelling the other foot to Calla and Davey. "Oooo, hello sailor!" the pair laughing like school kids.

"You sad pair of shites!" The rest of the lads laugh and point as the tannoy announces the gate opening. "Reet lads, there's away." Jonny stands up and gulps down the last of his beer!

Straight through to the gate and onto the plane troop the lads with their hand luggage, being greeted by the stewardess, Calla stopping as she speaks. "Excuse me luv, I'm wondering if you could organize me a cold can of beer as soon as you can? I need to take my tablet."

The stewardess quizzes, "Sir, would water not be better?"

"No it gets stuck in my teeth!" showing a toothy smile.

"Ok sir, can I get you anything else?" the confused stewardess asks.

"Yes, can I have a can of beer with that?" with a wink.

"Certainly sir, leave it with me!" she replies with a snigger.

All passengers are finally aboard and seated as the plane taxis down the runway. "I hate flying!" chirps Chuck!

"Better than crashing!" quips Willy as Chuck puts his hand over his face as the engines ramp up and the plane slowly lifts off the ground and steadies at altitude. Calla no sooner clapping his hands and saying 'tin time' when the stewardess hands him the tray with the 2 Budweisers and glasses, Calla shaking his head. "But I said I didn't want water?" He laughs and takes them from the tray. "Keep the cups, I'm a cowboy!" pinging the tin, spraying Jonny.

"Orr Cal, ya twat!" The rest of the lads order their beers.

Jimmy and Davey are snoozing to be woken by the announcement 'due to heavy fog in Orlando we are being redirected to Miami; we will update you further on schedule for Orlando – in the meantime sit back, relax and enjoy the complimentary drink service'. Jimmy looks at Davey. "Miami? Why the fuck Miami, surely they would head back to Myrtle beach?"

Davey shrugs his shoulders. "Maybe it's got something to do with fuel, marra."

Willy leans round the back of the chair in Calla's ear. "Maybe it's got something to do with you having spending money for Don Revie in your Skyrocket, big man!"

Calla pushes his big head through the gap of Jimmy and Davey's seat. "This is a fucking disgrace, Jimmy, we are supposed to be back to graft tomorrow!"

Jimmy turns round in his seat. "These are the times I am glad to be on the same jolly because there is no way I'd've believed you if you didn't show tomorrow."

Jonny chirping up. "Wouldn't the fucking fog be a clue like, Jimmy?!"

Jimmy raises his eyebrow. "That's a funny way to hand your resignation in, flanja."

Jonny sits back down in his seat as Calla presses the call bell. "Bong Bing, more hops for Callaghan!"

The plane lands in Miami waking Willy, his glasses landing on the floor. "Fuck me, have we been shot doon!!"

Calla picks his glasses up and puts them on. "It's a miracle! I can see the moon!!!"

Shouting at the top of his voice Willy says, "Giz them back, daftarse, I canny see the fuckin runway never mind the moon without the bastards!!"

Another announcement: 'Ladies and gentlemen this is your captain' – Calla interrupting: "Please pretty please with a fucking cherry on top, give us some good news!!" –'We have received weather reports from Orlando and we are rescheduled to depart at 7am tomorrow morning. Please contact the desk in arrival'

"Fucking get in there, captain, ya fucking bastard, legend!!"

Calla stands up in the aisle and does the moonwalk as the stewardess comes up the aisle. "Sir, could you please sit down as we are still taxi-ing to the stand."

Calla sits down. "We'll be taxi-ing straight into town in half an hour, pet!!"

Davey nudges Jimmy. "What we going to do, bigun?"

Jimmy cool as a breeze: "Same as we have always done, hamma, book into a hotel and gan oot and get pissed."

Davey settles back into his chair taking his little black book out. "Lend us your phone – I need to send our lass a teletext."

The usual hustle and bustle ensues as everyone begins getting their belongings from the overhead lockers and switching on phones to contact friends and family, Calla's smile beaming as he sings and dances all the way down the plane, planting a smacker on the stewardess's cheek. "So I'm in town for a while, can I stay with you?" as she blushes.

Davey nudges him off the aircraft. "Behave, lad." Calla looks over his shoulder, pulling a sad face towards her. "Davey man, I was in there – she was wetter than Rod Hull's roof!"

"You've more patter than a month of rain, you lad – she was having none of it!"

Willy walks down the air walk adjusting his glasses and smiling at the ground crew who are smirking at him. "Howw boys, a reckon Miami's Willy's town, the hens can't keep their eyes off me!"

The lads all sniggering, Tom nudges Jonny. "Say which one of y'all drew on his face?"

Jonny pointing at Calla – "Well Thomas I will give you a

fortnight and a million guesses!"

Tom chuckles to himself as he glances at Willy who is winking, waving and smiling at every woman who walks past, big black lines across his cheeks and a knob drawn on his forehead in mascara. Jonny nudges Tom. "That's why I never catch a snooze on a plane or drink with Callaghan – the man just can't be trusted!"

The lads all queue up in front of the carousel, Davey wandering across to the information desk to get the full story, Willy still beaming with the attention he is getting looking across to Jonny. "Here man, am nay kidding, the birds are all over me! I can't wait to get a splash on an head oot!" Willy just leaning back and catching a glimpse of himself in the outside window, squinting, he steps towards it noticing the marks and bellowing at the top of his voice: "CALLA, YOU WANKER!"

Dropping his hand luggage he takes after Calla who is already running, laughing, shouting, "Calm yasel, knob head!"

Willy is trying to kick Calla as Davey gets between them laughing. "Lads, lads! Hey now settle! It's like having frigging bairns! Reet we're here for the neet, choice of 2 flights in the morn – one's at 7:45 and the other is at 12! I want us all on the sharp flight so gan canny the neet!"

Willy, already spitting on his hand and rubbing the makeup off, smearing black across his face, Calla pushing him –"move, ya scotch scruff!" – picking his bag off the carousel!

"Right lads I've booked us into the Marriott, it's 10 minutes away, pool and gym and it's a bar and canny restaurant apparently so we can have a few jars and get our heads down!" Jimmy announces.

Calla stopping a group of young lasses in their late twenties, asks, "Ladies, where you lot staying?"

A tall blonde one replie, "Hey we are staying in the Clevelander on Collins Avenue downtown!"

Calla saluting her spins round. "Cancel my room, Helmet, I'm booking a room at the Clevelander! You think I'm coming to Miami and staying in some kip near the airport!"

"You haven't come to Miami, Cal, it's just for the night!" Jimmy tries to get hold of Calla who is walking past.

"Tough shit, Jim! We are here so I'm deeing a bit of sightseeing! I'll get a cab in the morn back here!" Shouting towards the girls – "Hey how far is it?"

One replies, "Maybe 20 minutes' cab ride!"

Calla rubs his hands. "Fucking bingo! Haweh, who's coming?"

Davey shakes his head. "Wyy it makes sense we all stick together then we can check in together and, Jim, at least we will be able to shake the soppy cunt in the morning when he sleeps in!"

Jimmy heaving his bag up and dialing back the Marriott – "That's if the cunt actually comes back to the hotel!"

Out the front Calla, Jonny and Willy are throwing their bags into the trunk of the big blue Crown Victoria, Calla putting one foot in the car and hanging across the door pointing straight ahead. "South beach! The Clevelander!"

Jimmy nudges Davey. "There's nowt booked y'nar, we might not even get a room!"

Davey standing on his finished tab replies: "We'll sort something out man, dinnit worry!"

The cab ride is like who wants to be a millionaire with Calla quizzing the driver on where to go and where not to go, the driver reeling off a number of hot bars and clubs and explaining some having strict dress codes.

"Are you trying to say we are some bunch of T-Ramps like?" Willy nudging the cabby's arm.

"Heyy man, no way man! Just don't want you guys getting disappointed!" The cabby laughs. "Hey guys, if you look to your left you will see Star Island – this is where anyone who is anyone lives, some of them properties are worth more than 100 million bucks, hey you know who...."

Calla cutting it short – "Shut the fuck up!" turning towards Willy – "You got owt to wear?"

Willy shakes his head and looks toward Jonny who is shaking his head aswell! Calla spins round. "Well it's a quick shop afor we head out! Howw, Tommy tour guide, where can we get some decent clobber for the neet out?"

The puzzled driver clicks on to what he means. "Oh you guys just walk down to Lincoln Road, y'all have all the designer shops there!"

The taxi rolls up to the front of the Clevelander which is bustling with people everywhere, sitting in the booths on the sidewalk eating and some around the open pool bar listening to the Live DJ in his booth spinning records. Calla flicks the driver 30 dollars and jumps out in front of the second cab with the rest of the lads dancing and singing 'Welcome to Miami' by Will Smith, Davey climbing out, Calla shouting: "Some fucking spot this, Jim me old china, there's fucking blart everywhere!"

***

Calla and Willy are in a Hugo Boss, Willy at the checkout with a pair of jeans, socks and a new t shirt giving the girl his credit card as Calla walks up and dumps a shirt, trousers, belt and shoes on the counter. "Alright, fannyballs, get them and I'll square you up when we get back to the hotelalero, hambone."

Willy looks at Calla with a worried look on his face. "Calla, these are about 500 dollars, I want the dosh, mind!"

Calla walks out onto the main street without even answering Willy. At the reception desk Jimmy is checking all the lads in as Davey leans on the desk. "We were going to be staying in the Marriott next to the Airport, pet. Our flight with American was diverted y'na so when we got to the Marriott they said it was full and I was wondering is this to be paid for or is it paid for by the insurance for our flight??" Davey winks at Jimmy as the girl picks the phone up.

Chuck and Tom are outside smoking as Calla walks over to them. "Are we checked in yet, lads?"

Chuck shakes his head. "Nope, that's what happens when you leave the bosses in charge, bud."

The girl on the desk puts the phone down with a smile. "Yes sir, if you have your flight tickets, there will be no problem whatsoever."

Jimmy smiles at Davey. "You old fox."

Davey nudges Jonny. "That's free dinner with drinks on the company tonight, son as we have saved them a fortune!"

Davey and Jimmy hand the room keys out to everyone. "All with our own rooms?! Fucking happy days. I thought I'd be sharing with onion feet!" Willy takes his card and jabs Calla in the ribs.

"Ooof, ya jock bastard, where's me bag of playing out gear?"

Willy holds the bag tight. "Meet me in my room in 20 minutes with green then you might make the Miami scene!"

Calla and Jonny laugh. "Not only do you sound like the mask, you need one, you horrible little cunt." Calla tries to grab the bag. "Nae chance of you getting the swag, spitballs, 2919 in 20." He walks off with Chuck and Tommy.

"Tell me, Willy, would it not be sensible to have a quiet bit of food and an early night tonight? You Brits do like to burn the candle." Tommy with his arm around Willy.

"Listen Tommy, we can all stay in and act sensible at some stage in our life but tonight we are in Miami probably getting drink paid off Fat Jimmy's company credit card."

Chuck laughs. "Well I guess I will have to turn my underpants inside out and join the party, William."

Willy is all ready, shaved and sitting watching the racing from Belmont Park as there's a knock at the door. He goes to the door. "Yes, password please." Silence. Willy peers through the spy glass. "Calla, is that you?" A waiter appears with a tray as another knock is placed on the door. Willy opens the door. "Aye, what's this?"

The waiter walks in. "4 bottles of Heineken and a bowl of fries, sir, you did order them?"

Willy munching a chip – "Aye a suppose I did, where do I sign?"

The waiter opens the door with the bill and 5 dollars as Calla runs in with just the towel wrapped around him and his wallet in hand. "Ha-ha, four eyes, you sold your soul for a bottle of beer and a plate of chips!!"

Willy laughs as he hands Calla a bottle. "I'm anybody's for a tattie and a pint, Calla." Calla opens the wallet and gives Willy 500 dollars. "Here you go, pal, where's the bag?"

Willy necking his bottle, stuffs the money in the pocket and points to the bag. "I'll see you in the hotel bar in 10." Willy picks up the hotel key and walks out.

Calla picks a beer out of the bucket and shouts after Willy. "By the way that Stevie Wonder looks like he picked the fucking outfit you have on!!"

Jimmy and Davey are already in the bar as Willy walks in. "What do you want, son?" Jimmy points to the bar. "There's a tab open"

Willy rubs his hands together. "Just a voddie and coke, bigman."

Davey laughs. "Your fucking kidneys must be like bits of coal, young'un."

Willy smiles. "Any crack from the job, Jimmy?"

"I spoke to Mally before. Everything went well with the shutdown and there's no big push until Tuesday so we are ok to be honest, might even get us paid for tomorrow if you all behave!!"

The barman answers the phone. "Hold on sir, is there a William Wallace in the bar?"

Davey bursts out laughing. "That Calla doesn't stop, does he!"

Willy walks over. "Aye this is William, is that Andrew Longshanks of Shitsville!?"

Calla's voice booming down the phone. "I suppose you think it's funny, you little twat. There is no way I am going out with a skin tight silver t shirt with YO written on the cunt!!"

Willy spits his drink all over the floor. "Well the shops are shut noo, bigun; try my bag it's full of wet stinky tops – wear one of them," as he hangs up and walks over to all the lads, Jonny and the 2 Texans at the table now.

"What's up with Calla? He is normally first down," Jonny asks Willy.

"Aye he tried to act clever up in Boss, throwing a full kit out on the counter as I was paying and walking out, so I swapped his nice white shirt for a gay design you wouldn't wear in a shit fight!"

All the lads are pissing themselves, Jimmy wiping his eyes with the laughter. "I think this is the first time in 10 years I have looked forward to seeing the big daft twat."

Calla strides over towards all the lads sitting at the bar as they all hold their thumbs up and shout YO! Shaking his head Calla shouts to the barman for a pint of Heineken. "And tonight, Matthew, I'm going to be...David Bowie!" laughs Willy as Calla takes a long swig of his pint.

"Ay laugh it up, Willy. I bet I still bury you with the fanny the neet! By the way I want my 500 chips back!" holding out his hand.

"No way, that cunt still cost 180 dollars and jeans and belt make it up to 500! Hey, at least it fits!"

"Ay like a fucking sticker!" Jimmy laughs. "I'm not paying 180 bucks for this shiny shower of shit, Willy. Haweh play the fucking white man!"

Willy shakes his head. "No way, Phil Oakley! You have worn it now and as it happens I think you look very nice!" bending his wrist and putting one hand on his hip, Calla shaking his head and necking the rest of his beer, pointing to the barman then his empty pot.

"Here, sparkles, you're not going out like that though, are you?" quizzes Jimmy.

"I've no other choice – the rest of my clobber is all dirty! Anyhow, clothes don't make a man!"

"Nor but they can mak him look a tiny touch homosexual! You do realize Miami is rife with vagina decliners?" states Jonny.

"Any fucking pillow biter comes near me and I'll punch their railings down their necks!"

Davey strokes Calla's arm and in a feminine voice says, "Or now haway pettle, play nice!"

"Ay bollocks to all yee, anyhow Davey, what's that on your lip, does Martin Luther King narr ya have his tash on!"

Davey sitting back and stroking his moustache replies: "I think I look areet, decided to leave it on when I was shaving!"

"Well you don't! You look a complete tosspot!"

Willy jumps up rubbing his stomach. "Reet boys, I'm away to fill the pan with dog eggs! You lot decide where we are getting some bait!"

Calla kicks Willy up the arse as he walks off, Willy hunching over. "Arrrr Calla, ya bastard, you caught me reet on the sweetbreads!" turning and holding his balls.

Calla laughs. "Ay good, ya jock twat! Hope ya shit kecks!"

Davey leans over to the barman. "Excuse me, bonny lad, where would we get something decent to eat, none of this fancy shite!" pointing to the waiter carrying a plate of sushi to a couple in the booth. "Well sir there is any amount of restaurants but 2 blocks down you have O'Reilly's, they do some good Irish pub food! Be just like your back home!" winking then picking up the glasses.

Davey frowns. "I'm from the north east of England, ya silly sod!"

"Oh I'm sorry, your accent, thought you were Irish?"

"No but the Irish boozer sounds good to me. Call, you might not get in wearing that mind?" Davey jabbing Calla.

"Har fucking har, Magnum. Bet that Irish bar sells polo shirts! They all do! I'll buy one of them and poke this cunt down William von dick feature's neck!"

Davey standing up – "There was no need fetching the tash into it, son!" – smirking then wandering off for a piss.

The lads wander down the street laughing and joking and they get alongside the Versace Mansion. "Here, that's were Mr Versace was killed! Shot on them very steps! Capped for being a homo! Calla, you wanna watch what you're doing round here!" laughs Jonny.

Calla shoves him into the iron gates then points to a bar up ahead. "That cunt's busy – look!"

"Bit cock-heavy though!" says Willy adjusting his glasses as the lads get alongside and notice it's a gay bar and the blokes are all staring at them. "Or fucking hell, lads, get a wriggle on, my fucking plums just shrunk! Dirty queers!" Tom nudges the lads forward.

"Get yasel in, Cal, you and YO t-shirt will go down a storm," laughs Willy as Calla pushes the back of his head and grinds his body against him. "Ooooo William, you make me sooo hard!"

Willy squirming out of the hold – "Get fucked, Calla, ya bender... Get off me, for fuck sake!"

At the Irish bar Calla storms to the bar. "Seven pints of the yeasty golden goodness, luv and one of them green t-shirts, extra-large!" pointing to the O'Reilly's t-shirt.

"But sir, they aren't for sale! We have a promotion on Guinness – 8 beers will get you one of these black ones or a key ring!" the barmaid replies.

"8 Guinness then, a t shirt!" whipping off his t shirt, rolling it up and throwing it into the bin behind the bar!"

Willy shouts: "Whoa there, that cost 180 bucks, ya crazy bastard!"

Calla ripping the plastic with his teeth then sliding on the polo shirt – "I couldn't give a monkey's fart! Arr this fits like a glove!" Looking at the white and gold stitched harp on the left breast – "Hey man, this shirt's a fucking belter!"

The lads make their way to a table next to the sidewalk so they check the talent walking backwards and forwards and look at the menu. "This cunt on the firm like, Jim?" Willy quizzes.

"Ay lad, al square the bait off!"

"Rib eye steak for me then!' Willy slams down the menu followed by the rest of the lads agreeing and doing the same.

"So it's 7 bastard rib eyes? You are some pack of greedy cunts! If a bowl of dogshit was the most expensive you wad all be having that!" shaking his head.

"Willy has that cunt every day! Have you smelled his breath!" laughs Calla as Willy cups his hand and takes a sniff.

"Doesn't!"

"Does!" quips Calla.

***

"I'm fit to burst!" Davey leans back letting out a massive burp.

"Do you want these chips, father?" Calla shoves his plate into the middle of the table.

"Aye gan on, son," Davey's big mitts taking the majority of the chips off his plate as Calla shakes his head.

"You'd eat 2 more tattys than a pig."

Jonny comes back from the bar with a tray of Guinness. "Fancy a game of pool, lads?"

Willy stands up. "Aye, let's play doubles for an hour to let the cow digest and we'll rock on round town." All the lads stand up and make their way over to the pool table area, Davey stopping to ask the girl behind the bar if they sold peanuts.

Two blokes around the 30 mark are playing as Willy walks over. "Any chance of a game of doubles after your game, lads?"

The smaller of the 2 looks a bit pissed, chalks his cue. "Nope!"

Willy shaking his head and slams his pint down as Jimmy steps in. “Easy tiger, so any chance of that game or what lads?” the lad looking at his friend who shrugs “Sure guys if you want a whopping”

Willy and Jimmy break off and Jimmy cleans up. “7 balls, fair fucking play, bigman.” Willy shakes Jimmy’s hand.

“It’s best of 3 guys, is it?” The little gobshite drinking his bottle swaggers over as Calla steps in. “You nearly got the sentence right, fannyballs, best you fuck off in 3 or I’ll rip your bastard head off.”

The bigger lad pulls his friend away. “You must be that mad bouncer everyone goes on about. Sorry for the trouble, sir.”

Calla looks confused. “What the fucks he on about, the daft cunt?”

Jonny laughs. “Must be the O’Reilly’s t shirt, hamma: – Gary Glitter by day, Patrick Swayze by night!”

“Fuck off, ya bell-end.” Calla picks a cue up. “Right Davey, are we doubles?”

Davey drinks his pint. “Aye, when I’ve finished me nuts.”

All the lads look over. “You just had half a fucking bull for your tea, man,” Chuck says in disbelief as Davey empties the packet down his throat.

“Eat to live, son, Eat to live.”

“You’ll live till you’re 157 then, for fucks sake, you can pouch!” Calla puts the white on the table.

Willy puts 20 dollars on the table. “10 bucks a man, there’s mine and Jimbo’s.” As Calla smashes the balls – “No bother sad sack.”

Two girls walk over to the table. “Can we put a dollar down?”

Willy slides over. “Of course, hen. We are playing threesomes – are you interested?”

She takes the dollar back off the table and walks away. “Charm of a fucking rattlesnake, ya cunt.” Calla pots a stripe as Tom comes back with the tray of beer. “Good shout, Tommy, bout time you bought a round. We have only been away four fucking days.” Jonny ribs the big Texan.

“Yep thought I’d get one in, John, keep us both even.”

Jimmy laughs. “So you have noticed Jonny is as tight as a squirrel’s arsehole, then?”

Chuck laughs. “Yeah me and Tom have been working together

for 15 years and he's never changed a damn bit. I said to him when we first met why the fuck are you saving 5 cent coins – he said 5 gets you a quarter, 4 quarters get you a dollar and 3 dollars gets you a cheeseburger."

Davey laughs. "Why there's nowt wrong with that train of thinking."

Jimmy taking his shot rattles the pocket leaving Davey on the black. "Aye go on, father. I will do all the donkey work and you take the glory."

Davey bending down rolls the black into the pocket, smiling at Calla. "Easy money, son, easy money."

Calla takes the 20 bucks off the table. "Good shot, raggytash, next!"

"Why don't we have a look elsewhere? There's nee clunge in here!" Jonny rocks back on a stool, Willy whipping the stool's legs with a cue, Jonny just grabbing a table as the stool skids from underneath him.

"And where do you suppose we go like, Casanova?" picking the stool up.

"Well what about that Sky bar the barman was on about? Then there is Lou's nightclub downstairs!"

"Way lad, fuck off, I'm not going to no night club!" replies Davey taking a long draw on his cigarette and thanking the barmaid who has brought over more Guinness. "Thanks pet!"

"Ay Trevor, you need to back for the 10 o'clock news!" laughs Jonny.

"For fuck sake, man, what's rang with the tash?"

Calla lays an arm around Davey. "Take no notice, Saddam, the tash is fantastic! More hair less fyass I say!"

Davey pushes Calla off, Chuck stepping forward and smashing the balls potting 2 stripes, Willy chalking the cue. "Your some spawney yank bastard!" as Chuck smashes 2 other balls in, screwing the white back for another easy pot.

"What about this little trick I learned in Nam" – potting the next ball while not even looking, smiling at Willy as the white comes to rest on the back cushion snookering himself.

Willy laughs. "Arrr that'll learn ya!"

Tom studies the table while chalking the cue. "100 bucks say I make this pot, then double the black?"

Willy leaning forward to the table for a closer look considers this to be impossible. "I'll take that bet, big man! And I'll run round the table with me boawby out anarl for the 7 ball trick!"

Calla pipes up. "Miss the cunt, Tom. Al half ya bet – you don't want Willy's little stinking cock flopping about, it wad mak a pig sick!"

Tom chuckles back as he holds the cue vertical and screws the top of the white ball bending around one of Willy's balls, bouncing it off the pocket and nudging the last stripe into the pocket to the cheer of the lads.

Willy scratches his head. "Best give yasel a pocket shuffle, son, give you half a bar on!" quips Jimmy as Tom slaps the stick on the nearside pocket and proceeds doubling the black straight in, throwing the cue on the table, holding his hand out to Willy. "Show me the green man!"

Willy flicks off 5 twenties and unbuttoning his jeans and underpants then waddles around the table like a penguin, all the lads crying with laughter, Tom shaking his head. "Crazy motherfucker!"

Calla shouts: "LOOK AT HIS PUBES! LOOK AT HIS MINGING FUCKING PUBES! Willy, you dirty bastard, do you not trim the privets back?"

Willy shuffling his strides back up and doing his belt a little, blushes. "Trimming's for poofs! Oh natural, me sunshine!"

Calla knocks the rest of his Guinness back. "Haweh, fuzzy plums, there's away!" putting his arm around Willy and wandering out of the bar followed by Davey, Jonny, Tom and Chuck, Jimmy shouting, "I'll get these then lads don't you worry!" as he squares off the barmaid then jogging to catch the rest of the lads up.

Walking down the street their sidewalk cuts between the bar on the left and tables where people are eating on the right. Calla is unable to resist pinching a chip from a bloke's plate. "Heyy!" the man calls. The lads keep on walking. A little way down the street and Davey stops a young couple to ask directions to Sky bar, the man explaining what street and avenue it's on and saying it's a small cab ride so the lads whistle down taxis and jump in.

The taxi pulls up in front of the hotel and asks for the fare, Jonny quizzing "This is a hotel, you berk!"

The cabby explains the bar is on the roof. The lads all get out and head into the lobby where they are directed to a lift where they all squeeze in to the top, Davey squeezing out a fart. "Hallow miss!"

All the lads shuffle around as the smell gets unbearable. "Davey, you horrible cunt! That fucking stinks!" says Jonny as he pulls his t-shirt over his nose, Calla wafting the smell – "Someone break my fucking nose! Jesus Christ!"

Davey holds his guts as he rocks back booming out a laff! The lift door opens and the 7 spill out onto the roof bar coughing and laughing, Tom jabbing Davey – "Man, you need to get that checked out, a say something's died up there!"

The bar is open cast with large upright timbers offering slight cover with lights and televisions supported on them, 8 or so flat screens showing basketball, hockey and baseball, small groups of Americans dressed sharp, watching the games, as the women stand looking bored. Call marches to the bar. "Seven pints of beer please, garcon and you can switch this shite off!"

The confused barman leans in. "Sorry sir, we only sell bottled beers and this is the quarter finals of the regionals!" – getting all excited – "Bulls versus the LA Lakers!"

Calla shakes his head. "It's Bull areet! Get me 7 beers, then!" swivelling round to check the place out, the barman tapping him on the shoulder.

"Sir, you want Coronas!"

Calla spins around and leans in. "No, I do not want a Corona – I wouldn't wash my feet in it!" Looking beyond the barman – "Hey, that brown ale?"

The barman smiles. "Yes sir, that's Newcastle Brown Ale – you want 7 of those?"

"AY and seven bottles of Smirnoff Ice and 7 of your largest glasses!"

Rubbing his hands as the barman puts the bottles and glasses down, Calla passes the 2 bottles and glass to each of the lads all looking bemused.

Davey quizzes: "Wayy lad, what the fuck is this?"

Calla pours his: half with Brown ale, half with Smirnoff. "This, fanny chops, is what's commonly known as a schmog! Twice as strong as Dog and gans down like pop!"

Tom's already poured in and he is drinking it without question. "Fucking schmog? You are some can of piss!" Willy barks!

Calla defends himself. "Had on, look around, it's fancy pants shiny bollocks in here. A mean, look at that daft cunt," pointing towards a guy around 6ft 5 with a silky grey shirt only held by 3 buttons, gold chain round his neck and his long black hair slicked back, puffing on a cigar, holding a cocktail glass. "Every cunt's swigging cocktails, even Pablo Escobar's got one! So, as my father used to say, when in Rome!"

Taking a large gulp then topping it back up, Calla squeezes past the lads towards a group of women stood round a tall table sharing a bottle of wine. "Now...let the cat see the monkey!" Davey shakes his head.

Calla is getting on famously with the girls along with Jonny and Willy. Davey and the rest of the lads are laughing at the bar as the barmaid prepares a mad looking cocktail with umbrellas and straws. "What do you call this again, hinny?" Davey tastes it.

"Miami misfortune," she says, Davey's face twisting.

"Send it to Middlesboro's misfortune, over there – that lad would drink turps." The waitress collects the drinks and carries the 4 over to the lads.

"What's this?" Willy takes a drink.

"It's off the gentleman over there, he said he is your younger brother."

Willy laughing raises his glass to Davey. "Tell him he looks like he ate my younger brother!"

Calla asks the girls if they would like a drink. "Do you like cocktails? You can sit on my knee and I'll tell you a few if you like," as Jonny spits his drink out through his nose laughing.

Willy puts his arm around the smaller of the girls. "I used to work as a cocktail waiter – what would you like? Can I recommend a very nice one?"

Calla butting in – "Fuck off! Cocktail waiter! The only cocktails you can make are buckfast and lucozade with a dash of Carlsberg special brew!" The girls wonder are they nice.

"Take no notice of him – he knows zero about cocktails."

Calla takes his wallet out. "Righto, sad sack, ask the lasses for a cocktail and I'll not only give you the fucking ingredients I will tell you how to mak it!! Then you, captain, can pay for them!! Deal?"

Willy drinking his cocktail wipes his mouth. "Deal, right hens, name a cocktail so Tom Cruise can work his magic!"

The gorgeous blonde puts her arm around Calla. "My favorite is a French mojito."

Calla rubs his hands together. "Fucking piece of piss soaked cake, I used to have them cunts with ice cream in them when I was 12 and me mam was at the bingo."

Jonny shakes his head. "Gan on then, daftarse, what's in it?"

Calla holds his hand to his chin. "Right, 50ml Remy Martin VSOP cognac, 1 lime cut into wedges, 2 spoons of brown sugar, 8 mint leaves, soda water. Pour slowly into a tall glass, lightly mix the lime with sugar and the mint, add the cognac, crushed ice and stir with a long spoon then top with soda water and garnish with a mint leaf!! And that, my friends, is a French mojito!!"

All the girls clap and Calla scores for a big kiss off the blonde girl as Willy stands speechless, shaking his head and wandering off to the bar. "And I'll have a bag of salty crisps n all, you little specky twat!!"

Willy slaps Davey on the back. "Areet Garry Bushell, what's on the telly the neet?"

Davey wiping his mouth and shirt where the schmog had spilled snaps: "You bastard heed if you do that again, you nearly knocked my bastard teeth out!"

"Wayy it won't be lang afor they fall out anyhow, you're a white one short of a snooker set!" Rocking back laughing as Davey gives him a jab to the ribs deflating Willy.

"Ooowf, ya heavy handed bastard!" Willy leans against the bar catching a breath. "You bastard winded me!" Catching a breath he shouts over to the barman: "Here, get me 4 French mojitos and 7 brown ales and 7 Smirnoff Ices!" He looks at Davey. "It's not a bad drop is that – seems Callaghan o halfwit has something right!"

Jimmy leans forward. "Don't be so fucking stupid, they are bastard awful; don't get me another one!" – as Willy stands the 2 bottles in front of him – "Too late, Gorbachev!" paying the waiter and asking him to bring the rest of the drinks over leaving Jimmy puzzled.

"Fucking Gorbachev? What's that silly sod on about now?"

Davey half turning and looking at Jimmy's head, shakes his. "You've had ketchup on your forehead since the Irish bar!"

Jimmy spits on his hand and wipes his head. "Thanks a lot, Davey! You're some cunt – why didn't you say!"

"Couldn't be arsed!" Davey quips. "Look at them 2! Tom and Chuck! They couldn't give a fuck – they just float about getting gracefully arseholed!" as Tom wanders over and picks the 2 bottles up and pours them into his glass. "Whatsupp!" then walks off with Chuck's drinks, Davey and Jimmy laughing.

Back at the table Willy's chatting away to one of the lasses. "So Louise, you live down here?"

"No I'm from Minnesota – me and Valerie are here for the weekend! Julie and Amari live down here!"

"We have a few pals working there at the minute – what's the place like? Here Calla, where's Screamer and Anto staying?"

The girl interrupts. "Anto and Screamer? Anthony Nicholson? I know them!"

Willy spits his drink into his pint as Calla pipes up. "Orr norrr, Willy's stirring Anto's porridge! Hahahaha!"

"What? What does that mean?" Louise replies as Calla takes a long sip and leans forward.

"Well I would happen to bet that Anto has either took you out? told you lies? or slept with you and considering you know of him I'd say all 3?"

Blushing slightly – "Well as it happens we just broke up last week! Said he couldn't handle long distance relationships and he flew back to the UK last week."

Calla laughs. "Here Jonny, pass your phone," taking the phone from Jonny Calla flicks down to Anto's, puts one finger in his ear and lets it ring. "Areet Sosaj broth! What's going on?" There's a pause as Calla listens "Ay! shut up! Nor here man we have had a weekend away for a game of golf! Class time! Just having a few scoops with the lads. You still in Minnesota? How long for? So you will be in Minnesota for another 7 weeks?"

Willy nudges Jonny as they both start to snigger, Louise stood with her hands on her hips staring at Calla. "Anto, do you know a bird called Louise? Ay! Yes! Honest! You didn't!" Calla laughs and palms Louise's hand away as she tries to take the phone. "Here Anto, I've someone who wants a word with you! Had on!"

He hands the phone to Louise who immediately bursts out with: "You lying bastard! Louise! Louise Moore! What? Anthony, Anthony! He hung up!" Slamming the phone back to Calla who is smirking.

Willy puts his arm around her. "Listen, Anthony is a worthless Irish sack of shite! Not all of us are like that, in fact we were pleased to see the back of him, weren't we lads!" Calla agrees and turning Valerie around slightly, says: "Now Val, tell me the truth! Screamer! You and him haven't?"

Valerie bursting – "NO WAY!"

Calla straightens his hands. "Fucking champion! Who wants a tequila!" He hands the phone back to Jonny who reads the text from Anto. "Calla you cunt!" Jonny laughs and shows Willy who sniggers then gets back to rubbing Louise's arm who is still foaming!

Calla gets to the bar and slaps Davey on the back spilling his pint again. "Or man will you lot fuck off!" wiping the beer off his shirt.

"Soz like, itchy lip! Here see that bird owa there! Anto has only spent the last few weeks rooting her! Ha-ha, gave her the flick and told him he has shifted back to the UK! So I just rang him in front of her, hahaha, he put the phone down on her!"

Davey shakes his head. "And you think that's funny?"

Jimmy leans forward. "We will be getting Anto back for a few weeks afor the job's finished ynar – he will tip you upside down!"

Calla ordering 11 tequilas – "Ooooooh am shitting my pants am sure!" Slamming down one of the tequilas – "Wooos!! That warms the cockles," putting the lime in his mouth and making a daft smile to Davey.

"Piss off back owa there will ya! You're deein my heed in!" Davey pushing Calla.

"Or. Hawch man, Hitler, lighten up, I've got you a tequila!" sliding the four tequilas in front of Davey and Jim then goose stepping back towards the table.

"That lad's got fucking shit for brains!" Davey slams his tequila down as the barman puts the receipt in front of Jimmy.

"He isn't that daft!" Jimmy holds the bill up and slides across the credit card. Louise and the rest of the girls have loosened off and are now enjoying the banter between the lads, Amari now perched slightly on Jonny's knee and Valerie and Julie sat listening to Calla.

"We are just popping to the toilet," Valerie says, winking at Calla then turning and walking away, Julie smiling at him as she slips away dragging the other 2 girls.

Calla leans in. "Or lads, a reckon am on for a threesome here like!"

Willy shakes his head. "Work away, son, you'll need the extra person to lift that Julie on! Fuck me, she must weigh 15 clem!"

Jonny laughs. "Ay I wouldn't like to clean her cage out!"

The lads laugh. "Laugh it up lads, she has massive tits!" Calla snaps.

"She has massive everything, Calla, to be fair" returns Willy.

"Ay well in all my horrible contracting days I have never followed a fellow work mate onto the job! Especially not a pipe fitter! Them cunts are riddled!"

Swigging his pint Willy takes off his glasses and breathes on them then wipes them on his shirt. "Call son, I do seem to recall you followed the pig onto that lass in Northwich!"

Jonny laughs. "Who's the fucking pig?"

Willy is laughing as Calla interrupts –"He never fucking shagged her, he was lying!"

"Yes he fucking did, Call. Tell the truth now, come on, no porkies!"

Jonny jumping in –"Who is the fucking Pig?"

Willy shouts over to Davey. "Howw Tash fyess!"

Davey's eyes rolling –"What!"

"Tell Jonny about the pig!"

Davey laughs. "Brian Wilson! He was the horriblest man I have ever laid my eyes on! His nose was turned up and it looked like a pig's snout! Here Jimmy, remember when he fell off the ladder and sprained his wrist and the word went round site he was off work because he had hurt his trotter?"

Jimmy and Davey nearly fall off the stool laughing. Calla jumps in. "Ay but he never slept with that bird – he was in the same room but nowt happened, honest!"

Jonny turning to Willy – "So go on, you tell me what happened!"

Willy sits up on the stool as the lads form a circle. "So me, Calla, the pig and the 2 Krankies were out on the piss..."

Jimmy interrupts. "Stevie and Alan Naysmith, the 2 4-foot brothers from Scotland?"

Willy nods. "Aye, tweedle dum and tweedle dee!"

Davey interrupts. "I had them 2 cunts working for me in Norway, 2 canny grafters but they loved the sauce and were both fucking mental."

Willy is getting annoyed. "Do you want to hear the fucking story or what?"

Calla laughs. "Not really, you boring little bastard."

Willy pokes Jonny in the head. "Get the drinks in, Hymey you Jewish Mackem!" as he continues: "So we are in this bar and the pig is chatting to this bird, bit of a heavy waller but, like daftarse says, shaggable cos of the breast department although she had an arse like champion the fucking wonder horse! So the pig fucks off with Red Rum and me, Calla and the 2 midgets try and get our hole with some gang of lasses on a hen. After an hour of trying and failing mainly due to Stevie Naysmith coming back from the bog with nowt on, just a bog roll wrapped around his head like a turbin shouting poppadum poppadum and us getting kicked out, we headed back to the digs. We opened the pig's Grey Goose and started playing cards. Daftarse here decides to let the fire extinguisher off in the pig's bedroom forgetting it was a powder canister! We just heard the big parfft noise from down the corridor and all 3 of us went to look as Calla was running back towards us pissing himself laughing. The pig and Shergar came out the bedroom, bollock naked, covered in white powder! The pig's face was perfectly clear as was the horse's fanny!! He must've been down there looking for bait when the powder went off!! He was fucking raging. We were all pissing oursels laughing. So to make things worse he came up with his shorts on and we were all into his goose!"

Chuck interrupts laughing. "Fuck me, William, this sounds like Animal Farm – pigs, horses and geese!"

All the lads laugh. "So how did Calla buck the horse then?" Tom asks Willy.

"So the pig fucked off to bed in the huff, with his goose! And the horse came up in the sitting room for a drink – she was like Marley's fucking ghost, all the white powder on her. Calla being the gentleman that he is said the dust was toxic and she should get a shower. He took her to his room so she could get washed! After 20 minutes we heard the latch go in his door and the next morning he had that big stupid grin on his face saying they just cuddled!"

Calla finishing his drink off nudges Jonny. "Haweh mingebag get the fucking shmogs in, and by the way the pig said to me the next day her fanny tasted like scampi fries! He was more upset half of his vodka was gone than the fact he never got to shag Desert Orchid!!"

All the lads wipe their eyes as the tall blonde girl walks over. "There's a VIP section with a free table for you guys, courtesy of the manager."

Willy jumps down off his stool. "He must be a brave man letting us fuckers in there!"

The girl laughs. "No, he said you guys have spent more money in the last hour than people spend in a full night! It's not too busy but it gets busy in an hour, so enjoy!"

Davey climbing down pulls his strides up. "Is there free bait as well, pet?"

Chuck sits down and Calla puts a few dollars in the jukebox and starts smiling. "Fuckin great tuneages on this by the way."

Davey takes a massive mouthful out of his pint and lets a massive burp out turning the heads of half the bar. "That fucking Gary Sprake has got me full as an egg."

Willy laughs. "I thought it was another 9/11. Fucking hell, Davey, you're one pig."

The lads congregate around a pool table as Bruce Hornsby and the Range comes on to the delight of Calla. "Classic, fucking classic! Right Chuck, who's your partner and get your money down while I tank your Texan arse."

Tom does a little jig while pouring his schmog as Jimmy takes a phone call and wanders off. Willy comes back from the bar with a round of shots. "Right, sadsacks, get these bastards down the neck." Jonny taking no second nudge to pick his up. "Whoa woah, captain, had on a wee minute, I am going to do a toast."

Chuck potting a ball and laughing – "You getting all sentimental on us, William?"

Calla flicks a nut off Willy's glasses. "Ay he's fucking mental alreet."

"Where's fat lad gone?" Willy putting everyone's glasses down in front of them as Jimmy comes back in the bar. "Do you want the good news or the bad news, lads?"

The lads are all listening up. "Good." Davey climbs off his stool. "Always good first cos if you get the bad first it always spoils the good."

Calla chirps in. "Ay well said, magnum, wipe your privet! Anyways, Jimmy, what's the news?"

"Well that was Tony Hamilton, CEO on the blower, he has been in with the client today – who are delighted with the progress by the way – and want us to do the second trip after our leave."

Jonny scratches his head. "So is that the good news?"

"No Jonny, that's part of the good news. We are getting another trip and we are going home on Friday this week instead of doing the full 3 month trip."

Willy, who has been absorbing all the information, totting up figures, going through the militant handbook of militancy handed down to him from his welder of a father, says: "Right o, so it's Sunday today, we fly back to Orlando tomorrow, so with that news we go back to blighty Friday, 2 weeks short of a 12 week jolly?"

Jimmy nods. "Ay that's right, Taggart."

"So," Willy continues, "2 days at graft left?"

Jimmy nods again. "We just have to wrap everything up, tool and plant because the shutdown means phase 1 is now live so the Yanks have to commission that. When we come back after the 2 weeks we move into the next area."

Chuck wanders closer, his big barrel chest pushing Jonny out of the way. "So that's a month with no pay?"

Davey shakes his head. "No Chuck, it means you just go home early, have your 2 weeks off and come back! Two weeks without pay."

Jim nods. "Ay he's right, lads, no big deal is it?"

Calla laughs his head off. "The good news sounds fucking areet to me, hammas. I just hope I dinnit get a fucking dose of the pox the neet or it'll not clear up in time before I buck the ex! on Friday neet."

Davey shakes his head. "Is that all yay think about, lad?"

Calla putting his head back thinking. "Erm, aye pretty much, oh and drinking lager and bucking lasses."

"So Jim, what's the bad news?" Jonny having his two penneth in.

"Well lads, according to the big cheese, some of us won't be back – they have a big job starting in Shanghai next month. Don't ask me who's coming back and who isn't – all I know is everyone will be grafting."

Davey agrees with the news. "Ay hinny, as long as we are all making a few bob, that's all that matters."

The next song comes on, If you're gone, by Matchbox Twenty, Calla laughing. "Fucking some tune for the occasion this cunt!!"

Willy laughs. "Ay raise your glasses, lads."

Everyone grabs the shot glass. "What's the big speech, William?" Chuck's big mitt dwarfing the glass. "Well it was going to be a thanks for the weekend, let's get fucking splattered and watch Calla make a cock of himself, but since Jimmy's splitting us up I might have other words to add."

Jimmy shakes his head. "I never split anyone up, you daft jock twat, I am the one making sure everyone gets yem with a bundle of dosh and the chance of another merry go round to play on after the fortneet off."

"He's only winding you, bigun." Davey puts his arm around Jimmy.

Calla stands on the stool. "Before my good friend William, 'the four eyed, dog breathed, onion feeted midget of a Braveheart cunt', rattles on I would like to have a few words."

All the lads are pissing themselves laughing. "Haway then, daftarse!" Jonny throws the chalk at Calla.

"I am not one for soppy speeches but here gans, Jimmy! You fat Geordie bastard, please don't send Chuck and Tom to Shanghai! The last time the Yanks went owa there they got their fuckin holes opened wide !! However send me and I will reverse the misfortunes and open the holes of the slanty eyed!"

Willy swipes the stool away as Calla lands on the floor in a heap not spilling a drop of his shot shouting "get it down ya's" as he necks the drink, all the lads following suit.

The 4 girls wander over to the entrance of the VIP and wave Calla over, marching over to where the doorman has stopped them. "Yes ladies, how may I help?"

The girls all laugh. "Come on, Andrew, stop messing – tell him we are with you!"

Calla puts his hands on his hips as the doorman glances back at him. "Well now, that depends on whether any of us are bucking any of you tonight?" The girls all sigh as Calla lifts the rope and nods to the doorman. "They are with me, fanny chops – at ease." Turning to Valerie –"Look, sweet cheeks, I'm only here one night and I don't believe in wasting time – life's far too short for what ifs and could've beens? So I'm going to be straight up with you, no lies, I haven't had sex since my wife ran off with my best friend 18 months ago and I've finally convinced myself there isn't anything wrong with me and I am able to..."

Val holds a finger over his lips. "Save your bullshit! You're coming back to mine tonight!" then kissing him on the cheek she wanders over to re-join the girls.

Davey stands up and gives her a kiss on the cheek. "Hello, bonny lass, I'm Davey."

Willy licks Calla's cheek. "What's up with your silly face?"

Calla smiles. "Eer that Val didn't even listen to my patter, she said she just wanted bucking!"

"No-one listens to your patter, you berk, you normally only get your hole because you get them pissed beyond the making the right decision threshold! But respect son, get her bucked," digging Calla and walking off.

Jimmy shouts to Willy who is back at the bar, racking up more shots, "What happened to your speech, lad?"

Willy prompted then bounds over with the tray of shots. "Tak one and pass them round. Right everyone, my few words – Calla, shut the fuck up!"

Calla holds his palms out. "Haweh!"

Willy pushes his glasses back onto his nose. "Lads, it's been yet another great trip away from yem! As always a bless to work with a bunch of blokes who narr exactly how to gan on with pipes! Whether it be welding steel ones, drinking from rubber ones or fucking flesh ones! You're all pros in my book and should it be east or west for the next spell I hope I'm with you lot! Raise your glasses, lads! To the best drunken fucking grafters in the land! Hoorarr!"

The lads all cheer then sling the shots down their necks, the girls following a little puzzled by what noises the funny little man just made as Calla picks Willy up and squeezes him giving him a big kiss on the cheek. "I love you too, you horrible little jock cunt!"

Davey stands up and claps. "Well lad, I'm sincerely touched," leaning in to give Willy a hug.

"Fuck off, bog brush lip!"

Davey sits back down. "Bastard, charming that!" The lads laughing, Tom and Chuck patting him on the back.

"Hopefully you'll be stateside for phase 2. Hand me over your email and if you're not we can keep in touch!"

Willy shakes his head. "Tom, Chuck, what I've came to learn in working the world over is you meet new guys, you drink with them guys, you get on with them guys and you create stories to tell about them guys and them guys become remember hims and that's and thens – let's be honest, you never keep in touch, them contacts end with the contract and them asking for your email is a polite way of saying fuck off, I don't like goodbyes! Right or wrong?"

Tom gets Willy in a hug head lock and squeezing. "Right on, man! Stick your email up your ass!"

All the guys laugh again. "Is it fucking speech day or what?" barks Jonny easing Amari off his knee.

"No, it's go to the fucking bar to buy the round in, you tight duck lipped mackem mongo faced twat!" snaps Calla as Jonny stands up shaking his head.

"Nee need for the duck lips, like!"

Jonny brings the drinks back to rapturous applause from everyone, even the bouncer coming over to shake his hand. "Well done Jonny, you tight err mak erm mother fucker," after Willy had tipped him the wink.

"You're a right shower of cunts the lot of you, I pay my way – you lot just drink too fast."

Calla puts his arm around Jonny. "Jonny man, we are only pulling the piss, we are here nearly a week now and we just thought it was time you broke your second fifty, that's all, hamma."

"Fuck off Calla, you gormless mong."

Calla turns back to Davey. "Haweh, bit strang that, father. So, father, fancy a little night cap back at the hotel? At least you have me yem?"

Davey smirks at Calla. "I thought I heard this good looking lady say she was taking you back to hers tonight?"

Calla grimaces. "Nowt wrong with your fuckin lugs, is there badger lip! What I said was I would go only if she bought me a nightcap." Calla nudges Val.

"I will of course buy you all a nightcap. I will just call my driver to pick us up."

Davey looks shocked. "Driver? Why you must be from the posh hooses, hinny, are ya?"

Val looks blank as Jimmy chirps in: "He means you must be well off to have a driver, pet."

Val looks a small bit embarrassed as Calla snuggles her into his chest. "Dont tell these lot anything, gorgeous – you just make sure the driver stops off and gets me 20 tabs and 8 cans before he get us, will ya?"

All the lads laugh. "How many can you fit in the car, Valerie?" Jonny is starting to sway a bit as Willy grabs Calla by the balls. "She'll fit pencil features in no bother, Jonny – he's hung like a fuckin moose! And that's a hoose moose not a Canadian moose."

Calla swipes Willy's glasses from his face. "Ay, is that right, goggles for eyes! See how clever you are trying to walk down the stairs like Stevie Wonder, ya Scottish ball bag."

Willy pleads with Calla. "Give me them back or I'll start swinging."

Jimmy nudging Davey. "Bowt time we got the bairns home before it ends in tears."

Davey agreeing and shouting up – "Haweh lads, the beers are on Jim back at the hotel."

Jimmy shakes his head. "I never said get them back for more beer, Davey."

Davey laughs as he swills his shmog down and pulls his face like he just sucked a lemon. "Fuckin arful, them bastards. Why Jimmy man, it's like telling the bairns the cartoons are on back home, they'll always fall for the free drink chestnut."

The lads make their way downstairs to the foyer with 3 of the girls in toe, Calla going through the automatic doors as Val looks for her car through the glass. "Can you get us 2 taxis, Jackie!" Calla shouting over as the concierge signals 2 cars. The doors then open and Calla rides through the doors on a pedal bike with the concierge chasing him. Calla waves over to the lads with 2 hands, as Davey and Jimmy shake their heads while the lads piss themselves laughing, pointing. "Why don't I give you a backer back to mine for a buck Valereeeeeeeeeeee!" Calla sings. As he rides round, a group of Japanese tourists all start clicking away, taking photos at the madman on the bike.

Calla then peddalling off along the sidewalk and trying to pop a wheelie only to flip the back and land in a heap with the bike on top of him, the concierge lifting the bike off and inspecting it for damage then wandering back to the foyer wheeling the bike back. "Well I'll just help mesel up will a!" Calla clambers to his feet looking at the graze on his elbow and walking back to the rest of the lads who have paid no attention and began getting into the taxis. The Japanese tourists all still snapping away and laughing at Calla's stunt bike riding as Calla pulls his zipper down and flops his knob out, puts his arms around the tourists and poses for a photo, one shuffling back to take the snap. Calla shouts, "Warr grasshopper and squints his eyes!" the tourists still unaware Calla's loin is on show. He walks off from the applause and bows as he links Valerie's arm and helps her into the Lincoln town car as Julie and Amari get in as well. "Clevelander, me old china, pronto and where is my tinzzzz and tabzzzzz!?"

The driver swivels round. "Excuse me, sir?"

"Never mind, Parker, just step on the gas. I don't want Jimmy the minjbug nipping me oot the round! Nice car this, Val! What's this all about?"

Valerie dabs the excess lipstick from her mouth. "We always get one. Amari's dad owns the company and it beats hanging around for taxis!"

Calla nods approvingly and taking the striking red lipstick off Valerie before she can put it away he applies it to his lips then smudges it up either side of his face like the joker. Cocking his head he turns to Amari and in a gravelly voice says: "This city deserves a better criminal and I'm going to give it to them." He lets out a cackle.

"What's wrong with you?" Amari quizzes raising an eyebrow at Valerie who is equally as bewildered! Julie turning round to speak with the rest as she catches a glimpse of Calla's tackle that's still on show – "Oh Jesus, dude! That's not cool!" spinning straight back around as Valerie, Amari and Calla look at his crotch.

"Oh dear, it appears my weapon is no longer concealed!"

"Andrew, put it away for god sake! What are you doing?"

Calla lifting his bum off the seat and beginning to tuck it back in his jeans – "Aya aya aya arrrrr me fucking clems have got caught in the zip, driver STOP STOP the fucking car quick!"

The car skids to a halt at the side of the road as Val jumps out, followed by Calla who is crouched over beginning to gingerly try and prise the loose skin from the zipper.

"Are you ok?" Valerie puts her arm around him, Calla shrugging it off.

"Course I'm not alright, you sackless bitch! You think these are tears of joy? My fucking left plum is caught!"

Valerie is miffed by this. "Well what the hell did you think you were doing? It serves you right!"

Calla, annoyed by this, half turns his head (sarcastically). "Well thank you for your comforting words, Valerie – have you considered writing jingles for sympathy cards?"

Valerie pushes Calla and gets back in the car. "You're an asshole!"

The door slams as the car screeches off, Calla yelling after it, "Wait wait, or fucking hell!" looking up at the distant light of the Clevelander 3 blocks away and unable to wriggle his bollock from the zipper which is now bleeding slightly. He begins to shuffle in pain up the sidewalk.

Back at the Clevelander pool bar Willy is getting on famously with Louise who opted for jumping in the cab with him, and the rest of the lads are chattering merrily at the bar when the Lincoln pulls up, Valerie and Amari jumping out, Valerie grabbing Louise's arm and pulling her from Willy, spilling some of her drink on Willy. "Howw man, steady on there, cheese tits! What's the beef? Where's Calla!"

Valerie looks past Louise. "Andrew the JOKER is limping along Collins with his testicle stuck in his zipper! He is a complete douchebag!"

Willy sprays both the girls with a mouthful of Jack and coke and lets out a roar of laughter, both girls wiping their faces not impressed. "You say, hen, he has done what? Could you not wait until yee got home? Randy bastards!"

Louise shaking off Valerie's arm – "What's gone on?"

Valerie shakes her head. "Look he is a prick and a clown! Then he went and spoke to me like shit so let's go!"

Willy lurches off his bar stool and stumbles slightly. "Whoa there, now, no-one's leaving. Let's all have a nice drink, relax and forget about Cally the Muppet!"

Valerie smiles and tilts her head. "William, I am leaving! Louise?"

Willy rubs her back then puts his arm around her, spilling some more drink on her shoulder.

Louise looks at Willy. "I'll stay, William, if you can tell me my name?"

Willy backs off from the two girls and salutes them. "You two ladies have a nice evening!" then staggering back to the rest of the lads to tell them about Calla. The girls storm back to the car; it then screeches off.

By this time Calla is about level with the gay bar on the corner totally forgetting he still has a large joker smile scribed on his face, crouching over cupping his swipe, still trying, rolling the loose skin between his fingers and his other hand delicately trying to prisedown the zipper...

Two rather extravagantly dressed lads are standing outside Danny's bar having a tab. "Do you need a hand with anything, sailor?" one of them camply shouts over at Calla. "No thanks. I have just got my balls stuck in my zip, that's all. Thank you, good night, fuck off." Calla points down at himself.

The 2 lads laugh. "Would you like us to have a look at it there for you?"

Calla walks over towards the door. "I tell you two queers what you can look at..."

"Excuse me, sir, are you ok?" A police officer on the beat interrupts Calla from the pavement to the side of the bar.

Calla, who rips the small bit of skin out of the zip with a muffled scream as he pulls the zip up and walks over to the cop, replies: "I am, officer, thanks, I am just a bit lost. I stay at the Clevelander hotel?"

The policeman smiles. "Are you from England? My family are from there, I am English."

Calla raises an eyebrow. "Well, the accent threw me a bit there, pal, which part you from? London?"

The policeman nods. "Yeah, how did you know?"

Calla shrugs. "Just a wild guess, mate, just a wild guess."

The policeman signals the car over the street. "C'mon, how bout I drop you off, buddy?"

Calla looks very surprised as he climbs into the car. "Normally

I am kicking and screaming at this point by the way, lads." The 2 cops laugh as they drink their coffee and speed off. The lads are by the pool bar as the cop car pulls up at the front of the hotel bar, lights flashing on the top, which Calla had asked them to do for his entrance, the car coming to a halt as it lets out a loud whoop whoop.

"Fucking hell they have found out you broke the second fifty. Jonny – they are taking you away for being cash crazy on the streets of Miami," Willy ribs Jonny who is drinking daiquiri looking worse for wear at this stage.

"Fuck off, Willy, you're as bad, you're as tight as a fucking camel's arse in a rainstorm."

Davey laughs. "A rainstorm? Where in the friggin desert!? Why, ya not piss wise, Jonny son."

Calla gets out of the car thanking the two cops. "Stay safe out there guys, it's a jungle," limping down towards the lads. "Areet strokers, where's the blurt gone?" Jimmy giving the barman the wink to get Calla a daiquiri.

"They fucked off after some tosspot got his balls caught in his zip and decided to be abusive." Willy with his glasses on his head now, always the same thinking he can see when he is pissed not realizing his vision is that blurred due to copious amounts of drink, him thinking his eyesight is normal.

"Aye well a little compassion was all I was after, and the stuck up bitch drove off and left me!! Luckily I was outside a gay bar and this copper came over and..."

"Whoa whoa whoa, hawldy hawldy, outside a gay bar with lip stick on, your balls out and a copper came owa?!!" Davey climbs off his stool as the lads and the barman fall about laughing.

"I suppose you're going to say it's not how it looked, are you?" Big Texan Tom chuckles next to Chuck who is asleep in the chair.

Jimmy hands Calla a daiquiri. "Get that down your Gregory Peck, Boy George."

"Aye he turned out looking like a fucking homo and returns confirmed as one!" Willy laughing his head off just as Calla pushes him in the chest backwards into the pool with a massive splash, soaking Chuck.

Calla takes the drink off Jimmy. "What the fucks this!? Now this is fucking gay by the way!!"

Calla tries to suck the daiquiri down in one pausing half way and rubbing his head. "FUCKING HELL BRAIN FREEZE!"

Jimmy laughs. "You need to own a brain afor you can freeze it, numb skull!"

Calla slams the drink down and uses both hands to rub his temples as Chuck, without opening his eyes, kicks Calla straight in the pool soaking Davey. Willy is beginning to climb out of the pool when Calla swims up behind him and yarks down his trousers revealing his white arse. Calla slaps the arse as Willy screams and flops back into the pool to struggle with his pants.

"Way man, you can't take these bastards anywhere!" Davey shakes his head and shirt of the water looking into the pool where Calla and Willy have begun wrestling. Tom and Jonny are laughing at Chuck who is now sound asleep with a bottle of Bud in his hand.

"Heyy motherfucker!" Tom nudges him with his leg, Chuck just raising the bottle to his mouth and taking a large gulp, eyes then rolling as he nods back to sleep letting the bottle rest gently again on his lap. "Hey motherfucker! Wake up!" Again Chuck just looking dazed out at Tom with eyes blank again raising his bottle and taking a large mouthful and again falling back to sleep. Tom lets out a laugh. "Well I give up, dude can spend the night there!" Tipping the last of his drink down his neck – "Yo barman, get me a round of Hurricanes in and a packet of them salty pretzels for Davey dustbin!"

Davey tips his drink approvingly. "These are the last ones for me, Davey. I'm off up the stairs – early rise tomorrow!"

Davey riving open the packet of pretzels and stuffing a handful into his mouth and washing them down with the hurricane – "Cheers! Well, you need to be necking that before these two dopey twats get out the pool, coz they will be wanting one for the stairs, the bathroom, the bed, the breakfast and the next day!" Davey quips.

Tom, shaking his head and slamming down the drink, flicking the dollars on the bar, replies: "Good advice, man," patting Davey on the back. "This is me and this is the back of me!" Tom walks off squeezing Jimmy on the shoulder as he passes.

"HOW, Larry Hagman, where you going? Not tapping out are we!" Calla shouts as he clambers out of the pool wringing his t-shirt into Jonny's drink.

"Twat face!" Jonny slams the glass on the bar, Tom not even breaking strides. "Yup! Timeout for now, Houston! Y'all don't be late now!" Tom not turning round and carrying on through the door.

Calla – "Fucking Yanks! Couldn't drink cold tea!" – turning to look at Chuck who is still fast on, digs him in the arm. "Wakey wakey Roy Orbison!" Chuck again just lifting his beer the same time as his eyelids and taking another mouthful then closing his eyes and lowering the beer. Calla rubs his hands. "I'll wake this cunt up! Seems Chuckles fancies a late neet dip!" grabbing the bottle and trying to prise it away, jerking Chuck's arm everyway, Calla pulls the bottle. "Has some cunt glued this to his mitt or what! How man!" Calla digs Chuck in the leg as Chuck snatches back the bottle and empties the last down his neck. Standing up and slamming the bottle on the bar, he picks up the hurricane off the bar drinking that in one, wiping his mouth he palms Calla in the chest who then flies backwards back into the pool. Straightening up with a stretch – "God speed y'all!" – then staggering off, clipping every chair and table on the way, he too disappears through the doors.

Calla clambers back out of the pool. "Some cunt!" – standing up and throwing the Hurricane down his neck. "By, that cunt's not well, why do they call them hurricanes?"

Davey turns. "Probs coz like Chuck you walk off like you are a bastard hurricane!"

"Right, one for the stairs, lads. Barman, get me 4 of your worst shots, 4 shots of tequila and 4 jack n cokes!" Calla slams a pile of soaked dollars on the bar and winks across to a couple of girls who are stood sipping cocktails! Davey shakes his head.

Jimmy is across to the bar and looks down at the drinks racked up. "Is this the suicide round or what? No wonder the yanky doodles tapped out!" slamming the tequila down his throat and chewing the piece of lime whole and swallowing it!

Davey winces. "Fucking hell, Jim, you're only supposed to suck it!"

Jimmy laughs then throws the murky red shot down his neck, nearly throwing up! Grabbing the Jack and coke and pouring it down his neck, ice as well, crunching through the ice and swallowing – "Fucking hell! What's in that bastard!" – grabbing the hurricane and drinking that straight over, he then grabs the leftovers of Jonny's watery daiquiri, Davey's tab quivering on his bottom lip as his mouth is wide open in shock. "What the fuck are you doing, ya simple twat?" Jimmy is a funny shade of white holding his hips breathing deeply. "You're not a fucking teenager, man and we have to be up sharp!" Davey rubbs his back.

Calla shouts, "Ay and that cunt's only the one for the stairs! We have the bathroom, bed and breakfast yet!"

Jimmy is holding his hand up waving – "What's in that fucking red shot you bastard!" – trying to talk while fanning his tongue. "It's fucking red hot! My lips are ahad!"

Calla leans in towards the bar. "How, you poisoning bastard, what's in that shot!"

"It's a flat liner, sir, they have chilli powder and tabasco in them!" the barman replies.

Calla pointing towards the barman says: "We asked for a drink, not a fucking curry, you gormless cunt!"

Jimmy gulps down a pint of water. "Calla, give that daft bastard a slap, will ya."

Calla climbing over the bar in a flash as the barman legs it through the hatch – "Come here, you little wop bastard, I'll strangle ya."

Davey tries to calm Calla down. "Andrew, get out from behind there, man, the bouncers will be over."

Calla now leaning against the bar asks the girls what they want for a drink.

"Right that's me, I'm away to bed. That fucking drink's knocked me sick! I thought I was ganna die." Jimmy gets up as Davey steadies him.

"Excuse me, sir, but you need to pay the rest of the bill," a voice shouts out from behind a tree as Calla gets under the hatch and tries to find the shitless barman. "Come here, ya little bastard."

Jimmy stands up with a big deep breath and a massive fart and holds the back of his pants. "Fuckin hell, I think I've shit mesel. I'll square the bill up the morra when we check out. Haweh Davey, there's away."

Davey pulling his strides up hanging round his arse and cramming the last of the pretzels in his mouth, washing them back with the tequila and a mouthful of Jack n coke, swerving the red shot, says: "Goodnight, godbless, lads, remember 6 o'clock rock down the dancers."

Willy is asleep as Calla draws a big moustache on him with a felt tip, Jonny sitting like a nodding dog bladdered as Calla skips away and joins the two lads. "Aye a think I'll have an early one mesel, in Willy's room though! The daft twat left his card when I got ready! Porn channels all neet for Calla." Davey shakes his head as they get into the lift.

By the pool the bouncers move over to Willy and Jonny. "Ok guys, time to leave."

Willy opens his eyes. "Ah for fuck sake I cannae see, where's ma gleggs?" The bouncer pushing the glasses from Willy's forehead over his eyes. "Ah, that's better, bigman, ta." He gets up giving Jonny a kick. "Righto, sleeping beauty, time for bed."

Jonny stirs. "Fuck off, Willy, leave me alone."

The bouncer shakes Jonny. "Time to move, sir."

Jonny waking up looks at Willy. "Ha-ha, where did you get the tash – Davey lend you it for the neet?"

Willy walking over to the bar looks in the mirror. "I'll fucking murder that bastard Calla." He tries to scrub the Hitler tash off with a beer cloth from the bar.

Jonny stretches. "Haweh Willy, let's fuck off, we are up in 4 hours for fucks sake."

"You fancy a nightcap at mine, Jonny son? There's a few Heineken going spare."

Jonny shakes his head. "Not a hope, Adolph, I'm wrecked, man."

Willy walks out of the lift. "Suit yasel, Jonny baby belly I villl drink all of ze bastads mesel."

Willy opens his bedroom door to find the plant has been moved from the window and is sitting in front of the telly with tissue paper stuck to the leaves with empty bottles of Heineken stuck upside down in the soil. "Whit the fuck?" Willy pulls a tissue off as it sticks to his finger, smelling it– "that smells of spunk, some fucking dirty bastard's been wanking on my plant."

Willy flicks the tissue off his hand as it sticks to the wall, Calla falling out of the wardrobe holding a bottle of Heineken laughing his head off with nothing on but his underpants and socks. "Ha-ha, Adolph you sshpecky von leetle focker, you dared touch the plant of wanks."

Willy is not impressed. "You're one dirty bastard, Calla – one for wanking in ma beed but to drink my Heiny is a cunt's trick! I was looking forward to them."

Calla picking up the phone "Ah quit your bleating for fucks sake, aye horse, 6 Heineken, 3 voddys and tin of coke oh and a 12" pepperoni! Ay as quick as you can." Calla puts the phone down and sits on the chair.

"Put some clothes on, Calla man, you're making me feel uncomfortable!" Willy is trying to put his clothes in his bag for tomorrow.

"Hawld on, I'm making you feel uncomfortable? You have a Hitler tasher and your hands smell of pineapple chunk! Some man to have a late night drink with alone in a room by the way!"

Willy laughs as he rubs his mitt across Calla's lips. "Err, you vile man!!"

Willy laughs. "It's your spunk, man, what's wrong with ya?"

Calla wiping his mouth and swilling his Heineken back, replies: "So this is the last night, hamma, back to graft for a couple of days then offski."

Willy flicks his shoes off and sits in the chair opposite Calla. "Ay wonder if we'll get back or is it away to the land of the slant?"

Calla shrugs his shoulders. "I couldn't care less as long it's away from chicken parmos and the Boro." The bell rings as Calla jumps up and opens the door. "Bout time, Stroker, entree."

The waiter stares at Calla, then across his shoulder at Willy who is now standing doing the zeig heil signal behind Calla's back. "Are you guys ok?" The waiter nervously pushes the trolley in with the pizza and drinks on.

"Ay we are just fucking starving, haggler, waiting for this, so get a shift on pal."

The waiter pushes the trolley over next to the table as Calla takes a Heiny off and opens it with a spoon. Willy leans forward and goes to take a piece of pizza as Calla raps the back of his hand with the spoon. Willy cries out and the waiter jumps back. "Ah you mad fucker, what ya dee that for?" Willy holds his hand under his arm with pain as Calla takes another slurp and picks a bit of pizza up. "Wash yer hands first, you dirty little bastard – they stink of spunk!!" as the waiter turns on his heels and closes the door behind him...

Willy wanders back in from the bathroom drying his hands throwing the towel over his shoulder knocking the lamp onto the floor. "Howw man, ya making the place a mess!" barks Calla, Willy stuffing a slice of pizza into his mouth whole, looking round at the wank plant and the empty bottles. He washes the pizza down with a big swig of lager, letting out a large burp. "You're making jokes, ham, it looks like some cunt has rolled a shrap grenade in!" Folding another whole slice of pizza into his mouth he begins to chew as Calla digs him in the ribs. Willy rolling over he coughs out the lump of chewed pizza –"owf ya cunt!"

Calla jumping up heads into the bathroom laughing as Willy stuffs the chewed pizza back in his mouth hearing Calla raking about. "What you doing in there, ya daft cunt?" Calla wanders back out with a shaving foam beard on and Willy's spare glasses, beginning to do the robot dance. Willy laughing stands up, stepping on an empty bottle, twisting his ankle, sending him flying into the trolley, knocking it over. He tries to catch himself, grabbing the wall mounted TV, ripping it off the bracket then rolling over the table sending the kettle, cups and glasses and magazine rack flying he lands in a heap on the plant with the TV landing on his back letting out a faint whimper. "Oww, me ankle!"

Calla stands rigid watching in total shock, the devastation unfolded in front of him, starts to laugh. "Fuck me, Stroker. How quickly things can gan rang!" Wiping the foam from his face and throwing the specs onto the bed, he steps over Willy (who is writhing in agony and trying to move the 42" plasma from his back), grabs a bottle and a slice of pizza, wandering out of the door, shouting back, "You always spoil everything,Willy! Ya mad scotch twat," as the door slams.

Willy finally gets the TV off his back and pushes the trolley away. He clutches his foot and swivels round on the floor getting the ice bucket and scraping the ice back in. He pops the top off one of the beers, pouring it down his throat, Heineken spilling everywhere as he gets onto the bed and slams his foot into the ice bucket, adjusting his glasses as he looks at the telly. "Orr for fuck sake!" Swigging his beer he dusts an upside down piece of pizza from the deck and rams it in his mouth, tilting his head to his right and sniffing up. "Arrr fucking hell!" peeling off a piece of tissue with Calla's sex piss on it.

Next day Davey bangs on Jimmy's door. "Jimmy. Haweh, Jim! Jamesey! How man, wake up!"

A bleary eyed Jimmy opens the door. "Areet captain crumb, haway in, al not be a minute!" Davey follows Jimmy in who wanders off scratching his arsehole, squeezing out a fart. "There its!" Jimmy sniggering to himself as the egg smell wafts Davey's way.

Davey gagging, he shakes his head, turning back around. "Bollocks with this, ya smelly arsed bastard, at least you're up. I'll go and wake the rest of them!" leaving Jimmy who is wandering round packing while whistling 'Piper to the end' by Mark Knopfler. Davey raps on Willy's door, the door unlatching, Davey walking in looking around. "Fucking hell, son, you've been burgled!"

Willy limps off, scratching his head, with his toothbrush hanging out of his mouth. "Davey man, I think I've broke my ankle!" spitting the toothpaste into the ice bucket and gargling with the dregs of a Heineken.

Davey rolling his eyes walks over to the bed where Willy's stretched out, looking at the ceiling. He examines the black lump on his foot. "Holy shit lad I'd say you have! Your foot's black as arseholes. How the hell did that happen?"

Willy sitting up, sighs. "Me and silly bollocks had a night cap!" He smiles and holds his thumb up.

Unimpressed, Davey standing the lamp back on the bed side notices the damaged television lying on its front. "You've totally wrecked the place!" Laughing, he notices the bed hasn't been slept in and is still all folded neatly, only spoilt by the crazy jock resting on it. He walks over towards the TV. "Willy son, Jimmy's going to go jalfrezi! He covered the incidentals on his credit card." Davey lifting the plant back onto the window sill peeling off a tissue –"what the fuck?" –taking a sniff, throwing the tissue on the floor in disgust. "You fucking dirty jock bastard – is that spunk?"

Willy sighs and lies back down. "Nor, big man, that wasn't me! It was Calla! Divunt worry about Jimmy, I'll explain it was mare accidentals than incidentals!"

Davey, shaking his head, walks towards the door. "Wey, get your arse into gear and square the room off – they might not notice it!"

Willy sits up. "You think?"

"Don't be soft, lad. Stevie fucking Wonder on a galloping hoss wad notice the fucking state of this place. Downstairs in 5!" Davey standing in the corridor letting out a big sigh begins to bang on Calla's door which is slightly ajar. He then pops his head in and finds no Calla and no sign of his bag. Puzzled –"where's he at now?" –scratching his head, he starts to wonder if he has the right room. As he wanders out he notices all the pictures and lamps are turned upside down. Tutting and rolling his eyes –"Ay Davey, ya not going daft, this was the brainless smog's room alreet!" –he wanders across the hall to Jonny's door and begins to bray on it as Jonny opens it, dragging his bag out.

"Morning Davey, everything alreet?"

Davey holds the door open. "Weyaye lad, the two morons are awake anyhow, although one is AWOL and it looks like the other has broke his ankle."

"Eh?" Jonny stops, Davey shaking his head and leading Jonny down the corridor. "Al tell ya owa a tab, young'un"!

Downstairs Jimmy is at the counter settling his bill, the clerk explaining the incidentals will be held until the rooms have been checked. "Ay areet, so lang as wet sheets and messy rooms isn't deductible!"

The clerk hands Jimmy a printed bill. "Had on there, mate, room service? Nor nor nor, just my room number, pal, the other monkeys will square that off."

The clerk adjusts it then hands him the bill. Jimmy takes it off him and inspects it! "Woah, what's this! Breakfast? You have made a mistake there as well, bonny lad. I've came straight down from the room, haven't been in the restaurant!"

The clerk checks. "Well sir, it says here that your room number has had a large breakfast?"

Jimmy shakes his head as Davey joins him. "What's the bother?"

Jimmy leaning on the counter hands Davey the bill. "Some bollocks has had a 38 dollar breakfast on my room!"

Davey is checking it as Jimmy turns back to the clerk to argue. Calla wanders past, Davey looking up. "Andrew...you just had a breakfast?"

Jimmy turns round as Calla stops. "No Dave! Couldn't stomach a thing. I'm not hungry!" rubbing his belly and pulling a sick face.

Davey looks over the top of his glasses. "Then why you walking back from the restaurant area?"

"Just wondered what was over there, Davey. Was up with the larks so I've just been killing time!"

Jimmy looks Calla up and down, shakes his head. He spins around taking the bill and signing it. Picking up his bag, he begins walking to the door. "Cal, you owe me 38 dollars, you cunt!"

Striding after him, Calla shouts, "Eh hawld on, it wasn't me, Jim! Honest!"

Davey laughs and puts his arm around him. "Calla son, cough up, you have bean juice on your trainer!"

Calla looks down at the red dribble on his shoelace; he begins to smirk. "Bastard beans!"

The lads are standing in the foyer as Chuck and Tom walk out. A porter is pushing the bags with Willy sitting on the top of them. All the lads start laughing. "Ay, very fucking funny! Crack on, my foot's in bits through that daft bastard."

Calla shakes his head as the porter starts loading the hotel minibus with the bags. Calla gets the trolley and pushes it back into the foyer with Willy clutching the bars. Calla points. "He's like a fucking monkey, look, with glasses!"

Jim and Davey climb in the front of the bus as the porter pushes Willy back to the van. As he is climbing off the trolley Jimmy shouts through the door, "Your bill hasn't been paid yet, little fella."

Willy gives the porter the wink to push him to the desk. "Watch his face, lads," Calla nudges the lads."I watched 2 pornos, we had 12 Heinys, voddie, chips, pizza and the daft cunt smashed the telly!! I hope he's got his credit card."

Tom roars with laughter. "I hope your 2 friendly cops who caught you sucking guys off last night help the manager out when he gives William that bill, Calla!"

Calla turning round gives him a stare. "Haweh!"

Willy's arms start waving and he starts pointing out towards the van. "Haweh Jimmy, let's fuck off, he's embarrassing."

Jimmy climbs out of the van, walking into the foyer as the lads continue to stare. Willy gives Jim a hug and Jim helps him out into the back of the van. "Calla, you owe me 400 dollars, that's 4 for you and 4 for hop along." Jimmy puts his seat belt on.

"Eh? What for – 12 Heiney and a couple of bits of pizza!! I'm fucking not coming back here by the way!!"

Willy shows all the lads his foot. "So, I was busy laughing at robonob doing his dance with me gleggs on and I stood on a bottle and done a fosbury flop into the food trolley."

Jonny is in bits laughing. "I wish I had come back now for the pantomime!"

Calla sighing looks out of the window. "I will live here one day, me."

Davey leaning back over the chair – "Beats Stockton for a neet out, son, eh?"

Big Chuck puts his arm around Calla. "You can count on us to come and see you, Calla – the entertainment value is the best in the States."

"Ay Calla, how much of your winnings have you got left?" Willy tries to put his foot up on the chair.

"Nen of your business, onion toes."

"Haway, big man, how much?"

Calla leans over to his bag. "Well I honestly forgot about the bundle I won and stashed." Calla pulls a rolled up ball of shite out of his bag resembling dollars.

"Fuck me, Calla, you expect to get through life holding onto the green like that." Tom moves across as Calla straightens the money out on the seat. "All fucking franklins too." As Calla continues to count them. "Shh, tomcat, I will lose my count."

Jonny can't believe the amount. "Calla, I thought you won once?"

Calla squares the pile up. "No, hymey von hymessen, I won a few times – forgot I'd stashed this, like." Calla taps Jimmy on the shoulder. "There you gan, Jimmy, 1200 dollars for me and Willy, me flight owa and whatever else I've ticked on while dancing with the devil."

Davey nudges Jimmy and whispers, "Dont care what anyone says – he always pays his debts and is the kindest lad on the planet when he's flush."

Willy sits up straight. "Cheers Calla, you're a gentleman."

"Lad, give owa!! You owe me 400 ok man, jock fyass, I squared him cos you were crying your gammy eyes out back there at the hotel, so I guess your pink lint." Calla rounds the rest of his money up, stuffing it in his jeans pocket. "And still over a grand left for my last 2 days' holiday!! Ye har!!"

Davey laughs as he lights a tab up. "I don't know, we come owa to work, graft our balls off and mak a few bob and he still makes everything sound like we are deein nowt but having a good time."

The bus slowly pulls up to the front of the airport and the lads all clamber out, Calla darting into the airport and returning with a wheelchair. "Hop in, William, your carriage awaits!"

Willy hops across and settles in, Calla throwing both their bags on Willy's lap and spinning him round and rocking it back into a wheely, pushing it, making car noises, Willy screaming. "Calla man, pack the cunt in, am ganna fall!"

The lads all laughing, Jonny shouting, "Well what do you expect, getting in anything controlled by that idiot? Evel Kneivel would've bottled that!"

Calla is now running behind the wheelchair towards the check in counter which is empty. "Good morning, treacle." He slides over the 2 tickets and passports. "What's the jack palance we can get an upgrade or a seat with more space? My retarded work colleague has injured his foot while playing hopscotch!"

Running the tickets and passport into the system the woman leans out of her chair to look at Willy who is looking back with his lip turned up. "Erm I'm sorry to hear that. We actually have a bulkhead seat available for your friend but unfortunately you will have to sit elsewhere!"

"I'm sorry but that won't do – as you see I am his carer and I need to be sat next to him in case he needs a hand getting to the bathroom or opening a can!" Calla kicks Willy's ankle, Willy dropping the bags and leaning forward letting out a moan. "See, he is in great pain. Surely there is something you can do? I mean this injury only occurred because your airline made us stay a night in Miami which we weren't happy about!" – kicking Willy again who grabs his leg and starts squeezing it.

"Calla, knock it off, it's nay joke, it's frigging killing me!"

"Now now, William, watch your mouth in front of the nice lady!"

The woman blushes then taps on the keyboard. "Ok, Mr Callaghan, I have upgraded you and Mr McLaughlin into business class. You should find yourself with more leg room there and it is only a short flight! Hope your foot gets better and you two have a nice flight!"

The woman sits back down and takes Davey's passport and ticket from him. "What have you two connived?"

Calla throws the bag back on Willy's lap. "Mind your business, Davey, see you at the bar!"

"Orr, Calla man, nee mare drink!" Davey calls after him as he watches Calla pushing Willy through the airport like a slalom shouting, "Weeeeeeeeee!" Willy's muffled "slow down, ya crazy bastard, slow down!"

Davey shaking his head smiles towards the woman. "Nice day, hinny, isn't it!"

The last through is Jimmy who flops onto the chair at the table.

"I'm bastard sick of the drink, me!" as Calla returns with a tray of lagers. "Calla son, you can boil your heed, am not drinking that–" pushing the pint into the middle of the table.

Calla takes a large slurp and wipes his mouth. "You come out with some shite, now get it down you else I'll tell your mam!" He pushes the pint back towards Jimmy.

"She's been deed 4 years, Calla, you moron!"

Calla lifts his glass. "Let's raise our glasses for Jimmy's mother! Gone but may she never be forgotten!"

The lads all raise their glasses prompting Jimmy to join in sniggering while taking a sip then pulling a face like he just ate a lemon. He looks at Calla – "Tit!" then continues to pour it down.

"It's like drinking green paint, that bastard!" Jonny's best effort leaving the beer with barely a mouthful missing, Tom and Chuck both shaking their heads as they take a sip.

"You lot are a right ungrateful bunch of twats!"

Davey walks over to the bar and wanders back with a can of sprite pouring a top onto his beer and passing it to Jimmy. "Get a splash of that in, mate, takes the edge off it!"

Jimmy pours it into his beer and passes it to Jonny. "Shandy? That's what them homos swally!"

"You calling my pint a puff?" quips Jonny, the lads all laughing.

Davey checks his watch. "Time do you mak it, Jim?"

Jimmy looks at his watch. "We have an hour to kill before the off!"

Davey stands up and stretches. "Well I'm going for a sandwich and some bullets to suck on the plane – anyone want owt?"

Tom stands up. "I'll take a walk with you. Chuck, you want a buffalo sub?"

Chuck nods approvingly, the rest of the lads chirping to Davey to get the round of buffalo subs in, all them with different orders, shouts of "nee onions, extra cheese, howw get me jalapenos, get me..."

"All yez, shut the fuck up!" Davey waves his hands. "I'll get the subs in and you fussy bastards can pick the shite off you don't want. That's not good enough – get off your arse and go yourself!"

The lads collectively – "OOooooooooooo!"

Davey walks off, the others shouting after him to get bags of crisps as well. Willy lifts his foot up; by this point it has doubled in size.

"You should get that strapped and iced, son!" Jimmy looks over at it, Calla leaning and poking it.

"Ow, you stupid cunt, man, it's sore, Cal!"

"Well I think it's just badly sprained and you're being a baby!" Calla finishes the rest of his pint, burping. "Haway, Willy, it's your round – go on, hop it!"

Jonny pipes up. "Leave him alone, Calla, he is sick of being pushed around by you!"

The lads laugh. Willy uses his hands to wheel the chair away from the table and rummaging through his pockets he throws the dollars on the table. "Calla, go and get them in, man – honestly my foot's like toothache!"

Calla stands up and counting the money out. "Willy, you minj, there's only 24 chips here! The round is 45 chips!"

Willy pushes himself back to the table. "Wey chalk it on, snaggle tooth!"

Calla goes to walk off then stops and feels his teeth. "Eh! Nowt rang with my railings?"

"Fucking will be if you don't make sharp with them beers!" snides Willy. The beers are being placed on the table as Davey returns handing out the sarnis and throwing a couple of packets of crisps to share on the table, all the lads rubbing their hands, unwrapping the foil. "That's fucking tremendous, that, good call, Tom!"

Tom licks his fingers and tips his glass to Jonny as he takes a swig.

"What else did you eat on the way up, Davey?" Jimmy quizzes Davey, looking up, thumbing a piece of chicken in his mouth.

"What you on about, lad – you think I'm some sort of giss?"

"A packet of Cheetos, bag of nuts and a steak pie!" replies Tom, Davey spinning round. "Sket gob!" looking back at Jimmy –"you needn't talk, you're nee racing snake!" taking a bite of the sandwich again using his finger to stuff it in.

Jimmy wipes his mouth. "You must be at least 17 clem, Davey, haway you're fatter than me! You're bastard massive!"

Davey sticks his hand in the packet of opened crisps, his huge hand splitting the bag and his pork sausage like fingers stuffing the crisps into his mouth. "Wey, wor lass will have me on one of them diets when I get back, man – nee drink except for a Sunday, bastard ryvitas for my dinner with cottage cheese and cucumber then tuna salads or grilled chicken and veg for me tea! Back to being the hungriest man in Gateshead."

Jimmy laughs. "Well what I've seen you trough this weekend wad see a grizzly bear into hibernation for the winter months, nee bother!" (A tannoy announces the gates opening for the flight!) Davey and the rest of the lads stuff the last of the sandwich into their mouths and swigging the last drops of the beer Davey is first to his feet. "Reet nyuck time, then all aboard the big bord yem!" tucking his shirt into his jeans and dusting the crumbs off he wanders towards the toilets.

Jimmy looking towards a puzzled Tom "Nyuk is a toilet Tommo!" Tom lifting an eyebrow then shrugging.

At the gate the announcement for first and business class is greeted with cheers from Calla and Willy. "That's us, me and ironsides are in the dear seats."

Jimmy looks in disbelief. "Are you two fucking winding me up?"

Willy makes a rowing boat imitation while Calla whistles Hawaii Five-0 as they sail past Jimmy. "Enjoy the cheap end, bonny lads."

Willy and Calla get to the door of the plane and the air hostess gives Calla a nice smile. "Hello miss, are you looking after business class today?"

The girl nods. "I am, sir. Does your friend need any special assistance at the other end?"

Calla smiles. "Yes he needs a wheelchair and someone to change his nappy."

Willy stands up. "Take no notice of him, pet, he's the worse carer I have ever had! Tried to feel ma balls in the bath this morning."

Calla nudges him forward. "I was in there, dopey hole!"

Willy limps to his seat. "Ah this is the life, Calla pal, free champers all the way for Willy boy."

Calla sits down, flicking his sambas off, odd socks on, as a business man looks over in disgust. "Aye the matching pair is in me suitcase, kidda."

Davey and the lads all shuffle on and get seated, all of them immediately positioning themselves for a snooze all the way back, Davey gaining valuable elbow room with Jimmy. "Big day tomorrow, marra, back to reality."

Jimmy leans against the window. "Ay see what bullshit Mally has – no doubt stitched us all up so we head to fucking Shanghai!"

Davey laughs. "Won't be the first time; he's more dangerous than a chimpanzee with a machine gun, the cunt."

Champagne or orange juice is offered to the lads up front. "One of each for me, gorgeous – bucks fizz, have you heard of them!?" Willy takes two glasses off the blonde hostess.

"I am not sure," she replies as she offers the tray to Calla.

"Well make your mind up, like making your mind up," Calla sings as Willy spits his champagne all over himself. "Do you stay in Orlando or fly back?" Calla asking the million dollar question.

"I fly back but I have a sister in Orlando I visit regular."

Calla winks at her. "Well anytime you need a chaperone I will give you my number," as she smiles and walks off. "I'm ganna bucks fizz her all owa, ya little gimpy bastard. What d'ya think about them apples?"

Willy sips his drink. "I think they are canny. Bet you that 400 dollars you don't buck her before we leave at the weekend."

Calla swills his drink. "Ok, I'll bet you, captain sad sack."

The blonde gives the lads the menus as Willy leans over to Calla. "Am I allowed to eat beef? You made me eat chips all last week in Miami."

Calla shrugging his shoulders smiles. "Take no notice of four eyes, petal, he hurt his foot rolling over a Heineken bottle and his moaning got us into business class. Anyway I will have the Charles Dickens and there's my mobile number. I am Andrew – for the next time you're in town."

The trays are laid for the lads. "I can't take numbers I am afraid, Andrew. Red or white wine?" Willy sniggers as Calla kicks his ankle.

"Red for me please, Sharon."

Willy rocks the electric seat forward, back, up, down. "Howw, Calla, look at this man!" tilting the seat nearly completely flat. "This is the way to travel, man, no worries of the bloke in front spilling your wine when he rams his chair back!"

Calla adjusting his back and sipping his red wine replies: "Arrr yes this is a touch, wonder how the cretins are getting on back there with the goats?"

Willy laughs. "Who cares!" as the service light comes on at the side of his chair.

The stewardess wanders along and leans over. "Yes sir, how can I help?"

"Can you get two more bottles of the red!"

"We are about to start serving the food, sir, so if you could wait just a moment we will serve you a drink with your meal!" half smiling then walking back to the galley.

Calla leans over. "You're useless you, Willy, can't even get us another splash of wine."

Willy snaps, "They are about to serve the bait – we will get it then!"

Calla looks over the top of his seat. "Ay but we are 5 rows back here, that's 20 people they have to serve afore us! And that's nee good. That 400 chips I can get us 2 bottles of wine before the meal?"

Willy rocks his chair up and looks towards the galley where he can see the stewardess preparing the trolley. "You're on, big man! You love giving that money away!" All smug Willy nestles back down into his chair flicking the screen out of his arm rest starting to clart on with the features. One of the stewardesses from the back walks past towards the front of the plane and pulling the curtain back chats with Sharon. The 2 appear to be discussing trolley contents. Calla leaning out of his chair slightly trying to get a glimpse sees the stewardess thank Sharon, closes the curtain and begins walking to the back of the plane as Calla catches her eye. "Excuse me when you get a minute could I please get 2 bottles of red wine please?"

The stewardess smiles back and turning round heads back to the galley and leaning through the curtain asks for the 2 bottles and wanders back handing them to Calla. "There you go sir!" and continuing her walk back to economy.

Calla jingles the 2 bottles and toothing a grin at Willy – "Willy, me old china bull frog, it's all about the way you ask them! None of this get me this, I want that! Pleases, smiles and thank you go miles! They aren't pieces of meat to be gawped at, they are real people carrying out a service!"

Willy leans over snatching the bottle. "Get fucked, Cal. Service and people? You just wanted to buck one of them before!"

"Had on! Nowt's changed, I still do! All I'm saying is to secure what you want sometimes you have to be a little bit nicer!" (in condescending voice)

Willy pours the wine in one into his glass. "You do come oot with some shite sometimes, you boy!"

Calla pouring his wine delicately replies: "Well maybe I do, William but this nice bottle of..." turning it round to read the label, "Chateaux de Puis will taste all the more well considering William Mcfuckwit bought it for 400 dollaroos!"

Willy spins his head. "Hawld on! That's nee fair – you used a different lass!"

Taking a sip and smiling, Calla responds: "William, the bet was to get 2 bottles of wine before our meal, nothing more nothing less. I could've pulled them out my arsehole and would still've won the bet! So you owe me 400 spuds simple as it it goes, hey diddly dee robbing the scotch will dee for me!" Calla is all smug.

Willy kicks back, smiling. "Well considering I owed you 400 for the trip anyway I would say I have lost nowt and witnessed you crashing and burning with Gloria Estefan up there."

Calla frowns. "Gloria Estefan?"

"Ay ginger cunt's a ringer for her, look!"

Calla leaning out watches Sharon hand out the in-flight meals. "She isn't ginger, you muppet!"

Willy adjusts his geggs and squints. "Is!" settling back.

Calla looks closer. "Looks dark brown to me how, hard to tell with it tied back!" Sitting back – "Anyhow nowt rang with an angry fanny! And was Estefan a ging?"

"Ayy, big man, she sang that song, man, ginger is going to get you tonight!" whistling it then laughing.

Calla shakes his head. "You're about as funny as a mouthful of ulcers! And the flight isn't over yet so our little 400 dollaroodle bet might still be on. See who's laughing then!"

This dawning on Willy as he sits up in protest – "Nar, big man, you asked! She burned you! Bet closed!"

"No, Willy Mcjamjars, I said I bet I buck her before the WEEKEND!"

Willy scratches his head and realizing this Calla laughs. "Yes William, never throw your betting slips away until the bookies have paid out! I remember once standing in a Ladbrokes on York Road watching the big race at Ascot. Tell ya, Willy it was jampers and every cunt's screaming at the telly, most had backed the favourite which was romping it the whole race then it clipped the last hurdle and another hoss leaped in front of it; as they both charged for the line the race finished with what looked like the favourite in second, ha-ha, all the blokes going bonkers, crumpling up their slips and throwing them on the floor, about to leave when your man on the box announces a photo finish! Ha-ha, it was like the O K Corral, there was fucking hell on! There was bodies all owa the floor, cheeky digs flying and all sorts! I just stood there with my slip until the announcement of the winner. I just stepped over everyone rolling about and collected my 240 quid! Che ching!"

Willy laughing takes his glasses off and wipes them on his t-shirt. "Ay class! Well we have to see if the outsider wins the bet and gets a buck, won't we!"

Calla leans over to Willy. "Howw man, ya balloon, it was the favourite that won!" leaning back and winking at him as the trolley arrives.

Sharon smiling at Calla as she presses the brake on the trolley. "Chicken or Beef, sir?"

"I'll have the beef please, Sharon, the chicken is foul." Calla takes the tray.

"Very witty, Andrew, would you like another wine?"

Calla nods. "Why not, it's the last day of my holidays."

Willy smiles up. "Same for me, hen."

Calla stops to sip his wine. "It's chicken, Willy – I've told you before the hen lays the egg!! Sorry Sharon, sometimes I do feel like his helper."

Willy starts again. "Beef please."

Sharon smiles. "Wine?"

Calla again interrupting: "Oh he whines areet, you should've heard him last night when he sprained his ankle! And this morning when he had to pay his dirty film bill in the foyer."

"Er that was your bill by the way." Willy takes a bottle of red off the air hostess as she moves down the aisle.

"Sharon, you have lovely eyes, by the way." She blushes to the next row of chairs.

"Slavering cunt." Willy sticks his fork in the prawn cocktail and eats it in one.

"Calm down, Willy. Fucking hell, we are in business class y'na, shovelling prawns in like some prawn shoveller."

Willy swallowing the lot – "Says he, sitting with fucking odd socks on!!"

Calla wiggles his toes. "I lost the other pair in the pool." Both of them laugh as they clink glasses.

The trays are taken away and Calla jumps by. "Gan for a piss, ham."

Willy presses the chair into sleep mode, letting out a sigh as he pulls the blanket up. Three minutes 25 seconds later the chair is on its way back up and Calla is standing smiling with 4 bottles in his hand. "Oh were you asleep, fannyballs? Ah well, you're up now, nice kip? Here are 2 bottles of chateaux blow your fucking brains oot before touchdown!!"

Willy straightens himself, taking the bottles. "How long was I asleep for?"

Calla sitting down looks out of the window. "About 5 minutes."

Willy puts the bottles on Calla's tray. "Funny that, it only felt like 2!! I'm away for a tab."

Calla smiles as he gets up and the hostess comes over and puts a paper coaster down and 2 more bottles. "I probably will be in town

Friday," she smiles as she walks off and Calla turning the coaster over sees her number. Taking his mobile out he turns it on and puts it in, putting it in his pocket as Willy sits down. "Fucking hell, Willy you stink of smoke!!"

Willy takes his bottles off the table. "Aye, bigman, that would be because I had a tab!"

Calla laughs. "You could've sprayed some of that perfume shite on you, for fucks sake."

Willy takes a gulp of his wine. "I could say the same to you about your socks!"

Calla wiggles his toes again. "Don't get me started on socks, onion feet, just don't go there!"

Jimmy wanders up towards Calla and Willy, Sharon stopping him. "Excuse me, sir you're not allowed up here, please return to your seat!"

Jimmy points at Calla. "Look love, I only want a quick word with Andrew – I'm his gaffa, will only be 2 minutes!"

Sharon allows Jimmy to squeeze past, Calla hearing this as Jimmy taps him on the shoulder. "Calla son!"

Calla stops him short. "Jimmy, what have I told you! Now is not the time to discuss a wage rise!"

Jimmy leans back. "What?"

Calla turns to Sharon shaking his head then back to Jimmy. "Go and sit back down, Jim and me and my partner William will get you into the office first thing tomorrow to discuss this. Here's 2 bottles of wine for you and Davey as a small thank you for behaving yourself this weekend – I hear you have to pay back there! Go on now, run along!" tapping a confused Jimmy on the hip, Sharon easing Jimmy away.

"Eh! Hawld on...just wait...Calla man, I need a word...Calla...Calla son, you some bollocks!" as he is led back through the curtain.

Calla stops Sharon as she passes. "Apologies for that, Sharon, I'm deeply embarrassed!"

Touching him on the wrist, she says, "It's quite alright! He said he was your boss – that's why I let him past!"

Calla letting out a fake laugh rolls his eyes towards Willy. "Yes,

Sharon, of course he is, he is that good of a boss he puts us up here and takes a pew back there! Really? ha-ha."

"Oh I thought you said you got these seats because of Willy's foot?"

Calla agrees. "Yes of course, we would've gladly sat back there so we could chat with the rest of the lads but with Willy's foot being like that we decided to pay the extra so he had more space!"

Sharon is embarrassed. "Again then, I'm sorry for that. I assure you I won't let you be disturbed again!"

Calla smiles. "Don't worry. Get us a couple of wines and I'll forget about it!" winking at Sharon who smiles and walks to get the drinks.

Willy leans over. "Some cunt you, smoglet brains, I thought Jimmy was ganna start to cry!" Both the lads laugh.

At the scruff end of the plane Jimmy is sitting back down pushing the bottle of wine towards Davey who takes it and cracks the lid. "Cups, young'un?"

"Y'nar what that sackless twat just did?"

Davey taking a mouthful from the bottle – "By that's sharp! Wee?"

"Calla: gets up front to tell him that about the wages and he maks on he is my gaffa and makes the bord throw me oot!"

Davey coughs a laugh. "Ha-ha, class, is that wee sent these boozes back?"

Jimmy tries to stop a stewardess for 2 glasses as she ignores him, darting past. "Ay he said it was a thank you for behaving ourselves in Miami! Some cunt!"

Both blokes laugh as Jimmy joins Davey drinking out of the bottle. "Ger by there, Jim, need a slash!" Davey half standing up so he doesn't bang his head, dragging his weight up by gripping the chair in front, jerking it backwards violently.

Jonny turns round! "Davey man, ya ham fisted cunt, I was asleep there!"

"Shut ya twisty mackem hole!" clipping Jonny on the top of the heed as he passes and wanders to the bog.

"A thought a we had crash bastard landed there, Jim! The cunt!"

Jimmy laughs. "Ay, wor Davey's got some tactile! He is as elegant as a mouse! A fucking hippopotamus!" The stewardess asks Jimmy to take his seat as he is blocking the aisle. "Is this pick on bastard Jimmy day or what! Have I done something rang to these lasses?!"

He flops back into his seat as Jonny turns round. "Ya did say the bait was a bag of shite to her!"

"Way it was! The beef was tuff as bulls' lugs!"

"I got the chicken, wouldn't know, and how, a didn't notice any left on ye plate!"

Jimmy crosses his arms. "Wey man, am sat aside Davey pedal bin! It was all fat n gristle, half chewed and he clagged it between his bread bun and dipped it in my gravy!"

Jonny laughing turns back round.

Davey taps Calla on the shoulder.

"Howw, dopey hole!" Calla turns around and in a loud voice "David, I have told James I will deal with you the morn!"

Davey grabs Calla and squeezes his arm, Calla wincing in pain as Davey's vice grip digs in. "Arrr Davey man, a bruise like a peach!"

Davey smiles at Willy. "Mare wine, son! And crisps. It's costing a fortune back there."

"Nee botha, Davey, ay, just let go!"

Davey leaves go and rubs Calla's hair making it a mess. "Knew you would see sense," rubbing his hands as Sharon walks down. "Sir, I'm afraid you will have to go back to your seat!"

Davey smiling holds his hands up. "Oh I'm sorry, bonny lass. Andrew here is my son – I'm just checking he is alreet, pet, that's all!"

Sharon smiles. "Oh ok, can I get you something?"

"Aye, pet, I'll have a couple of bottles of that white wine and have you any snacks?"

Sharon nods and wanders off, Davey leaning in towards Calla and Willy. "You will have to rise early to beat me young'un!" laughing. "Listen, Jimmy wasn't ganna say till we got back but I told him to get his arse up and tell you 2! But you sent him packing, ya silly sod!"

"Tell us what, big man!" Willy pulls his seat up.

"Mally called afor we boarded! Apparently he wants to pay us the weeks owed next month!"

Calla turns around angry. "Fuck that, Davey. We will be yem then and I don't trust that horrible cunt!"

Davey pats him on the shoulder. "Calm down, calm down. Jimmy says he will have a word and get us our bit but he can't be sure! He just wanted to prime you, that's all, says you will defo get it but they may hold it as a retainer to make sure they can get you back for the next gig!"

"Arr fuck that! Mally can get stretched, boy, he will keep it on and we will be working a month in hand or he will give us it when we get back knowing we will then have money, he will save the expenses and the dropsy we got at the start of this job – no way! Hate the cunt; he can fuck off!" Willy shakes his head that hard his glasses nearly come off.

Davey stands up straight to take the wine and salty biscuits from Sharon. "Cheers, love! Look lads, it's just a heads up that's all – forewarned is fore-armed!" Tapping Willy on the shoulder and turning and walking off – "See ya when we land!"

Calla stretches out and sticks the red wine soaked paper coaster on the back of Davey's chinos.

"Mally must think we have just landed from the moon. Firstly he can't retain a retainer not in our contract and secondly he can't retain owed wages as a retainer in case we don't come back to a job we might not be coming back to! Do you honestly think the next gaffer in Shanghai is going to pay money off the last project?"

Calla looking through the end of the wine bottle at his feet says, "My socks look the same colour through this, Willy, look!"

"Did you listen to a word I said, daft arse?" Willy nudges Calla.

"When I look at you it's like picture nee sound, I kind of press the Willy is a mute button in my head!"

Willy leans back. "Well, I am just saying."

Calla opens his wine. "And I am just saying I will pull his fucking head from his shoulders if he even mentions the word retainer."

Willy slides the coaster with the number into his hand. "So did you get ginger Shazza's number?"

Calla pretending to pat his pockets – "Aye it's on a coaster somewhere?"

Willy smiles. "We defo still on for that bet, Friday midnight you have to buck her?"

Calla smiles at her as she walks past. "Chalk it down, moon man."

"This wine's going to my head like, I feel a bit pissed." Willy blows like he just done 20 press ups, looking at his glass. "Aye, will we have a short?"

Calla agrees and presses the bell. "Ay, I'd say that will pull us round." Both the lads laughing as Sharon attends.

"Yes boys, last orders. We will be descending into Orlando in 15 minutes."

"No problem, could we have 2 gin and tonics each please, we need to rehydrate?" Calla passing the empties back. "And Sharon, I will have us booked into the best corner of the best Irish bar in Orlando when you're over."

Sharon laughs as she walks away. "You smooth talking bastard," Willy winks at Calla.

"Aye Willy, no expense spared! I might even let you sniff my tissues again after I wipe my jizmo off her fyass."

Willy makes a vomit impression. "I can still get the waft of scampi patty off the last time, you horrible twat."

Two double G&T's arrive as the lads taste them and both pull a sour face towards each other. "Ah lovely," both at the same time.

"Are we out the neet, Willy or what?" Calla moves his chair back and crosses his legs out straight.

"I think we should, just to test the water with the lads in case they have heard owt."

Calla agrees. "We'll go home first for a power snooze and a shower."

Willy nods. "Fair play to Jimmy, getting us paid for drinking booze like, we should repay the favour by asking him if he wants to come out."

Calla shakes his head. "No way will Jimmy be out, he'll be in the cot for 20 hours and in at 5 tomorrow! There's no better man for getting back in the swing! Davey might come out though after he's spoke to the queen."

Seat belt signs come on as Willy and Calla neck their drinks. "We'll have a 10 minute snooze eh?" Willy putting his chair up as

Calla leans against the window.

The plane touches down and taxis to a halt at the gate, the usual bloop of the seat belt sign as Calla and Willy both stand up stretching, a businessman catching Calla with his case out of the overhead locker. "Howw pal, be more careful eh! What's the rush like, the airport's not ganna tak off! Will still be there when the doors open."

The businessman straightens his suit up and stands waiting to disembark staring straight at Calla who is kicking his shoes on, bending down to pull the backs from under his heel. Willy taps the guy on the shoulder and whispers, "Free advice, boy, take yee eyes off me, pal! He catches ya he'll poke them oot and I'll jam that Samsonite briefcase up yer arse for not listening to me!"

The businessman holds his hand up, stepping backwards to let Willy out and looking at the floor as Calla turns round and looks, "Reet, hop along! Haweh, there's the door open!" Calla puts his arm across the seat allowing Willy to hobble out, Calla walking behind getting level with the door where Sharon is wishing everyone a good day. "Safe travels home, thank you!"

Calla mimes the 'call you' sign, stretching out into the air walk as the businessman storms past. "Ay Willy, briefcase wanker didn't have much to say!"

"Ay bigman, I had a word in his shell when you were putting your sambas on like a monkey with a deck chair! Told him to stop staring at ye please!"

Calla nods. "Ay, he areet with that like?"

Willy shrugs. "Wey, am sure he didn't fancy his eyes poked oot and that lovely case jamming up his arse!"

Both lads laugh as they are greeted by the airport help with a wheelchair, Willy pleading with the young lad to push him as Calla flops the bags on his lap and snatches the handles. "Ger by!" and running again with Willy screaming, "Calla man, am ganna tip oot!"

Davey and the rest of the lads wander towards the exit where Willy's lying on the floor, wheelchair on its side and Calla doubled over laughing. "What's going on?" Jonny asks, standing the wheelchair up.

An unimpressed Willy is holding his foot being helped up by

Tom. "That stupid cunt pushed me and let go! Tried to slow down and put my foot down and it went under the wheelchair and flipped me oot! Foot's worse now! Calla, you fucking idiot!"

Calla gasping for breath, drying his tears. "Orr, stop being a cry baby, ya soft twat! Davey man, you should've seen him – he went flying towards some aard wife and he panicked, sticking his gammy foot down! The old woman nearly shit her pants as well and hit Willy with her handbag, calling him an idiot!"

Davey walking past shoves Calla. "One day, son, you will grow up!"

The lads follow, Jonny pushing Willy who dumps Calla's bag at his feet. "Dick!"

Calla is still sniggering. "Haweh!"

Jimmy returns with the van and Tom insists on driving as Jimmy had a few wines on the plane. "Nor Tom, am sound man, honest!"

Tom pulls him back from jumping in the driver seat. "No way, Jim, not happening!"

Jimmy admits defeat and jumps in the back of the bus, Calla giving him a pretend strangle from behind. "Arr, haweh son, you dee as ya telt and sit back here with the bairns!"

Jimmy shrugs Calla off as Willy slides into the front seat, the rest of the lads climbing in the back, Davey next to Calla. "Howw Davey, look what I've found!" Calla holding 2 cans of lager, Davey palming it away.

"No thanks, son, it will be warm!"

Calla presses it against his face. "Nor, it's areet – feel!"

Davey touches the can. "Nor son, honest, am areet!"

Willy shouts from the front. "Pass it this way, Cal, haway!"

"Get lost, ya twisty scotch twat! Says on the back for over 21s only, ya baby!"

Willy turns the radio up. "NOB!"

Calla pings the can, it spraying Davey before Calla gets his mouth right over it, Davey wiping his face giving Calla a stare, Calla gargling a sorry then handing the other tin to Jonny. "Reet wee's on for heading straight to Nelly's?"

A collective groan as Jimmy pipes up: "Yem n bed for me! Busy few days ahead!"

Chuck says, "No way, y'all, I'm having one off!"

Jonny nods. "I'll come down for a bit!"

Tom shouting back he will, Davey sighing, "Wey I'll probs pop doon for a gallon after I've spoken with our lass!"

"What about you, Willy Mchuffhuff? You and your minging foot making an appearance?"

"I'm there but I'm going to have to get this strapped first!"

Davey shouts. "You need to get that looked at, man, it could be broken!"

"It's not broken, man Willy, take nee notice! You would know if it was broken!"

"Ha-ha, or ay, listen to Dr Callaghan back there! What do you know about being a doctor!" Jonny crushes the can and stuffs it in the ashtray.

"Never missed an episode of Casualty!"

Davey shakes his head. "Willy son, get down to the quack's and get it checked. You're covered with the insurance. You might make it worse – mebbies needs a pot!"

"Ay, that's a great idea that, Davey, enjoy your 4 hour wait in casualty for a tubey grip and a box of cataflams! Or and being told to keep off it as much as you can!" Calla frowns.

"I'd say Calla's reet like, Davey, al get some anti inflams, a tube sock and 10 pints of Kronenbourg and it will be grand for a 12 hour Shift, the morn! It's been worse than this after 5-a-side, honest!"

"That's my boy! You have redeemed yourself!" Calla pelts the empty can off the back of Willy's chair.

Jimmy putting his head in his hands says, "That's reet, son you listen to Dr bird brain!"

Tom waves to Earl on the gate as the minibus pulls in, Calla's arm out of the window flicking Earl's cap off. "Missed us, big man?"

Earl's deep voice – "nope" – sounding back as Calla is like a child trying to get off a ride.

"Calm down, for fucks sake, Calla, you moron." Jonny pushes Calla back into his chair, Calla springing back up opening the sliding door as Tom pulls up.

"Haweh father, we're yem."

Davey shuffles out of the seat. "What's the plan, Jim?"

Jimmy gives the famous chestnut.

"I'll buzz you in an hour. If I fall asleep I'll see you at graft in the morning, pal." Davey opens the door to the digs.

"See you in an hour at Nelly's, Willy." Calla waves the bus off, walking in behind Davey. "Are you ringing your lass?"

Davey nods. "Ay a quick ring, let her know the Bobby Moore about the weekend, I didn't ring her over the weekend. She's just back from Haggerston Castle with the bairns."

Calla slumps on the sofa. "Bucked a lass in the band up there one summer."

Davey shakes his head as Calla goes on. "Nearly knocked her through the side of the caravan, although not surprising – she was built like a brick shithouse! Anyway, Davey, say nowt to your lass about the wages until tomorrow in case Mally the rattlesnake has his way."

Davey puts the kettle on. "Ay, you're reet, d'ya want a cuppa, scad son?"

Calla turns the telly on. "Na, I'm going to throw my bag on my bed and get a David Gower I fucking hum, shave n shite, then head down to Nelly's."

Calla turns his phone on and a rake of texts start coming through.

"Week's teletexting you, Calla," Davey shouts from the kitchen.

"Ah every cunt, I had it off for 4 days." Calla's first text is to Sharon as he smiles and throws the phone on the table.

"Well?" Davey comes in with a cuppa tea and a sarnie.

"Ah I texted young Kris to meet him after graft, see what's happening. Davey, that ham's been there 2 weeks mind."

Davey opens the bread and smelling the ham then takes a big bite. "Tastes areet, son."

Calla jumps up kicking the table, spilling the top out of Davey's cuppa. "Clumsy twat." Calla grabs his bag and jumps up the stairs. "Bag's on your bed, father."

Davey is dozing in the chair, big snores, as Calla comes down the stairs, pristine pink stripey Lacoste t shirt, jeans and gazelles, checks his hair in the mirror, ties Davey's laces together and picks his mobile up before heading out the door. Nelly's is very quiet when Calla walks in, only Rosie behind the bar and an old bloke sitting at the end of the bar. "Afternoon, gorgeous, bottle of Heiny and a can of sprite please."

The barmaid smiles from ear to ear on seeing Calla. "How was your weekend away, Calla?" as Calla pulls up a stool and salutes the old timer with a wink before leaning over and kissing Rosie on the cheek.

"Quiet enough but we were diverted to Miami because of fog so that was a mad night."

Rosie pours the sprite into the Heiney and puts it in front of Calla. "There you go. You smell lovely – what you wearing?"

Calla smells himself. "Aqua di gets intyapanties by Armani," taking a big drink out of his pint glass, "ah that's fucking tremendous," flicking a 100 dollar bill on the counter – "get yourself one and put whatever Steptoe is having on me as well."

Rosie drops a single measure of bourbon up to the old timer who raises the glass to Calla and carries on reading his paper.

Calla's just about to take a sip of his lager when Keegs slams 2 hands on his shoulders. "Boo!"

Calla nearly jumps out of his skin, spilling the lager down his shirt, standing up and wiping the shirt. "Keegs, ya cunt!" digging him in the chest then shaking his hand, Keegs moving in towards the bar as Calla shakes Kris's hand then "areet custard cock!"

Then Mark puts his hand out, Calla smacking him full force in the arm – "Areet ferret features!" Mark rubbing his arm "Knob!" then Calla giving him a big hug. "I've missed ya, youngn, I've fetched you something back!" Mark's eyes light up as Calla reaches in his pocket but pulls out his two fingers, Mark shaking his head and taking the lager off Keegs. "We were hoping you would fucking die while you were away!"

Calla coughs into his pint. "Fucking bit strang, lad!"

Keegs pats him on the back. "Now then, how was the weekend?" As the lads take a seat at the bar, Calla tells them about the whole trip.

***

Back at the apartment Davey stirs, stretching his arms out and letting out a roar and a tiny bubbly fart. “Hyup best get to the shit hoose!” Standing up, clenching his arse and holding one hand over the back of his trousers, taking one step forwards, his shoelaces jarring him back, sending him stumbling forward into the coffee table, rolling over it and banging his arm on the chair, his glasses flying, sitting himself up, he looks down at his shoes pissed off. “I’ll have that bastard!” He looks at the lump on his arm and putting his geggs back on he begins to untie the knot when a half brick flies through the window sending glass everywhere. “What the fucks going on?”

Davey stands up as Willy’s head pops through the broken window. “Sorry there, big man, didn’t want tee limp over the road so thought I’d tap ya windee with a styen!”

Davey picking the half brick up throws it up and catches it. “No smaller stones like, you idiot?”

Willy blushes. “Sorry man, didnae mean to break the window!”

“What did you think would happen to a window if you threw a rock at it?”

“Sorry man, I’ll pay to get it fixed, dinae worry!”

Davey shakes his head. “I really can’t believe you lot sometimes, a really can’t!” He throws the rock back out of the window and wanders into the kitchen for a brush and shovel.

Willy scratches his head, looks at the mess and pulls a face, shouting through the hole, “Say Davey, I’ll wander down to the gatehouse and see if Earl can call the maintenance!”

Davey lights a tab and starts to sweep up the broken glass, still bleary eyed from waking up.

Willy reaches the gatehouse still hobbling although he has had a handful of anti inflams and strapped the ankle up tight and managed to squeeze it into an old Samba. He shouts into the gatehouse, “Hey, Earl!”

Earl’s head pops over the top of the newspaper. “Yup!” He folds the paper and standing up opens the door.

“Some cunts threw a stone through Davey’s window – do you

know anyone that can fix it?"

A shocked Earl replies: "Say what? I'll catch that bastard – wait here, let me scan the surveillance cameras, this just happen?" He turns round and flicks switches on the control panel, thumbing through the footage on the screen.

"Ay man that cunt was me!"

Earl stops and turns to Willy with a puzzled look. "You? Why the hell you do that?"

"Look! Was an accident, didn't mean for the window to smash, do you know anyone?"

Earl shakes his head and lets out a sigh. "Yah, I'll call the grounds maintenance, they should be able to patch it up, maybe not get the glass until tomorrow now but I'll see what he can do! What number?'"

Willy smiles and leans in, giving Earl all the details, Earl dialling the number and calling maintenance. After a brief chat he puts the phone down. "Ok he will be here in the next half an hour. Is somebody going to be in or you want to leave the key with me?" as Davey walks up behind him.

"Sorted Willy or what!" – still angry, giving him a dig in the arm. "Areet, Earl! Sorry about this but shit for brains here thought it would be a giggle to break a winda!" Lighting a tab he offers the box to Earl who takes one, letting out a booming laugh, stepping out of the cabin to get a light from Davey, taking a big draw.

"Surprised it aint your man who lives with you!"

Davey coughing rubs his chest. "Even he isn't that sackless!" giving Willy another stare.

"Davey man, will you give owa, it's only a broken window. Earl's sorted it – the bloke will be here in 20 minutes!"

Davey finishes his cigarette and scrubs it under his foot. "That's not the point, Willy. What with dopey hole tying my shoelaces together, me nearly setting my neck, then you lobbing bricks at the place! Can you not just grow up and use a bit of savvy for once in your frigging lives!"

Willy leans against the cabin scratching the back of his head. "Who stole the jam out of your doughnut!"

Davey shakes his head. "What's the story then, Earl?"

Earl straightens up. "Jeff is on his way! Say a half hour, you want to leave your key with me?"

Davey shakes his head. "No pal, I'll be in!"

Willy pushes back off the cabin. "Haweh, Davey man, gets doon the boozer. I'll buy you a pint!"

"No! Not in the fettle!" Davey not looking pats Earl on the arm. "Cheers pal!"

Earl waves. "No worries!" – turning to Willy – "Seems you're in the weeds there, buddy!" climbing back into the cabin.

"Arr fuck him, I didn't mean it! Can you call me a cab there, Earl?"

Earl shakes his head and smiles. "Say, why don't you just hop right on my back and I'll run you down!" pressing the numbers for the cab and making an immediate booking. Sitting back in his chair Willy in the one beside it.

"How long you done this for, big man?"

Earl rocks back. "16 years now!" letting out a sigh.

"Like it?"

Earl shakes his head. "Seems to be the only thing a 6 foot 5 inch black dude can do around here!"

Willy frowns. "How old are you?"

Earl laughs. "Your momma never tells you! You should never ask another dude's age! I'm 44!"

Willy does the maths. "Leaves some unaccounted years there, pal, a mean from leaving school or college, whatever you call it!"

Earl slides open a drawer and takes out a water, taking a big gulp. "Was in college football team, could've went pro but I had problems with my knees! Tampa Bay Buccaneers had all but signed me up when I took an injury!" Shaking his head and looking down, he continues: "Got myself fit again but their docs said my knees were a liability! Breaks you man, really does! Playing football was all I knew!"

Willy leans forward and patts his shoulder. "Ay man, share ya pain! I was a YT at Rangers! Best right winger you have seen! But that all went tits up!"

Earl quizzes, "Injury?"

Willy shakes his head. "Naye man, got myself pished one night and decided to go for a spin in me mate's car!"

Earl quizzes again: "Oh god! You have a car accident, that what's up with your face?" laughing, Willy joining him.

"No, ya cheeky bastard, police pulls me over for not having my lights on!"

"They breathalyse you?" He takes another drink from his water.

"They didnae need t, I had nee clathes on!"

Earl rocks back on his chair laughing, drying his eyes. "So you got arrested for drink driving?"

Willy sighs and peers out of the window at the taxi at the gate. "Ay man, banned for a year which meant was a bastard coz I was nearly an hour from the ground meaning it was a nightmare to get to training! Still, managed nee botha for a while but I was still drinking and turning up late!" – putting his head in his hands – "Remember Souness saying to me I was one of the best he had seen but I had to decide either Rangers or the swally?"

Earl leans back. "So then what happened?"

Willy stands up and stretches, "Well Earl, I fell in love with a wee hen called Wendy and she loved the swally mare than me! Told Souness to shove Rangers up his arse! Me father got me a start at his place on an apprenticeship and here I am!"

Earl stands up. "Regrets?"

Willy limps out of the cabin. "Narr, regrets are for poofs! See ya, Earl!"

Earl waving him off closes the door – "Damn fool!" – laughing.

Willy limps into the bar to a cheer from the lads, Calla standing up – "I'm Jake the peg de de de de de derr de derr de de! With my wooden leg!"

Willy pushes him back onto his stool and gives Keegs a hug, shaking the other lads' hands. Keegs looks down at his foot, laughing, Calla pushing Willy. "Did you not give Davey a shout like, ya snakey jock twat!"

Willy shouting in a beer: "Orr dinnit ask! He is in a reet mood! Nearly broke his neck after you tied his laces together!"

Calla cuts in, "Wasn't me!"

Willy shakes his head. "Orr aye, Davey did the cunt his sell and fell asleep!"

Calla laughs. "He is getting ard y'nar! The old squash between his lugs isn't as sharp!"

Willy shakes his head. "Wey that and someone slung a brick through his window!"

All the lads gasp in shock!

***

Davey lets Jeff in, who looks at the window. "I spoke with Earl, we'll patch it up for now and replace it in the morning.

"Nee bother, son." Davey's big digits are trying to press the buttons on his mobile while he puts it to his ear. "Ah balls, no credit left – these phones cost a fortune. Gone are the days when you could put ten bob in the phone and talk to your lass for an hour, eh Jeff?"

Jeff boarding the window up nods not understanding a word of Davey. "Well that will hold up for now, sir."

Davey gives Jeff 20 dollars. "Here are, son, thanks."

Jeff not wanting it – "No way, this is my job, I couldn't take it."

Davey laughs as he pats Jeff on the shoulder and sticks it in his shirt pocket. "The 20 bucks is for running me down to Nelly's, son. I'll take one of the lads' phone off them for a minute until I get credit."

Jeff picks up the tools. "No problem, van is just outside."

The lads are all still talking about the stone throwing. "Honest? Who the fuck done that? Does Earl know? Did you get him to check CCTV?" Calla is all concerned as Davey walks over.

"Thanks for the concern, CSI Miami, but we managed to find out who done it, when I got up from nearly setting my fucking neck because some arsehole tied my laces together and the brick coming through the window at a moment I already had 5 bar on me clacker valve! I was surprised I never shat me kegs, only to peer through the 8 inch hole to see this dozy bastard waving at me!"

All the lads are pissing themselves laughing. "I thought you of all people would've known not to throw stones, bearing in mind you look through glass houses, ya specky little cunt," Calla nudges Willy.

"Between the pair of you I'd struggle to gather any sense at all, all weekend wiping both your arses," Davey complaining as Kris puts his arm round Davey.

"See you should've took me with you, father, nee problems with me."

Davey sips his pint. "Na, son, we had to leave the youth and brains behind to keep an eye on Fitzsimons, the lanky streak of piss. What's the craic with the job?"

Keegs jumps in. "Ay it's been a mad couple of days, nae bother with the install. Client delighted, all tests off on time, just they moved the goalposts yesterday saying they need to commission the first phase and it will take at least 2 weeks."

Willy shrugs his shoulders. "So what's that got to do with us?"

"Well phase 2 is on hold until they are happy with phase 1! So they aren't going to want to pay us for 2 weeks of doing nothing."

Davey chirps in, "By the time we move all the kit, get the drawings and material sorted out it will be 2 weeks though, son."

Just then Mally walks into the bar with 2 other blokes, pulling up a stool at the bar and ordering drinks.

"Look at the cunt, never even looked over. I'm gan owa." Calla stands up as Davey pulls him down.

"Sit down, daftarse, they look like the 2 lads looking after the commissioning – he might be sounding them out and anyways let Jimmy deal with what's next tomorrow."

Mark agreeing – "Mally was areet yesterday, just said if phase 2 is on hold only half the squad will be needed but no panic as there is a massive shutdown on a job in Shanghai for the rest."

Willy smiles. "Ay, what's the deal though, need to find out what the dosh is, like."

Keegs disagrees. "I think they will let us know who they want here then they will tell the deal once you are yem – if it's a better deal imagine the uproar here."

Davey calms all the gossip. "It'll be the same deal, lads, simple as it gans. Right, who's up for killer?"

Calla jumps up. "I'll get the round in," heading over to the bar. "How's it going, Malcolm?"

Mally is in conversation. "Sound, Calla, I'll be over in a bit, just having a pint here with the two lads."

Both lads smile at Calla as he winks back and shouts Rosie over. "Right, beautiful, usual shout for the reprobates over there and whatever the lads are having here."

Mally spins round to Calla. "Cheers son, got some good news for the lads but we'll do it tomorrow in the office with Jimmy."

Calla takes 4 pints in his hands. "As long as it doesn't begin in R and in R everything will be tickety boo, Mally." Winking at Mally then walking off as Mally looks confused getting back into the conversation. Calla drops the pints on the table. "There you gan, courtesy of zero."

The lads take the beers. "What did Judas from Billingham say, Calla?" Kris asking what Calla was chatting to Mally about.

"Fuck all really, just said I'm alright for a good while here and you're bagged cos you couldn't put a pipe in a snowman's mouth or summit like that."

Kris is going red as Calla takes a drink of his pint. "Eh honest? He fucking never said that yesterday, he said–"

"He's winding you up, you fucking halfwit," Willy interrupts, giving Kris a 10 dollar bill. "Here, gan put some tunes on, my ankle's snapping."

Willy flops down in one of the booths. "I'm fucked! Calla, I'm still pissed!"

Calla chalks the cue. "That's areet! And you have good right to, we got through some wine on the plane!"

Keegs laughs. "Davey, lads reckon they blagged themselves an upgrade and even got Jimmy kicked out?"

Davey sips his pint. "Ay that's reet, the pair of twats! Here, Calla son, pass me your phone. I need to call our lass!"

Calla breaks the balls. "It's nee credit on it, Davey, honest!"

Mally just joining the lads holds out his phone. "Cheers, Mally son, back in a jiffy!"

"Playing Killer like, Mally?"

Mally waves it off. "No Calla, just having five, heading back over there in a minute! Everyone ok? Good weekend?"

"Great weekend, Mally. The diversion to Miami nearly spoiled it though; being laid up there for a night was a bit of a bastard!" Willy winks to Calla.

"Ay weather was terrible down that way I hear. Not busting your balls, lads, but you're not having a mad one today, are you? Big meeting tomorrow to discuss commissioning hand over and we need to address the temporary de-mobbing and where you lot will be placed! Some of the big cheeses will be there!"

"Dinnit worry about us, Mally lad and if you want to start telling us what to do in Nelly's I'll be getting you to sign my timesheet for owe time!" Calla swigging back his pint, Mally shaking his head.

"Not at all, Cal, just tipping you the wink is all – you've scored an extra day being away and we have a busy few days; tomorrow's most important, could cause you botha if you cut a shift!"

"Mally, will you pipe down? It's not even dark! We'll be in bright and early! Probably before you, now piss off back to the squares at the bar, you're fucking up my chi!" Calla smiles then nods to Willy.

"Reet o, catch you lot later!" Mally turning and walking away, shaking Tom's hand on the way over who has just come in.

"Hey y'all!" He flicks a dollar on the pool table and crashes next to Keegs, squeezing his leg.

"Now then, big man, how was the jolly with the chaos twins?" quizzes Keegs.

Tom rocks back laughing. "Yup was interesting to say the least! These guys seem to attract trouble. I could write a book about our 3 days. Say, Cal, how's your left nut?"

Calla straightens up rubbing his crotch gently. "Ay Tom, it's a bit tender, squirted a bit of savlon on it!"

Keegs asks Calla what that's about; as the story unfolds Keegan, Kris and Mark are in hysterics as Davey wanders back in. "What's tyou lot laughing at?"

Keegan explains it's over Calla's bollock incident, Davey laughing and wandering over to give Mally his phone back, Calla flopping down next to Mark, flipping his phone out, catching him with the cue. Mark tuts and pushes the cue back. Calla reads the text from Sharon who is asking how he was and how it was lovely to meet him. Calla texts back about a grand farewell at Nelly's on Friday night, asking her to get down and she is welcome to bring some friends He smiles and winks at Willy who looks back confused.

"What? Who's texted you?"

Calla laughs. "Your sister says she can't wait for me to get back!"

Willy shakes his head. "Fuck off, Cal, our Sandy's got eyes! She wouldn't go near you."

"She wouldn't have to if her geggs are the same strength as yours, ya cunt! She'd be able to see me from outer space!" laughing back, digging Mark for approval who stands up and sits near Kris. Calla looking and frowning – "Narr, Willy, mind ya business!" snaps Calla sliding the phone back in his pocket, smiling at the reply (saying ok see you around 8 at Nellys s x).

"It's either your mam telling you your father's been bummed in the joint or your ex-lass asking when the wages are going in coz she's that pink lint the bairn's gone back on the tit!" Willy laughs.

Calla sitting up all serious – "Here, ya daft jock cunt! I hope the old man gets bummed every day, the cunt and the bairn is areet! Don't fucking start with that!" Calla pours the rest of his lager down, the atmosphere tensing up.

"You haven't spoken to him since you came out here, man! Ya fucking hopeless!"

The lads sensing Calla's anger, Keegs piping up – "Haweh, Willy give owa, man!"

Calla jumps up and slams the glass down on the table. "You keep your fucking nose out of it before I bite the cunt off, Willy, you fucking arsehole!" pointing his finger red with rage. "You know the story there! That slag won't let me near the bairn!"

Keegs stands up trying to calm Calla down, Willy stands up – "Bite my nose off? You smoggy twat! Don't you threaten me with that. Come on then, outside, I'll fucking bury you!" waving his hands towards Calla who is trying to push Keegan out of the way, Kris holding Willy, Mally and crew looking over as Davey wanders back. "Lads, lads, what's going on?"

Calla, furious now, pushing Keegs out of the way, pushing Tom back and grabbing Willy by the throat, throws him against the chair. Willy losing his footing and going over on his bad ankle with a yelp, tries to stop Calla grabbing his face, scratching his cheek as Calla lies on top of him squeezing his throat, Tom and Keegs unable to pull him off.

"Bury me, bury me! Don't you talk to me about my boy! I'll fucking KILL YOU!"

Willy, unable to get him off, stops wriggling as Calla holds his fist ready to punch him. Davey yarks his hand back and drags him off, the barman and other bystanders all around now pushing Calla back and pointing to the bar. "Fuck off over there, Andrew!"

Calla protesting – "Narr, Davey, I've had..." Davey cutting him off – "Bar, right now! Cool off!" pushing Calla towards the bar. Calla is still staring at Willy who is being helped up by Tom straightening his shirt; Calla pushes the bar door open and storms outside, Davey turning back to Willy – "What was that all about!" – the bar returning to order, Keegs standing the knocked over chair up, Willy straightening his glasses and pulling his top down "Wey Davey, he is number one for giving banter but he can't tak it!"

Keegs chirping in – "Ay but. Haweh, Willy, ganning on about him not speaking to the bairn and calling him hopeless was a bit strang!"

Davey turns to Willy shaking his head. "Bad craic that, son! You know as daft as the lad is he is getting some shite off the witch back yem!"

Willy takes a mouthful of beer and rubs his neck. "Ay, mebbies didn't think he wad blar his top like that though!"

The lads all sit down except Davey who wanders to Mally who asks if everything is ok and what had happened. Davey explains it was all and nowt, just daft lads' crack and everything is areet! Shouting the beers in he wanders outside to see if he can see Calla.

Davey lights a tab and looks about, seeing Calla walking back from the shop over the road. He stands waiting until Calla is level. "Areet, son?"

Calla opening the box of cigarettes and sparking one up takes a big draw. "Al fucking kill that cunt if he mentions the young'un again!" pointing to the bar with his cigarette eyes welled up.

Davey puts his arm around him. "Anar ya will, son, anar! Think he is sorry, said it was just banter! Dinnit fall out, man! You've known each other years and we have had a class time; dinnit spoil it owa daft words!"

Calla finishes the cigarette and flicks it away, blowing the last of the smoke out. "Wey, there's some things you dinnit throw at people!"

Davey pats him. "Ya reet, there is! But if I blew at everything I put up with off all of you lot I'd be deeing time now!"

"Not the same, Davey – he knows the shite I get from the ex!" Calla lights another tab.

"Calla son, let it go! He is sorry for what he said and I'm sure after nearly having his eyes popped out his heed he won't be mentioning it again!"

Calla laughs. "Ay he fucking shit his pants, the fucking Jockroach!"

Davey laughs. "Ay so did Mally, wet noodle spine! I thought the skinny bastard was ganna jump owa the bar!" Both of them laugh. "Haweh son, there's away back in, shake hands and get a booze doon ya! He says 2 words and I'll banjo him mesel!"

Both walk back into the boozer, Calla walking over to the pool table.

"Look Cal, sorry for mentioning the bairn! I stepped owa the line!" Willy offers his hand.

Calla smiles. "Ay apology accepted! Not sorry about mentioning ya sis though, I'd still fucking podge her if I got a chance!" He drags Willy up on his feet and gives him a hug, both of them laughing.

"You ripped a button off me shirt and twisted me ankle, ya bawbag!"

Call, leaving go, points to the scratch on his face. "Wey look at me chevvy! Looks like I had a go at Freddie Krueger. You need to get ya nails cut, ya fucking tramp!"

He sits next to Keegan who squeezes Calla's leg. "You're both soft as shite! I've seen bairns fight better!"

Calla smirks. "Ay, Captain America, you did well restraining me!

You're weaker than mead!"

Keegs, finishing his beer and wiping his mouth – "And why would I want to restrain you! I wanted you to deck the cunt – he gets reet on my wick!"

Willy looks up. "Eh! fuckn Haweh, Keegs!" All the lads laugh.

Mally waves to the lads as he and his two comrades leave, all the lads waving back. "See you tomorrow, you fucking fifty faced cunt," Calla whispers at table level.

"He's away yem to climb into his saxophone case for a lie down, the bent bastard," Davey slipping a dig in as he stands up. "Away for a gypsy's kiss, lads."

Mark laughs. "Why does he always tell everyone when he goes to the bog?"

Calla shrugs his shoulders. "He always has – he's some man."

Davey is in the toilet when a tall, middle-aged bloke comes in to wash his hands wearing a nice pinstripe suit, looking hard at Davey as he turns round. "Davey Ramsay!"

Davey takes a second look. "Billy Hamilton, what the fucking hell you doing in Orlando, son? Let me wash my hands and I'll shake yours." Davey washes his hands as Billy dries his.

"I'm over to go through the demob plan with Jimmy tomorrow and sort the bonuses out for you lot."

Davey dries his hands and shakes Billy's with a big pat on the back. "We'll slide through into the lounge for a pint out the way, son." Davey glances round the corner as they scoot into the lounge and sit in the corner. "You in here on your own like, Billy?" Davey signalling 2 pints to the barman.

"I texted Jimmy to ask him where you would be. He had me collected from the airport and said he was cabined up tonight."

Davey takes the pints off the young barman. "Stick them on Calla's tab, youngn." Both the lads chinking glasses. "I knew you were up in the directors' box but I thought it was Middle East and Asia you were looking after, Billy. Fucking hell, we had 5 years on the Montrose together back in the day! Fucking north sea tigers! Think the last time we had a pint was at Geordie Burnet's funeral, bless his soul."

Billy nods. "Aye, that's right. I am vice president now, Dave, believe it or not so I was more than happy when I found out you were bringing a squad out here. Just sorry it wasn't the 6 months for the full squad although there will be a requirement of 6 lads for the next 16 weeks. We are ok with that. Between me and you it's fell at a good time because we need bodies, good bodies in Shanghai in a month, at least 8 lads and at least four 3-month trips."

Davey nods. "Sounds good to me, it's arl the same pipe, Billy, lad."

Billy laughs. "I was onto the main actor for the client here, good pal of mine, played a few rounds with him, says he was delighted with the install ahead of schedule so he is paying us the bonus we had tied in with the contract. With the news of a demob I had a word for it to be paid now rather than the end of phase 2 and he agreed."

Davey takes a massive gulp out of his pint. "So the lads get a bonus then, Billy, is that what you're saying?"

Billy nods and takes a drink. "Mally wanted to retain 2 weeks of the lads who we need back as an incentive – he is still in the Middle Ages, bless him! I have just sent the mail to accounts to release the month's bonus per man in this month's salary and there could even be another touch for the lads who come back if they finish phase 2."

Davey sits back, smiling, signalling another 2 pints. "Sounds good to me, the lads did pull their puddings out for Jimmy like. Mally is Mally – he'll never change! And Shanghai? Is the deal the same?"

Billy opens the laptop case and pulls out the contract. "150 dollars a day more for the bears, 200 for you with the same everything else and you get to pick the squad."

Davey takes the pints off the barman and gives him a fifty. "So I'm not coming back?"

Billy shakes his head. "Jimmy said you should run this one, Mally's level you know, Dave? Jimmy will be flying there in 2 weeks to set it up."

Davey shrugs his shoulders. "Ok Billy, but two things: one, don't mention the bonus until tomorrow and two, don't mention that I have to pick the squad for Shanghai."

Billy clinks glasses. "No bother me, old pal, so how's Marj?"

***

"Where's Davey at? He's been away ages! Mark, go and check on him!" Calla barks.

"Get stretched, Calla, I'm not raking about a blokes' bog!"

"Wouldn't be the first time, ya toucher!" Calla, shaking his head, stands up, pulling his jeans up and tucking in his shirt. "Well, tell you what, fairy features, I'll check on the old man! You get to the bar! It's that long brown thing owa there!"

Mark stands up and wanders with Cal; as Calla goes right towards the toilet he shoves Mark into a table. Calla, catching a glimpse of Davey's unmistakeable foot stretched out drawing circles in the air, creeps over – "Rarrrr!" – Davey and Billy both jump, spilling their pints.

Davey pats his shirt down. "Ya stupid howwer, ya!"

Calla laughs, Davey pointing to him, "Billy, this is Andrew Callaghan."

Billy stands up and shakes Calla's hand. "Good to meet you, son. Billy Hamilton. Heard your name mentioned once or twice!"

Calla firmly shakes his hand then leans back smug. "Yes Bill, you will've heard I'm a great hand and excellent with the fanny!"

Billy laughing looks at Davey who is shaking his head. "Yes lad, think that was about the sum of it – sit down please!" Taking his seat.

"No Billy, thanks but I'm only on a recce mission to see where the old goat got too! Went for a slash half an hour ago and even with his twitchy prostate he is nee langer than 10!" Davey gives Calla a stare. "Ay Davey, had me worried!"

Davey smiles at Billy and tips his glass at Calla. "Cheers son! Nice to know someone cares! Worried in case I'd rowled down?"

Calla pats him on the shoulder. "Nor Davey, I've forgot me key!" Laughing then wandering back off.

Davey looking over his glasses to Billy "Some cunt! Tell you though Billy as mad as bucket of frogs that lad but he was reet about the first bit he is a great hand!" Billy smiling as the two carry on cracking on

Mark is just slowly walking back to the table with a tray of drinks when Calla sneaks behind him and whips his strides down, all the lads cheering as Calla then jogs past laughing! Mark is like a road cone stood rigid with the tray of drinks as the lads start taking pictures with their phones. "Calla, ya bastard!" Mark waddles to the nearest table to put the tray down then drags his pants back up.

"So where's Davey?" Keegs asks.

"He's reet round the corner in the lounge bit talking to some gadgy called Billy or summit!" taking the pint off the red faced Mark. "Ta love!" winking at him.

"Billy who? Not Billy Hamilton?"

Calla puts the glass down and rubs his hands, top lip covered in froth. "That's him! Why?"

Keegs straightens up. "That's one of the main men, sound bloke but a ruthless bastard when it comes to hiring and firing! You didn't say anything did ya, Calla?"

Calla looks up at the ceiling. "Nope! Was polite as fuck!" picking his pint up. "Mark, gan get some menus, man, ya lazy shite!"

Mark looks up. "Calla! Fuck off!"

Call stands up and heads to the bar. "Nee need! Howw, Rosie, can we have some menus when you get a minute please!" Glancing at the old timer who is still sipping his bourbon and reading his paper – "Yo, Jackie!" The old man tipping his paper – "Yes son?" in a broad Irish accent. "Want another Bourbon?" The old man puts the paper down and looks at his glass! "Sure son! My father always says a trout in the pot is better than a salmon in the sea," wiggling his glass and smiling!

Calla shouts to Rosie: "And get your man another bourbon!"

Rosie hands him the menus and nods towards the old guy. "He is called Walter! Real character!"

Calla nods a thank you. "Walter! You get bored of that paper you get yourself over for a bit crack with us!"

Walter raising his new glass – "Maybe I will, son, maybe I will and thanks for the drink! What's your name?" taking a sip and laughing as Calla walks off. "Calla!" shaking his head, whispering, "Owld fucking minj, could've said he'd get me one!" throwing the menus on the table as Davey comes behind him, putting his hand between his legs and making a fart noise. Calla leaps up, Davey laughing. “Lads, this is Billy Hamilton. He is here to pump the lot of ya! Said Epsom Salts wadn’t work ya!” Laughing as the lads take turns shaking his hand, Calla kicking over a chair. “Get your arse down, Billy! Pint?”

Billy raising his glass – “Might as well, Calla, the last two were on your tab!” Laughing at Davey. Calla looks at Davey. “Some cunt!”

Rosie is over taking the lads’ orders, Davey ordering a sharing platter for 4 of ribs, wings, onion rings, garlic mushrooms and prawns. Rubbing his belly he looks over towards Tom. “Just a starter for me the neet! Had a ham sarni afor a came out!”

Tom sniggers. “Motherfucker, I don’t know where you put it!” the rest of the lads shouting up their orders.

“And another round please, Rosie, need a hand carrying them over?” Keegs offers as he follows her to the bar.

“Watch him, Rosie mind, he is only nice when he wants something!” Willy shouts after them, the lads all cracking on.

Well on through the food when old Walter wanders over and sits down next to Calla, all the lads frowning over while they stuff their faces, nodding to Calla in gesturing who he was? Calla licking his fingers and wiping his mouth on a napkin, says, “Lads, this is Walter!”

The lads all wait for more as Calla picks up his burger again, the lads all waving and saying hello. “Sorry to join when you boys are eating. Calla here said I’d be welcome to join you guys for a drink?”

Davey throwing a mushroom in and wiping his hands – “Of course you are, marra, everyone’s welcome at this table, my name’s Davey!” The rest of them follow suit introducing themselves, Calla nudging Walter. “What do you do then, Walter, retired?”

Walter, nodding and laying his stick against the chair, replies: “Sure am, retired down here about 12 years ago! Worked and grew up in Portland! Damn cold it is up there!! Good for your own being this sunshine!” waving his finger at Calla. “Live round the corner, head in on a Tuesday for a cheeky few. Used to meet my son in here but he took a big job in Calgary working up there with them folk!”

The guys all look at each other over their food, unsure what to say, as Walter waves Rosie over. “Yes honey?” she asks smiling. “Get the boys a drink in on me please dear!”

Davey throwing a skinned rib bone on the platter – "No no, Walter, most kind of you but there's a few of us, just get yourself one!"

Walter focuses on Davey. "You'll not be telling a man how to spend his money now!"

Davey holds his hands up. "Go on then!"

Calla puts his arm around him. "You get him telt, Walter, the fat shite! And here was me thinking you were a minjbag!" squeezing Walter who is looking at him puzzled.

"So where we headed next, Billy, any rumours?" Keegs asking the million dollar question with all the lads looking on.

"Not sure exactly. I've to speak to Jim in the morning. One thing is for sure, lads, is you all have a job for the next 6 months."

Willy nods his head. "That'll dee for me. If there is a split in the camp make sure I am on the other list to this balloon," pointing at Calla as the lads all burst out laughing all happy with the news.

"I'll speak to Jim tomorrow." Billy nudging Davey with his foot.

"That works for me as well by the way, I spend most of the day looking after the little jock twat anyway."

Willy standing up gives Calla a big kiss. "You love it you big English oaf."

Rosie drops the drinks down from Walter as he takes out a roll of dollars thick enough to choke a donkey and peels a hundred off. "Fuck me, Walter what did you say you did up in Portland? Rob fucking banks!!" Calla nudging the old man as the lads laugh.

"Well that would just be letting the cat out of the bag, Calla, wouldn't it!"

Davey leans forward taking the last wing as Calla's hand hovers over it. "Haweh, father, I had my eye on that" as Davey sucks the meat off in one chow and hands it back to Calla. "There yar."

Calla shakes his head. "You're like a fucking piranha, ya cunt."

Kris and Mark are setting the balls up as Neil Diamond comes on. "Now we are talking, pour me a drink and I'll tell you some lies." Calla laughing as he drinks his pint.

"What part of Ireland where you born, Walter?" Davey pours the dregs of his pint into his fresh glass.

"Place called Salthill, County Galway. Well my father was a

Limerick man and my mother was a good old Galway girl."

Billy nods. "Beautiful part of the world. I have had a few lost weekends at the races there. When were you last home, Walter?"

Walter takes a picture out of his wallet and shows the lads a picture of himself and John Magnier next to Rock of Gibraltar – the legendary owner and world record winning horse. "2001 at Coolmore stud was the last time I was home, son had me a paid visit to the stud as I like the old nags."

Billy looks closely at the picture. "When did it win the breeders' mile?"

Walter puts the picture back in his pocket. "He didn't, he won 7 out of 7 group one races and I backed everyone. When he came to Chicago in 2002 as favourite for the mile I had a gut feeling Domedriver, another Irish stud would upset the party."

Davey is intrigued with the crack – "and did he?"

Walter smirks as he raises his bourbon. "Here's to upsets and fairytales, boys." All the lads raising theirs to Walter.

"So what price did he win at – surely the Rock of Gibraltar must've been fave?" Davey still on the story, mad about horses.

"He won at a small price and I had a small bet on him. I'd won about 5000 dollars on Storm Flag Flying so I thought 1000 dollars on the hip for a few for the day and the rest on Domedriver."

Davey is still staring waiting for the answer as Calla chirps in, "Davey, man, who gives a fuck as long as old Walter has enough to get out on a Tuesday!"

Walter laughs as he gets himself up on the stick. "Count me in for pool, fellas," winking at Davey as he heads to the toilet leaning in – "28-1, David, 28-1."

Davey leans back, sipping his pint in disbelief, as Rosie walks over to collect the plates asking how the food was. Davey frowns and rubs his belly. "Rosie, I'd actually like to send my food back, it was that bad!"

Rosie looking down at the plate of bones and dipping sauce lets out a laugh. "You're sure, Dave? Seems you only left the bones!"

Laughing back – "Rosie, pet, if I had dog's teeth I would have eaten them anarl! Top snap, love, cheers!" winking as Rosie wanders off with the plates giggling, Tom stretching out. "Get tired on Tuesday!" He smiles toward Davey.

"Ay bigun, it's a tough old life this!" stretching his legs out and pouring half a pint over.

"See you have started taking it easy with the ale, Davey!" Billy laughs, shaking his head.

Davey straightens up. "Here, Tom, worked with this cunt at the Tyne Brewery on the new bottling plant, what year was that, Bill?"

Billy rubs his forehead. "99, Davey! Remember it well, it snowed like the clappers! Had you lot fighting over who was doing the set-ups! Tom, they had made a small lay down area in the waste yeast area and were using it as a small fab shop! Now this place stunk! The waste yeast was everywhere, smelled like dirty fannies!"

Tom squirts his beer into his glass, laughing, Davey chirping in – "Ay but it wasn't minus 5 and we could close the roller shutter! The bottling hall was wide open with them bastard forklifts bombing back and forth!"

Billy agrees.

"So, Tom, the job started with 12 of us, so the firm got us a steady away portacabin, small heater, kettle, sink, y'nar, good enough! Wey we would take our bait in there all sat along the 3 tables like school benches, plenty of room! The job manned up to around 35-40 blokes, Billy?"

Billy nods. "Ay Davey, that's reet!"

Davey continues. "Ay well, it was areet for this bastard being the gaffa y'nar, he had a small bench at the end facing down the row of tables at us full desk to himself. We were all hemmed in like sardines, wey we put up with it for a few weeks but it got beyond a joke so a couple of us complained to Billy about it. He said he would sort it! We thought there would be another portacabin delivered to site – little did we know he was planning a de-mob anyhoo as they had missed out on extra works! Friday bait time we are all sitting there fighting owa space when one of the lads barks to Billy about it being a joke and we needed another portacabin! Billy, calm as you like, putting his paper down and pointing at 14 different blokes – you, you, you and you – the guys looking at Billy who responds 'finish your bait then get yasels yem, you're finished!' picking his paper back up and curling in a corner – 'there's your space lads!'"

Tom coughs again. "No way! Motherfucker is some ruthless piece of work!"

Billy smiles. "Now Tom, there's no nice way of paying lads off but it's part of contracting and that killed 2 birds with one stone."

Davey rocks back laughing. "Tell him what happened next though! Tom, this bastard is also brass necked!"

Billy sniggers. "Wey Tom, the blokes all grumbled up, got their gear and stormed out to pack up their bags. They couldn't've been out the cabin 5 minutes when Peter Thomas came in and told me they had finally agreed on the extra work and they wanted it starting ASAP! Well this left me in a bit of a pickle – I'd just scattered 14 blokes!"

Tom shaking his head – "So what did you do?"

Davey jumps in. "He got up and wandered to the cabin door shouting after the blokes 'you, you and you', some even turning around, Billy replying 'not you, you, you' he continued getting 6 of the 14 to turn around 'you're all re-hired!' then sitting back in his chair and picking his paper up, curling in the corner again. 'Sorry boys back to snug dinners!'"

Tom laughs. "Wouldn't get away with that here; union would be up your ass in seconds!"

Davey shakes his head. "Ay Tom, like I said, some cunt!"

Billy finishing the last of his pint – "Could've been worse, Dave, could've shouted the full 14 back!"

The blokes are laughing as Walter gets back, rapping Calla across the back of his legs with his stick. Calla, about to take his shot, jumps round. "Howw, man!"

Walter sniggers as he eases himself back down, Davey straight into him about the odd flutter on a horse. "Well Davey, I've my own system and a good ear to the ground. There is a horse running Thursday at Arlington Park 3:45 called Bonustime sitting at 16-1 at the minute! I'll be sliding a few bucks on it!"

Davey kicks Billy with his boot and sits up. "That will do for me, Walter, think I'll have myself a punt on that! Same again, Walter?"

"No Davey, that's me!"

Davey insisting – "Come on, one for the road!"

Walter shaking his – "No,No, No, No. Go away now! no more!"

Davey insisting again – "Last one, Walter, let me get you a drink!"

Walter looking down – "One more now then I'm gone!" Davey smiles and sets off to the bar.

"I thought we were playing killer?" Tom shouts over at the lads.

"Na, Texas, doubles 10 bucks a man."

Keegs and Willy are playing Kris and Mark. "I'll go on next, me and Walter will play if he is up for it?"

Walter nods his head. "One game then I am off. I have a meeting with a man about a dog in an hour."

Calla nudges Davey as he comes back with the drinks. "Bet he's a right queer cunt by the way."

Davey agrees. "Winning 100 thousand dollars at the races is something to go off, I'd say son."

Calla gets up to go to the jukebox. "Aye canny touch, what d'ya want on, Walter? Any preferences?"

Walter takes his drink off the table. "Well, if you English lads don't mind an old rebel song! Come out ye black and tans – that's my tune."

Willy laughs. "Nee bother old one, I'll get it on for ya."

Calla – "Haweh lads, nee shite like."

Keegs shrugs. "Calla, the songs are from back in the day, still good songs, man."

Walter standing up to get on the table with his stick and glass of bourbon, singing before the song even comes on – "I was born on a Dublin street where the Royal drums do beat, And the loving English feet they tramped all over us, And each and every night when me father'd come home tight, He'd invite the neighbours outside with this chorus, OH, come out you black and tans, come out and fight me like a man!!"

All the lads are clapping – "Go on, Walter, you mad old cunt!" Willy laughs as Rosie comes over. "Get down now, Walter you silly old fart," as he jumps into Calla's arm.

"Fuckin hell, Davey, can we get him a start on the next job?"

Billy, smiling, stands up. "Well, lads, thanks for a great few hours. I have to get back to the hotel now to Skype the wife and kids. See you all bright eyed and bushy tailed tomorrow."

All the lads salute him off as Davey walks out into the lounge with him. "Thanks for that, Bill, thanks for coming out your way to see me."

Billy gives Davey a big handshake and hug. "Never forget good lads, Davey, and would never forget the best. See you tomorrow, pal."

Davey walks back to the bar where Walter has potted the black, leaving Willy and Keegs with 5 balls, taking the dollars off the side giving Tom 20 and walking over to the bar. "There you go, Rosie, tips from heaven."

Keegs shakes his head. "Willy, you're fucking useless!"

Willy looks shocked. "I just cleared up before, you dozy bastard. Didn't realise Walter knocked around with Paul fucking Newman anarl!"

Walter goes over to shake Calla's hand. "Enjoyed my few hours, Calla son, will see you next week?"

Calla shakes his head. "I am afraid this rodeo is over for a few of us Saturday, hamma, last night Friday – why don't you pop in?"

Walter putting on his old Portland Beavers baseball cap – "Maybe I will, maybe I won't, take care and remember to live life to the full."

Calla laughs. "Nice cap, pops, Portland Beavers – bet you had a few of them bastards in your time!"

Walter laughing and winking – "You bet your ass, boy."

Davey shakes his hand as he passes towards the door.

Calla stands, gripping either end of the cue, bending it behind his. "Y'nar, fair play to old Walt! Wanders owa, breezes straight in, few drinks, bit crack, then fucks off! Sound, that is fucking sound!"

Davey agrees. "Ay the old fella was a good egg and he gave me a tip for Thursday!"

"What's that Davey – shave off that stupid tash!?" Willy laughs.

Davey takes the cue off Calla. "So we're back to the tash, are we!"

Calla snatching the cue back – "It's my shot, Willy Thorne!"

Davey rolls his eyes.

"Anyhoo what's the tip? I'll phone in a bet!" Calla breaking just as

Tom lifts out the triangle giving Calla a hard stare. "Muthafucker!"

Davey smiles. "Bonustime Boys, 16-1."

Calla whips the phone out and dials the numbers. "Howw, ya queer cunt, you said you had nee credit?"

Davey grabs Calla's arm. "I lied!"

Shaking off Davey Calla holds the phone to his ear. "Ay, Hamma! Is that Brad? It's Calla...Ay hallo mate...spot on lad...Nelly's...Nor mate, away yem soon but get yasel out on Friday, it's our leaving doo...anar, but the job's been cut short...look man, fucking listen will ya! Can you get me a bet on? Reet, stick me a thousand bucks on Bonustime at Arlington 3:45!"

Davey pulls Calla's arm. "Stick me a ton on son!"

Calla putting his hand owa the phone – "Gan steady there, Davey I think you have a gambling problem!" moving his hand – "Ay make that 1100!"

Tom holding his hand up signals 2 fingers for him. "Hawld on, Brad! What's that, Tom? 20 bucks?!" Tom barks "2000", Calla shuffling over, resting the phone on his ear and shaking Tom's hand, nodding approvingly. Tom shrugs and Kris holds up 5 fingers. "Ay Brad, make that 3600!!"

Kris holds his thumb up, Davey looking around the room and doing the maths grabs Calla. "Or, had on, Brad! What Davey man, ya irritating cunt! You only want 50 on noo?"

"Fuck it Calla, lay me a grand on, anarl!"

Calla raises his eyebrows, digging Davey. "That's more like it. Ay Brad, look, tell ya father to put 4 and a half grand on Bonustime, it's 16-1...He fucking will pay it out, Hamma, else I'll have the two of ya in the car park and remember last time, son, you bastard caked yasel the thought of a tangle with big Calla!" Laughing down the phone – "Reet o, son! Meet me the morn in Nelly's at 8 for the crumb, just get the cunt on noo afor it comes in! Ta ra! Bet's on, lads, so get the green ready for the morn!"

Davey wipes his forehead. "Marj would dee her end if she fund oot!"

Calla hugs Davey. "Only way your lass will find out is if you rock up with new pearl necklace for her and ya arse pocket stuffed with fifties! 17 jib back, man Davey!"

Davey shrugging Calla off and taking the cue, replies: "It's got to win first, sackless hole! For all we know Walter might be some dozy old soul who tells more lies than Tom Pepper!"

Calla is confused. "Tom Pepper? Does he do the tips in the sun?"

"Mark Twain character, ya dippy twat!"

Calla is even more confused. "Does he do the tips in the Star?"

Davey walks off.

"Me and you halfs like, big man?" Willy shouts over to Calla.

"Are we fuck!"

"Haweh, Calla man! 500 apiece, half the risk!"

Calla leans against the pool table. "So that will be 900 dollars you owe me?"

Willy smiles. "No, because you ain't bucking Sharon! It will be 500!"

"Gan on then, Willy, you ugly little fucker!"

Willy sits up straight. "Eh! Bit strang that!"

The lads laugh, Davey jabbing Tom, whistling, "2 grand, Tom, fucking hell!"

Tom shrugs. "I either gamble it or mattress back in Texas gets it to spend in the Malls!"

Davey laughs. "You might win?"

Tom rubs his eyes. "Maybe I'll use it to put a bounty on her fat ugly head!"

Davey laughs. "So how long you been happily married for, son?" laughing and walking away.

"What's with you, Keegs, not like you to have a bounce on a tip?"

Keegan straightens up. "Changed man, Andrew, knocked all that shite on the head in search of actually saving a few quid!"

Calla frowns. "EH?"

Keegs laughs. "Nar pal, I've five hundred on Castle Rock in the same race! Favourite!"

"That's reet, another jock who limits the risk! Favourite, my fucking hole – how much do you win back? 28 pence and a curly wurly bar!" Calla snarls sitting on Mark's knee, Mark pushing him off, Keegs leaning in from his shot and glancing towards Calla as he pots his ball. "And my 500 back! Haweh!" both laughing, Davey stretching out, taking the cue from Calla letting a fart out in front of Willy who catches a mouthful of it, jumping up and hot footing across the pool area yelping, "Davey man, nee call cocking ya arse towards me! That's fucking awful!"

Davey laughing – "Howw, William the blacksheep with the Blackfoot! That's the fastest I've seen ya moves since your roll around with a Heineken bottle in Miami!"

Calla laughs and points, Willy takes another seat with his t-shirt over his nose, Davey straightens up for his shot looking down the cue then stands back up and holds it in front of him. "This cue's bent! Andrew, that's off you stretching it behind ya head you idiot!" putting it back in the rack and inspecting them, the other ones all of them with various damage to their tips etc. "Keegan son, pass me that cue over here!"

Keegan shakes it demonstrating a clunking noise in the butt. "This one's fucked anarl, Davey!" Looking at his watch – "Lads, it's quarter past ten. I say we retreat to the bar for a last beer and a handful of lies then bed!" Keegs throws the cue on the table and the lads pick their things up and loose dollars from table, wandering to the bar.

"Final round in, Rosie!" Davey rubs his hands.

"And a round of Tequilas, sweet cheeks!" Calla doing a little dance towards the juke box looking at Davey smiling. "What's wrong with you!" Davey shaking his head. "Wha?" Calla standing with his hands out.

"Get these rebel songs off before Willy paints his face again." Calla pushes 5 dollars in the machine as Rosie lines the drinks up.

"So wonder who gets the transfer tomorrow." Kris sits down next to Davey.

"Wait and see, son. Long as you have a job, that's the main thing." Davey swills his pint then stands up. "I think I'm cutting another tooth how–" Fingering his arse through his jeans walking off, Mark laughing. "See! Every fucking time, man."

All the lads laughing as Calla comes back. "Turn the jukebox up, Rosie please, my little rum truffle of trufflenesses."

Keegs shakes his head. "Why, what romance did you play?" just as Journey's Don't stop believing comes on.

"Nice song, Calla." Rosie smiles over the bar.

"Classic, kid, classic! Where's Davey?" Calla getting his tequila ready.

"He just advised us he was cutting another tooth and headed for the bog." Mark pours salt on his hand. "So we have 3 more days, I might stay in the morra and Thursday."

Kris does the same with the salt. "Ay, I'll probably do the same."

Keegs takes his tequila as Willy slaps him on the back of the head – "That's enough of the sissy talk for fucks sake, you two cunts will get contracting done away with."

Calla chinking his glass with Willy – "Well said, sad sack. Haweh, father, hurry up, we are doing a toast."

Davey tappy lappying over to his stool, climbing up, lets out a groan. "What's the matter with you, Dave?" Tom is all concerned.

"That wing sauce has got my arsehole like a fucking blood orange, Tom." All the lads wincing as he picks his glass up. "What we toasting?"

Calla asks Rosie to take a photo with his phone. "Take a photo, Rosie and I might take you back to mine for a game of Twister."

Rosie takes Calla's phone as all the lads pose with their drinks. "There's a message on the front from a Sharon, Calla."

Calla popping his head up – "Ah ignore that and press the camera – it's me sister." The camera phone flashes as all the lads are captured smiling, apart from Willy whose photo is of a gormless stare in Calla's direction. Calla getting the phone back and looking at the photo, says, "What a set of handsome bastards," shoving it in his pocket and raising the glass. "Get it doon yas," all the lads necking the shot with a scowl.

Willy checks the scrumpled paper coaster is in his pocket, sliding it out and looking at it confused. Calla points. "Ay, ya dodgy jockroach, knew it was you who whipped that away. Put the digits straight in my phone while you were pissing out ya midget cock!"

Willy shakes his head. "Mate, it's massive! What's the text say?" Calla looking at it on the sly. "Says she can't wait for me to burst her like a grape!"

Willy leans over. "Giza look!"

Calla putting it away – "No way, William, you snake!"

Willy sitting back – "Well, Calla, I'll be wanting proof you bucked her!"

Calla laughs. "You don't trust me now? Well, I tell you what, ya specky little twat, how's about you just give me the 400 chips back that I paid for you down in Miami instead and we won't speak anymore of it, eh?"

Willy blushing drinks his pint. "Let's not be hasty, man Calla, let's see how you get on on Friday eh?"

Calla is still staring. "Thought you would say that!"

Keegs digs Calla. "She fetching along any of her friends?"

Smiling – "Of course she will be, Trevor but Sharon will be told about the big scotch Jock who treats women like shite!"

Keegan turning to Calla – "Haweh man! Dinnit say that!"

Calla smirks. "Then Trevor, you will painted as a golden boy so long as you don't let the little Willy Krankie here fuck things up!"

Willy raises his eyebrows as Keegs holds his hand out to Calla. "You have a deal there!"

Willy shaking his head – "Fucking Judas!"

Davey stretching out cracks his knuckles and shouts to Rosie. "Is that chef still on? Ask him to send out a basket of nuggets and fries for the lads please!"

Rosie smiles and heads into the kitchen. "Reet lads, bit chow, ONE more pint, NO more tequilas then beed! Reet?" The lads nodding as Peter Gabriel's Sledgehammer comes on, Calla and Kris both shimmying their shoulders and miming the wail at the start. "You two aren't piss wise!" Davey laughs and stands up taking a deep breath and squeezing out a fart that lasts the full two lines of the song loud and raspy, trailing out into a gargling pop, all the lags gagging as Davey straightens up, Calla shouting "MUD!" from under his t-shirt, and "Fresh air time!" taking out a cigarette and lighter and heading for the door.

Keegs standing up and wandering with him, Willy shouting behind him, "Ay dinnit spark up in here the boozer will go up!"

Kris and Mark are wafting the smell as Rosie returns with the drinks, placing them down and rubbing her nose. "Man! You guys are really stinking the place up tonight!"

Calla with his shirt over his nose points at Mark.

"Nor, Rosie, it isn't me! It's Davey! Calla man!"

Rosie gives Mark a questionable look and walks off as the lads start laughing at Mark who has crossed his arms in a huff. Willy pulls his shirt down and plucks his bottom lip towards Mark but catching whiff of the smell quickly drags it back up. Kris getting up shakes his head. "Bet Davey's under crackers are in some order!"

Calla snaps, "Ay bet the backs of them are like Barry Sheen's Helmet!"

Tom stands up, rubbing his eyes. "Man, it's making my eyes water!" He wanders off to the bar coughing as Davey and Keegs walk back in laughing, Davey doing a little shuffle dance to the chorus then squeezing Calla on the shoulder as he walks past, Calla yelping, "Aya, man, ya cunt!"

Keegs looks owa at Davey. "Still stinks, man! Davey, get yasel for a shite!"

Davey cocking his arse and tensing, Willy digging him – "Cock it the other way, maan!"

Davey relaxing – "Nowt there yet, son! Shit shuttle's a good 20 minutes away from this stop!" He laughs and takes the basket of food from Rosie. "Ta love!" putting it down and throwing a nugget straight in his mouth. "Ooooo ooooo oooo that's bastard hot!" rolling it round his mouth, blowing out, the lads all tucking in, Willy blowing on a chicken nugget. "Fresh out the fryer might have something to dee with that, ya goon!"

Davey now beginning to chew it with a mouthful of lager – "Ahhhh! That's better and goon? Haweh!" Holding his hand out for a brief second before throwing another hot nugget in, doing the same routine again, all the lads laughing, Tom sitting back down taking a handful of chips as Willy goes to squeeze mayonnaise over the basket, Calla stopping him. "Whoa what you doing?"

Willy frowns. "Putting some mayo on the jockeys?"

Calla snatches the sachet from him and tosses it on the deck.

"Any of that shite goes in there!" pointing to the basket "and these bastards will be all owa the jockeys!" holding both his fists up. "It's red sarce or vinegar! Thems ya options!"

Willy picks the mayo up off the floor and squeezes it delicately onto a nugget. "Nay culture, some people!" shaking his head. Davey now throws 2 nuggets in at a time as he reaches for the basket, Kris holding his hand over it – "Haweh Davey, ya greedy twat, give us a gan!"

Davey leaning back licking his fingers – "Areet son! Fair enough!" Kris moving his hand not quick enough to put it back, as Davey grabs a handful of chips and a nugget, ramming them in his mouth, chewing them up, laughing.

Calla shouts the bill in, Rosie pulling him to one side. "Will I come round after work for a cuddle?"

Calla smiles. "Please do, I was thinking about your cuddles all weekend," looking down her t shirt, licking his lips as she digs him in the arm. "Cheeky, anyway Walter left 300 bucks for the bill in case he didn't get in before you head off."

"Some man, how much have we left?" Calla smiles.

"60 bucks." Rosie shows him the bill.

"Take a 20 tip, put a double in the till for Walter's next visit and spend the rest on tins for when you come over."

Rosie squeezes his arm. "Ok, honey."

Calla picks his pint up as the lads are singing bed of roses at the top of their voices. "Free neet, lads, Walter the ledge left some dosh for the bill in case he never got to see us before we piss off."

Davey winks. "Aye, some man for one man, old Walt."

Willy supping up – "Will we have one mare for the ditch?"

Calla supping up – "Not for me, flanja. I'm away for an early night to see what my destiny is, sober tomorrow."

Keegs laughs. "Horseshit, Rosie must be coming round."

All the lads laugh as Calla kisses Davey on the forehead. "I've been rumbled. Haweh father, I need someone to hold me hand crossing the road."

Davey slides off the stool dropping 10 dollars on the bar. "Na nite, flower, see you in the morning no doubt."

Rosie laughs as she tills up, the rest of the lads finish their drinks and head for the door. Suburban outside, Brad sitting at the wheel with another Mexican looking lad in the passenger seat. "You English! Here's your betting slip. My dad has put 2000 on it at 16-1 for you, he can't take any more of a tick on a bet than 2 grand."

Calla taking a wad out of his pocket – "Hold on," counting it. "Davey, lend us 500, Tom you can only have 500 on! Gist." Davey shaking his head gives Calla 500 as Tom hands over his 500. Calla looks at the betting slip, all above board – "even signed at the bottom for pay out, pops will just square it off elsewhere no doubt, no biggy."

Calla rubbing his hands together –"Roll on Thursday!!" opening the door, pushing the Mexican along into the middle seat, "move owa, handsome. Brad, will you drop us up?"

Brad laughs. "No problemo, you want to roll into town for a nightcap first? Nice titty bar and a cold beer?"

Willy leaning through the seats – "Count me in sunshine."

"Ay me n all," Keegs climbing in – "Fuck it we are in–" Tom.

Kris and Marky laughing as they climb in. Davey flicks his tab out and climbs in. "Hallo, Brad son, long time no see, youngn, you dropping us up, good lad."

Calla is trying to nudge Brad as Brad comes straight out with it. "I will after the titty bar and a nightcap," as he puts the motor into gear. "Hawldy hawldy hawldy!" Davey pulling rank –"drop me up that bank at yem, first. You daft bastards can do what you like!"

Brad laughs. "No bother, granddad."

Big Earl lifts the gate. "Hey Bradley!" A high five exchanged and an envelope passed to Earl as the motor drives into the car park and the sliding door opens. "Remember, lads, big hitters are in town and there could be some news on the horizon."

Calla laughs. "Aye reeto, ta ra."

Davey opens the door. "You, out. You're going nowhere."

Calla sitting back – "Eh? Haweh, Davey man, stop embarrassing me."

All the lads laugh. "Go on, fanny balls, get the fuck oot," Willy ribs Calla.

"I'll fuckin knock you oot in a minute, jamjars."

Keegs joining in –"Bye bye, Calla, sweet dreams."

Calla turning round – "Sweet dreams of your ma," as Davey climbs into the motor sticking his massive head in. "Look lads, I didn't want to say this but now we are away from the bar. Billy Hamilton is here tomorrow to give us a month's bonus for finishing this phase early – a month, lads."

All the lads are in disbelief. "Honest, Davey?" Willy shuffling over.

"Aye, on Marj's life, son."

Willy rubs his hands together. "Well let's gan fucking mental the neet, there's nee way I'll be able to spend a month's wages on lap dances the neet!!!"

Calla climbing out of the van – "And there's nee way I'm spunking a month's wages away sleeping in tha morn, early nite for me."

All the lads are in disbelief. "Haweh Calla, are you for real or what?"

Davey climbs out. "Just a heads up, lads, I know what Billy's like. Regardless of the hard couple of months' graft, if you turn in half-cocked tha morn he'll have no problem not giving you the brucey."

Silence as the lads think and Davey and Calla walk off –"Thanks for the lift, Brad, tell your father I'll see him Thursday."

Bradley gives Calla the captain's salute as the rest of the lads climb out the motor. "You chicken shits not partying?" Brad leaning back at the lads, Willy patting him on the shoulder. "Thursday, bonny lad, when the wages are in."

Calla and Davey walk up the path, Davey shaking his head at the piece of timber nailed across the broken window.

"Fucking hell, Davey, it's like being back at me ma's house in Redcar!" Davey laughs and puts the key in the lock, glancing round seeing Calla pissing in the bush at the door, Davey loafing Calla in the kidneys then pushing him head first into the bush.

Calla lets out a yelp as he crashes cock out head first into the pissy bush pushing the sticks away. "Davey man! What ya deein!"

"It will be bastard stinking tomorrow and the flies will be everywhere! It's bad enough having nee frigging windows but the place isn't reeking of your piss as well!" Pushing the door open and storming in, Calla dragging himself out of the bush, twigs in his hair, piss all over his jeans rubbing his side, walking into the living room, shutting the door, wandering past the kitchen past Davey making himself a cuppa – "Wey you're the one that spoilt the bush!" He slams his door. Davey looks up, screwing the lid back on the milk with a smile.

Davey flicks his shoes off and presses the remote for the TV – high speed chases is on. Flopping off onto the sofa he starts dipping Oreos into his tea mumbling "fucking custard creams are shite oot here!"

An hour later Calla wanders out of his bedroom twigs still in his hair but with only his duds on. Scratching his balls he shakes Davey who is asleep clutching his cup! He jumps up, spilling cold tea on himself. He rubs his eyes. "Wha?" swinging his legs round, lifting his glasses, rubbing his eyes again. "Time ist?"

Calla shakes his head. "Divunt nar! Still the neet! Has Rosie been round or not?"

Davey dropping his glasses back on shakes his head. "Nor!"

Calla wanders round then opens the door for a look – not a soul in sight, no approaching cars, nothing! Scratching his head he wanders back in. "Strange?"

Davey stands up stretching. "She's strange? Your bastard huggers are on the rang way round!"

Calla looking down then half spinning to see the buttons on the back lets out a laugh. "Ya reet!"

"Goodnight, dopey hole, I'm away to the scratcha – you in ya duds is ganna give me nightmares!" He puts his cup down and wanders off, Calla swinging a kick at his arse. "Ay piss off and goodnight, George Clooney!"

Davey laughs. "That's what they say youngn!" farting and slamming the door.

Calla flops on the couch with his head in his hand sitting wondering why Rosie hasn't turned up. "She was fucking gagging for it, the cunt?" rubbing his head, thinking about whether something had happened to her or heaven forbid nicked his cans, jumping up and knocking on Davey's door. "Davey man, I'm worried sick! Rosie said she would be round!"

The door opens and Davey's stood with just his duds on, brushing his teeth, trying to speak with a mouth full of toothpaste. "Calla son, Hamma and the little gringo probs headed back and gave her a lift yem! Get yasel to bed, man! She'll be areet!"

Calla looks at Davey concerned. "Them fucking undercrckers are the worst I've seen!"

Davey looks down. "What? What's rang with them! 4 pair from Matalan!" doing a half twirl – "Champion them!"

Calla shaking his head. "You're an awful sight, Davey!"

"Well piss off braying on my door then and you won't have to see me! Goodnight, Calla, sees ya in the morn!" Davey goes to shut the door.

"Ay, Davey though, she was fetching me some cans round!"

Davey looking at Calla dropping his shoulders – "So it's the cans you're worried on! You're some unit you, Andrew!" slamming the door.

Calla wanders to his room and gets his phone, reading a message from Rosie – 'home now! tired! spk tomoro x' Calla stots the phone off his bed. "AY! she's had her chance! Selfish cow!" He wanders to the fridge and finds a can in the bottom drawer next to an old onion. Slamming the door he pings the can, takes a big swig, leaning against the bench. "Ahhhhh!" then smelling the can, "Fucking stinks!" Shaking his head he wanders over to the sofa and flicks on channel 624 late night xxx some bird with big tits getting bent over a desk. Calla sits up and whips his undies down. "Time to pull the bald fella's heed off!"

Davey comes down all fresh faced at 5am whistling and singing bits of Bon Jovi's Bed of roses – "I wanna be just as close as, the holy ghost is, fucking stroll on, CALLA!" Davey shouts at Calla who is sitting upright with a can of Bud in his hand and his underpants round his ankles with the flickers of the telly blasting. Calla sits up looking down at his knob, tissue stuck to the bell-end. "Davey man, I was dreaming."

Davey in the kitchen putting the kettle on – "Aye well, I wasn't, son, I was having a fucking nightmare coming down the apples and pears seeing you balls naked."

Calla puts the tin down on the table and pulls his underpants up, sniffing his hands. "What time is it? Fucking hell I divnt nar what smells the worse – the can of onions or the smeg on me fingers."

Davey reaching in the kitchen – "Fucking dirty bastard. D'ya want a cup of tea or what before I'm sick?!" No answer as Calla goes upstairs straight into the shower whistling.

Big double hoot of the horn outside as Calla and Davey walk out. "Morning troops," Calla sliding the door back climbs into the happy bus, half the lads alive, half asleep. Calla looks round. "Fucking hell, has there been a gas leak?!"

Davey climbs in the front with Tom driving, Chuck shuffling along. "Morning son, how did Jimmy get in?"

Tom pulls away. "Mally took him. He texted me last night to tell me to bring you guys in."

Willy is sound asleep at the back in the corner as Calla leans over and nips his nose. Willy gasping for breath opens his eyes. "Fuck off, daftarse. I couldn't sleep last night – the rats were eating ma toes!! All the drink off the weekend's got me fucked."

Calla sits back in his chair. "Must be fucking hungry rats, hamma, eating the onion rings off your stinky crispy onion toes, you manky bastard!"

Davey chuckling in the front, Kris asking him what's funny. "Wey man, I came down this morning and–"

Calla butting in –"Anyway, fuck that Davey, what's the plan for when we get in to graft?"

Mark laughs. "Haway, Davey, what happened? Was Calla asleep, cock in hand, telly on?"

Davey turning round – "Aye, how did you guess, son?"

Willy sitting up puts his glasses square. "Cos he's a serial wanker! The bastard cost me 280 dollars, man, tossing himself off in my hotel room, like some sneaky perv that gets off wanking in the night!"

All the lads laugh as Calla sticks his fingers in Jonny's nose. "Does that smell of spunk?"

Jonny moves aside nearly spewing. "Ner, it smells like onions."

Calla laughs. "Aye exactly, I was massaging Patio Eyes off last neet."

Davey roaring laughing – "Fucking give owa, it's too early for laughing and carrying on. When we get into work I bet there's a big pow wow."

Tom pulls into the 7 eleven. "Yep so energy needed guys, who'sup for bacon and cheese bagels and Red Bulls?"

The lads all troop out of the van, Kris swiping Calla's legs, sending him flying through the electric doors, "Knob!"

The lads all self-serve at the hot stand, grabbing bagels and pouring coffee, Calla laying his food and coffee on the counter then dashing back to get some Gatorade and a bag of Cheetos. "Haweh, man, Calla!" Davey snaps, all the lads stood in the queue.

"Shut ya face, man! You know I can't operate without me Cheetos!" paying the girl and gathering his things, the girl handing out his change "Keep it!"

Calla wanders off. Davey, seeing the 20s and 5s among the change – "Hawld on, love! Give me that here!" Calla obviously using a 50 instead of a ten.

Climbing back in the van Calla is already halfway through his bagel.

"Here, dope stick!" – handing Calla the change, mouth full of eggy bagel, yolk dribbling down his chin. "What's this?"

Davey turns back around. "You gave her a 50! Not a 10!"

Calla wipes his chin and stuffs the money in his pocket, Willy flopping in beside Calla, knocking the other half of his bagel on the floor. "You clumsy bastard! Gan get me another one!"

"Fuck off!" Willy settles in, unwrapping his bagel and taking a bite. Calla bats it out of his hand onto the floor. "What ya deein, man!" Willy stares at Calla.

"Squits!" opening his bag of Cheetos.

Willy smacks them onto the floor. "I had more bagel left than you. Now we're squits."

Calla lifts his leg over and stamps on Willy's carrier bag repeatedly, Willy trying to move it and stop him. "Howw howw howw, man Cal, give owa man!"

Davey turns around. “For fuck sake, will both of you rap in!”

Calla stops and looks at Willy. “Cunt!”

Willy shakes his head, looking in his bag at the fizzing can of Red Bull and the squashed snickers and doughnut! Kris and Jonny are the next in, looking at the mess at the back of the minivan. “What on earth have you two been doing?”

Calla uncrosses his arms. “Speck fyess started it!”

Willy looks to his right. “You had one bite left!”

Calla is about to pipe up when– “FOR FUCK SAKE SHUT THE FUCK UP!” Davey spins round, glowing red, looking at them both. “It’s like having the frigging bairns with me!” Calla sits back, looking away then bending down and picking a cheeto up!

Mark and Keegs hop in followed by Tom and Chuck, sliding the door shut. “Halfs with ya bagel, Mark?” Calla rests on the seat.

“No way, fuck off, Calla!”

“Greedy cunt!” swigging over his Gatorade.

Willy is still shaking his head trying to stop the Red Bull spraying with his teeth.

The minivan pulls up on site and the lads all clamber out, Calla stretching. “Arrr nowt like getting to graft! Missed this place! Reet lads, I’m away for a half hour shite!” trooping off to the bog, Mark following him.

Davey calls out, “Boots n lids on, boys, straight into it!” bouncing the orders across to the lads who wander towards the fab shop. Walking through the doors they notice all the pipe stands gone and the orbital welding kits are gone.

“Davey where’s all the gear at?”

Davey scratches his head. “Have no idea, son! Maybe they have started shifting it?”

“It was here yesta, Davey?” replies Kris, lifting his helmet, rubbing his eyes.

Davey squints at his phone then presses the call button to Jimmy. “HALLO! JIM, CAN YOU HEAR ME?...Or reet, sorry! ay can you hear me?...Ay champion, listen we are in the fab shop and all the gear’s been shifted?...ay...reet...well I thought...nee botha, youngn, al send them down! Ay am on my way!” He squints again as he puts the phone down, "Right lads, the kit's been shifted to the next phase already. Seems Mally has done you a favour and got the night shift to wrap it up and move it! See he isn't that bad! That's the donkey work done!"

Keegs shakes his head. "What's the cunt done that for then, Davey? You get nee favours off that bastard – he has something up his sleeve!"

Davey shrugs. "Keegs son, ask me one on sport! I'm away up there anyhow so I'll find out!" walking off.

"Me and Kris will carry on from yesterday, Davey!" Keegs shouts.

"Areet son, so long as you know what you're deein. Jonny, you stop here and keep them 2 cunts out of mischief! Willy, start piling up the gash pipe!"

Davey gets to Jimmy's office gasping. "Morning, boys!"

Mally and Jimmy wave him in. "Mally, where's all the gear at? I thought you wanted the lads to pack it all up the day? What they going to be doing?"

Mally looks at Jimmy then Davey. "I got the night shift lads to send it down to phase 2 last night, Davey. I was thinking the lads could get a start down there? Three good days setting up, maybe even throwing a few brackets in!"

Jimmy looks up. "Mally son, them decisions are for me!"

"With all due respect, Jimmy, the lads would've used these 3 days for fun and games laughing and carrying on shifting the kit around site and we can't afford any of the gaffers seeing this! Best to get them down on phase 2, out the way, being productive!"

Jimmy shrugs his shoulders. "Makes sense like, Davey!"

Davey nods. "Reet, get me the drawings and I'll see what we can get on with down there!" sighing and stepping forward.

"Davey, I'm not squeezing the dregs out the lads, but you see my point! We are all under the watchful eye and the Yanks won't be laughing at the lads swinging the lead!"

"Anar, Mally, settle down. I'll tell them you're the cunt that's got them back on full steam for the last few days!"

Mally looks nervous. "Ay and tell Calla to put his Connect 4 back in the van!" laughs Jimmy, Davey joining in as the door swings open,

Billy standing there.

"Knock knock, boys! How are we all this morning!"

Jimmy standing up shakes his hand as do Mally and Davey. "Tip top Billy!"

"Mally, get Davey away with the drawings and we will all meet back here in half an hour! We have a meeting with the big boys in the main office at 10!"

Mally stands up with Davey. Jimmy and Billy wander out of the office as Mally and Davey go through the drawings.

"How's things, Jim?" Billy with a cup of Starbucks coffee as Jim lights up.

"Sound, Bill. Was a bit worried about the squad at first."

Billy laughs. "That's management for ya."

Jimmy continues. "Davey is the linchpin that keeps them all on their toes, commands respect – that's why I thought it was time he earned a few bob."

Billy nods. "You don't have to tell me, mucker, I have his bonus squared off for two months."

Jimmy smiling – "And mine?"

Billy winks. "Like peas in a pod, the pair of you, aye two for you and Mally as well; the rest of the lads get a month."

Jimmy flicks his tab away. "Let's get the lads and head into the head office."

Davey has the drawings rolled up as he gets to the door. "Give me half an hour. I have Keegs punching out snags and I'll give Calla and Willy these – it's only two days' graft but I suppose it keeps them busy." Jimmy holds the door open as Davey marches past.

Mally sits down as Billy takes a seat. "So, Billy, heard there's a job in Shanghai – who's running that one? This one's only going to be another few months aren't it?" Mally sits cross legged.

"Big Jim will go down in a week or so, followed by Davey."

Mally sits up. "Davey? What about me?"

Billy leans forward. "Mally, you have had 18 months here – you will get probably 2 years – what's wrong with that?"

Mally ever selfish – "Can't you let Davey finish this one?"

Billy looks at Jim. "And you get another good run? Don't worry,

Mally, we have plenty coming off." Billy throws his cup in the bin. "Ay mate, we'll no doubt need you in Shanghai in a few months."

Mally is not looking happy. "What will I do for the squad coming back? Hold their wages?"

Jimmy stands up. "Aye I'll go and tell the lads will a?"

Billy also standing up – "No need, Mally, I have arranged their bonus to be paid this month. They will be back, don't worry and I have arranged for your 1 month bonus to go in as well."

Mally smiles. "Sounds good, I'll get the car keys."

Jimmy looks at Billy. "Thought he was down for 2?"

Billy winks at Jimmy. "You can tell Davey in Shanghai." Both the lads laugh as they walk out of the door.

"Right lads, Mally has given us a mission; told him 2 days, I reckon 1. Get these brackets up, there's 10 and get the bagged and inspected pipe up on them with a hard purge next to the valve in phase 1."

Calla leaning against the bench – "Any news on the bruhooheeeeceeey bonusalero ball how?"

Davey shakes his head. "I have to go up now, so you know what you'd are at?"

Willy nods. "Aye, Davey, get yourself up to the orifice and sort the candy oot"

Davey gives Calla the drawings and heads off.

"We'll make a few shilling out the scrap as well." Willy points at the heap on the floor.

"Sound, I'll ask Chuck where the scrapman lives while you'd get the string line and get the brackets marked." Calla wanders off, drawings on the bench.

"Get them up on the wall, lads." Willy handing them to Kris and Mark.

"Fuck me, delegation city?" snatching the drawings and walking off.

"Dropped your dodie, youngn." Willy laughs as he opens the tool box.

Keegs walks over to Calla who is setting up a purge whistling away to himself. "What's happening, Calla, any word on the street?"

"Word is we are fucking away yem Saturday with a pocket full of dollars, hamma."

Keegs laughing – "Aye that's what it's all about, pal. No, I mean with the job? Who's heading where do you know?"

Calla shrugs his shoulders. "My guess is Mally will be left here with the golden boys, ie you, Kris and four eyes, possibly Hags when he gets back from his jolly with their lass."

Keegs nods. "Probably. Mally wouldn't keep you here, nor Davey, as you two are a threat."

Calla laughs. "I wouldn't be surprised if the twisted cunt keeps us here and paves the way for him and you lot to go to the land of the rising slanty eyed sun."

"Wouldn't be surprised either, Hags is back tonight. He texted me before asking if he was bagged you know how the pessimistic old bastard is!"

Jonny and the lads walk in through the plastic sheet carrying a load of brackets. "Well here's the numpty squad, billy bob and the bracketeers!" Calla winding the lads up as he finishes tightening the valve set up.

"Fuck off, Calla, you mong." Jonny bounces a washer off Calla's hard hat.

"Haweh," as he walks over. "What's the dimension on the west side of that run and I'll go down and mark it? You get this side marked and we'll lash a line down," as Tom and Chuck wander in carrying 6 lengths of stainless pipe.

"We'll be doing lines on Thursday if that horse wins, boy." Tom winks at Calla.

"Aye could only get 2 grand on it, you still have 500 on it though!"

Willy comes through the sheet in a scissor lift. "Out the way, sad sack," as he manoeuvres past Calla.

"Who let you on that, danger mouse!" Calla walks alongside Willy. "Willy, I was thinking..."

Willy looks down at Calla. "I thought I could smell burning wood."

Calla climbing on the side of the machine – "No man, listen up, I reckon they won't want us in the same team for Shanghai so why don't you strike before they make a call? Head up the office and ask Jimmy if you could go, say you have a cousin there or something?"

Willy stops the lift and leans over the side. "Is it more bread, like?"

Calla shrugs. "Fuck do I know but would you want Mally for another 6 months?"

Willy agrees. "Ay might fancy a few months of shagging a slanty fanny," as he jumps out of the lift and gives Calla the harness. "You go and mark up at that end, I'll go and see Jimbo."

Calla takes the harness and kicks him up the arse as he walks off. "Hurry back, fannyballs, I need a hand." Willy walks off holding 2 fingers up behind his head.

Jimmy, Billy and Davey are standing chatting as Mally pulls up in the car. They go to get in as Willy comes up. "Jimmy, got a minute?"

Jimmy wanders over. "What's up, Willy?"

Willy offers Jimmy a tab then lights both of them. "Look I know you will already know who's going where, but is there any chance of Shanghai? Tell you why, I have had a good run here and, well, a change of scenery would be nice."

Jimmy laughs. "Did you find out about the money?"

Willy smirks, his brain doing overtime. "Well I have kept it to mesel, bigman, but I did hear a whisper from other avenues, none here! You know I have my finger on the pulse."

Jimmy knowing in his mind Willy is as cute as a fox – "Look, Willy, we'll see what Mally has to say – he could be running it."

Willy blowing the smoke out – "Well, er, look I am quite happy here you know, but just..."

Jimmy roaring with laughter – "You can't kid a kidda, son, I know you want to be away from Mally. We'll see," as he walks off flicking the tab away.

"What did he want, Jim?" Billy asking as Jim climbs in the car.

"Just wanted to know if he had a job – you know what the lads are like when they see you in town, Bill."

Mally pulls off. "Mad as a march hare, the little fucker, but some hand with the piping."

Back down on the park the lads are flying along fitting unistrut brackets that Calla had marked off – he is up at height texting on his phone when Willy walks back under him. "Howw, fuckface, lower yourself down."

Calla doesn't stopping texting. "Fuck off, shit for brains, I'm busy."

Willy kicks the emergency stop button at the bottom of the scissor lift, walking off. "Suit yersel."

Calla goes to press the joystick to lower. "Howw, dipshit, turn it back on, will ya."

Willy marches on up to the lads as Calla makes himself comfy as a text comes back in.

"Willy, we'll be done here by 5." Jonny is up on another scissor lift marking unistrut.

"Maybe we should slow down. Mally will probably dock us for tomorrow, the cunt," Kris chirps in as he throws another bracket up.

"Fuck him, we'll finish it tonight, wrap all the gear up and tell him there's a bit more left tomorrow." Willy gets a few bits out of the box as Keegs comes back in with the two Texans and Mark. "That's the lot of the pipe, we are just waiting on you balloons to get the brackets up," the lads laying the pipe out ready.

"Where's Calla?" Keegs asks Willy.

"He's stuck up that scissor lift for being a cheeky Charlie."

Keegs laughs. "He'll not be giving two fucks about that knowing Calla."

Willy walks back along turning the key to light the power up on the lift. "Righto, bring her doon."

Calla stands up, lowering the platform down and sticking a blown up glove on Willy's hard hat without him even feeling it. "Right, ham, what did Jimmy say?" Calla jumps out.

"He said Mally might be running Shanghai! I think he was double dusting me though, the bastard, cos I tried to dust him."

Calla laughs. "Well at least you planted the seed. Let's go and grab the rest of that strut – we are all marked up here."

Willy walks down with Calla. "We'll be finished this today."

Calla shrugs. "Mally won't know. We'll tell him there's another few hours tomorrow."

Willy laughing – "Great minds, Calla."

Davey and Mally are sitting at the bottom of the table with Jimmy, Billy and three other blokes around the top half.

"Just to recap," Billy standing up flips the chart back. "Mally you will stay here for phase two and handover with Trevor, Gordon, the two young lads Kris and Mark, along with the two Texans and obviously the local labour who we will demob down to 30 tomorrow. Jimmy, you will give me a handover by Sunday so that I can send it to Head Office with a copy obviously sent to Mally – you then head straight to Shanghai Monday. Davey, you will fly to the UK Saturday for 1 week and head to Shanghai, 10 days for visas, getting the lads' accommodation and sourcing a few locals, then the lads will fly out 17 days later. Andrew, Jonny and Willy, with possibly 20 more from other sites."

Everyone nods, Jimmy leaning back. "We'd've loved to have sent the same squad but better all the eggs are in different baskets lads, I think."

Again all the lads nodding, the taller of the three fellas, the HR manager Tony Ball chirping in: "Yes, job well done. I'll make sure all the flights are sorted for Saturday."

The small fella, the accountant, closes his laptop. "40% margin with bonuses paid, not a bad 2 years' work, lads, that's without the few variations. This last few months push has been amazing and we are really impressed with what has been achieved."

Billy stands up. "Invite all the lads out to dinner tonight. Frank's steak house, 8pm free house until 11.30 on the company."

Jimmy stands up shaking the lads' hands. "Will do, will do, see you all later."

Outside the lads have a smoke, Davey is quiet. "What's wrong with you, Davey?" Mally asking the question.

"Ah nowt, son, just was looking forward to a fortneet off. Suppose our lass will be happy with a week and 5 grand bonus."

Jimmy laughs. "It's more than that Davey, 2 months' wages, like."

Davey smirking at Jimmy – "Ah na, son, the owld deeptank needs topping up though."

Mally laughs. "I'm pleased for you, Davey, bout time you ran the show somewhere. Bet you are sick of listening to me over the past few years."

Davey pulling his strides up and walking towards the car, replies, "I never started listening to ya, Mally, so it maks nee difference, son."

Jimmy chuckles as Mally flicks his tab away. "I'll get the jamjar."

Jimmy puts his arm round Davey. "And you get Willy in the squad, nice bonus."

Davey looks down at Jimmy. "Are you for real, him and fucking Calla? Why man, I'll be a nervous wreck by the time the job's finished."

Jimmy shakes his head. "Ay but the job will be reet."

Mally pulls up in the car and the lads climb in. "Nice touch with the meal, wasn't it?"

Davey smiles. "Aye proper bloke and proper company and I'll be proper fucking clamming by 8 o'clock."

Jimmy looks at Davey. "I hope they have a slack herd of cows round the back of the steakhouse – some hungry bears turning up the neet for a bit meat."

The lads are all in the site canteen having a bit of banter with the American lads, one of them, Jim Buxter, a 6ft 6inch gentle giant of a man with no brains and a big mouth, always one of the targets for Calla and Willy. Being the shop steward Jim would be the go-between for the lads with any grievances on site.

"Jim, when do we get our holiday pay?" Willy leans down the table.

"You guys made an agreement at the very beginning to be absconded from the system and remain consultants for tax purposes – I checked that out, remember?"

Willy eating his hot dog – "Oh aye, big man, I forgot about that."

With around 70 Americans on the books the British lads were outnumbered yet their expertise in the game made them better than any of the locals. Despite this the locals always had the attitude of an expert. "Hey, Willy, we hear you guys are bugging out," one of the locals, Tommy 'tomcat' Lamar shouts over.

Willy had shown Tomcat how to operate the orbital machine; now Tommy thinks he is a tradesman with more edge than a broken pisspot. "Bugging out? What the fucks that, Tommy!? Leaving you mean? Ay we have trained you donuts up and done most of the work so it's time to head home!"

Tommy laughs. "That's right, thanks for the memories, Willy."

Calla putting his hotdog down – "Ay we also have bucked all your women and took all the dollars. That's why it's time to move on, hamma."

All the lads laugh. "Oh yeah? You think you know it all, Calla, you know jack shit!" Jim Buxter is not able to keep quiet.

"Jack Shit? I think he is on the darts team in the local! I also know about geography, something you lot don't!! You still think the Boro is in fucking London man! Do you know what American state has the initials PH? I do. Gobshite!"

Jim shouting out straight away as Calla tucks into his hotdog again. "Phoenix."

One of the other Americans corrects him. "That's in the state of Arizona, Jim."

Jim is confused. "Oh shit, yeah, I meant Philadelphia."

Again corrected by one of his colleagues – "That's in Pennsylvania, dude." – Tomcat trying to help the big fella out. "Calla, there aint no such state, so shut the fuck up."

Calla standing up wipes the ketchup off his chin, leaning on the shoulder of Willy. "Yes there is, you fucking clown, Pearl Harbor!!"

All the Americans laugh as Tomcat stands up. "That's not a state, you dipshit."

Calla, just about to sit down, shouts back, "It was when the fucking Japanese were finished with the cunt."

All the lads are pissing themselves laughing as Willy shouts over. "Sit down, you fucking bell-end."

Calla hits him on the head with the rest of his hotdog. Tommy Tomcat sits down as Jim Buxter stands up. "Ok, good one, Calla. Hey are we having a beer to send you off?"

Jonny nods. "Come down to Nelly's tonight if you like."

Jim tapping him on the back – "Yeah we'll come down for a few, look forward to it."

The lads are back on the job as Davey, Jim and Mally walk in.

"Look out, the heavy artillery are in." Calla nudging the lads.

"Fucking hell, lads, it's not job and knock." Mally looking up at the job. "Fair play, you will have this done by 5 o'clock." Mally wanders down the rack as Willy sticks a blown up glove on Mally's hardhat. "Forgot to mention that, Calla you bastard! I was out having a tab, every cunt laughing at me."

Calla laughs. "They'll be laughing at that lanky streak of piss in a bit, pal." Davey leans against the bench. "So lads, who is up for free styek and beer the nite? Courtesy of the company."

All the lads nod except Willy. "Fuck that, what's the jackanory with the Brucie bonus ball?"

Jimmy laughs. "You're one miserable bastard, Willy, do you know that?"

Willy shrugs. "One miserable bastard if we don't get the bone eye in this week's packet."

Davey puts his arm around Willy. "One month's bonus in with your wages Friday, son."

Calla rubs his hands together. "Fish, chips and fucking mushy peas, hammas! Right, what's the dance for graft!?"

All the lads standing round as Jimmy takes stage. "Right, Calla, you, Willy and Jonny are heading to Shanghai for Davey in 3 weeks; the rest of you are coming back here for Mally, probably for two trips, then I'd say to Shanghai as well."

Silence for a small while. "Any retainer?" Mark chirps in.

"No retainer, everyone gets the well done bonus in their bank prior to going home this week."

The lads all seem happy as Mally comes back, Willy making the chicken dance as the big glove resembling a chicken's head bounces back and forwards on Mally's head, Calla joining Willy, elbows flapping around. "I feel like chicken tonight, chicken tonight!"

Mally shakes his head. "I'm pleased I don't have to put up with these two balloons for much longer!"

Willy laughs. "Balloons, aye balloons areet" as Mally nudges Jimmy. "Lads are flying, probably let them away early Friday."

As Jimmy and Mally walk off Davey shakes his head. "You're fucking mad, you two." Calla hugs him.

"Aye but you will love us long long time soon enough, father."

Davey pushes Calla off. "Get off, you big daft cunt" as he wanders through the curtain shouting in Keegs and Calla. "Lads, get the last of the pipe up and on purge and we'll head away at 4 o'clock. Pointless killing ourselves. We can square it off and tidy it up tomorrow leaving Mally no choice to let us away early Friday."

The bus is waiting outside, Willy driving for a change as the lads all swipe out, climbing into the bus, every one of them slapping him on the head as they get in.

Willy revving the minibus while gripping the steering wheel as Davey flops in – "Will you give owa!"

"Buckle up David!" Willy winks across as Davey pulls his belt on. Willy crunches it into drive, the minibus lurching slightly forward then bouncing as the wheels start spinning, the minivan on the spot, plumes of blue smoke wafting up.

"GO ON WILLY!" Calla shouts then waves to the Americans who are walking past, laughing.

"Fucking give owa, Willy man!" Davey snapping again as Willy eases the power off and allows the bus to move off, leaving the plume of smoke behind. Then gathering speed he slams a hard right, Davey banging his head on the side window, all the lads cheering, the van gathering speed as the security guard steps out from his hut to sign the lads out.

Calla laughs. "Run that cunt down, Willy!"

The security guard is waving in the middle of the road as Willy roars towards him. Davey tenses up. "Brake, Willy son! Brake! Brake! BRAKE!" as the security guard has to dive out of the way, the van roaring on through and bumping out onto the main road, all the lads laughing and spinning round, waving at the security guard who is dusting himself down waving his fist. Willy dries his eyes.

"You're not driving again!" Davey loosens off as Willy swerves into the far lane just missing a car. "For fuck sake'!" Davey puts his head in his hands. "I'm ganna be a nervous wreck sitting here next to Juan Fangio!"

The van hangs a left and heads towards the gates of the accommodation. "Hand brake the bastard!" Calla shouts.

Willy shakes his head. "We have no chance the handbrake grafting with you fat twats on!" Willy hanging a left and easing up to the gatehouse, big Earl wafting his nose as the van passes with a smell of the clutch and burning rubber. Davey takes his belt off as Willy eases the van up to the door but the last 20kph jams the brakes on sending Davey into the windscreen.

"You sackless sod!" Davey digs Willy in the leg and jumps out holding his hands up to the sky in prayer!

Calla slides from the back banging everyone. "Move, shift, look out!"

"Calla, why do you always sit at the back when you're first off?" Jonny pushing Calla's arse past.

Calla lets out a fart as Jonny pushing him out and slides the door shut. Willy leans out of the window. "Nelly's in an hour? For a pre steak warmer and an ass whooping at pool?"

Calla still on all fours kissing the pavement glances up. "Ay for the first part! But you haven't beat me at pool since the ill-fated summer of 2005 when I sprained my wrist falling down the stairs at Tall Trees!"

Willy laughs. And repeats the wheel spin, Davey wandering across the front of the van having to leap onto the kerb. "Weyhey!" Willy shouts as the van speeds off, all the lads inside getting slung about like rag dolls as it screeches round the corner.

"Get up, ya silly bastard! He isn't right in the head, that scotch bastard! I think he bought his licence!"

Walking up the path Davey looks up at the board still over the window. "Look at the state of the place, man!"

Calla shakes his head. "Anar, look at the privets, man! Tak years for them to grow back! Shit crack, Davey, shit crack!" Davey stops speechless with a tab in his mouth as Calla opens the door and wanders in.

Davey's in the kitchen making a brew and eating a bowl of cornflakes when Calla runs down the stairs soaking, with just a towel round him and a pair of socks on. "Davey lend me some deodorant man, I've ran out!"

Wiping milk from his chin and spraying cereal everywhere,

Davey replies: "Nor! I've only got roll on!"

Calla shakes his head. "So? That'll dee!"

Davey putting his bowl down – "Calla, ya not borrowing my roll on!" looking over the bench – "Howw, they are my socks!"

Calla looks down. "No they are not!"

"Get them off, Calla, they are my good ones. I got them off one of the bairns for Xmas!"

"What and you're the only one to have these socks like, Davey? Like they only made one pair in the whole wide world?"

Leaning on the bench, Davey picks the bowl back up and shovels in another mouthful of cereal, staring at Calla. "No son! But I'm sure I'm the only person in this whole wide house that has a pair!"

Picking his tea up and wandering over to the sofa, Calla lifts a foot up and inspects the sock's print – 'The Greatest Grandad in the world'. Calla stands on one with his other foot pulling it off and doing the same with the other one, grips them with his toes and flicks them at Davey, one landing in his bowl splashing him with milk. "Well, isn't that clever!" watching Calla troop back up the stairs, his towel dropping, showing his arse.

Davey's nodding on the sofa when there's a loud bang. Jumping up startled he hears the bang again on the board. Walking to the door he sees Willy on the other side of the street with a rock in his hand. "Areet, Davey boy! Gi Calla a shout!" juggling the rocks.

Davey looks at the board. "Have you got shit for brains or what? We have a door bell!"

Willy throws the rocks down. "Wha? I'm a canny shot, hen but don't think I could hit the doorbell!"

Davey shakes his head, walking in and slamming the door and sitting down, angrily muttering to himself.

Calla bounds down the stairs. "Is that William McGaylord asking if I'm playing out?"giving Davey a big kiss on the head and running out. "See you at Nelly's, ya owld git!" The door slams as Davey shakes his head and hears the faint sound of Calla's voice asking Willy where he found his shirt.

Willy looking down at his shirt – "Miami, what's wrong with it?"

Calla catching him up – "What's fucking reet with it!"

Willy looks at Calla's outfit. "Fucking chinos, denim shirt and fucking deck shoes!? Fucking hell, man, I wouldn't wear that outfit in a shit fight."

Calla pushing Willy across the fence into a hedge – "Bet you would now you've got privet stains on yours, foureyes."

Willy climbs out of the bushes, dusting himself down, holding his ankle. "That's it, sprained for the third time, you twat."

Calla laughs. "Fucking still on about the ankle! Change the record! Anyways I want to ask Rosie why she didn't come over the other neet."

Willy sniggering lights a tab. "Maybe she decided to call up to my door?"

Calla looking back at Willy limping up – "She would rather be raped by Mexicans, man, Willy, than have to chow on your stumpy little tadpole."

Willy laughs. "Well, maybe you should ask her instead of being paranoid."

Calla crossing the road to see Earl – "Paranoid, right o! Howw, Earl, any news on the street? We are off Saturday if you fancy a pint Friday?"

Big Earl smiling – "Yeah, why not, Calla."

Then Earl climbs out of the gatehouse as Calla stick his head in. "What's the smell in here? It's like Amsterdam high street."

Earl laughs. "A little medicinal cigarette, Calla, from Brad. He drops me in an envelope every 6 weeks to keep me sane."

Willy catches up. "Well, give us a blow, then." Earl offering the lads a pre rolled cigarette. "Cheers, bigun." The lads walk off down towards Nelly's.

"We'll give Mally this before the meal, tell him it's a class mellow, makes you feel proper calm and cool."

Willy laughs at Calla. "I hope it makes him shit his pants, to be honest." They both laugh as they walk into the bar. Rosie is sitting this side watching the TV with a cup of coffee. "Two pints of your finest please."

Willy walks to the bar as Rosie turns round. "Hiya lads."

Calla, ignoring her, goes over to the jukebox. "Turn the jukebox on and that shite off."

Rosie walks behind the bar. "What's up with him?"

Willy shrugs his shoulders. "Think he has the hump about you not going round last night."

Calla drops the quarters into the pool table as The Calling comes on, Calla whistling the tune as Rosie kicks him up the arse. "Hey you! I stood for an hour outside your door last night."

Calla shakes his head. "No missed calls."

Rosie kicking him again up the hole – "I had no battery, you clown!"

Calla shrugging – "And that's my fault?"

Rosie walks away shaking her head. "Ten minutes after you left the bar and I am at the door. I would expect you to be up waiting."

Calla setting the balls up as Willy comes back with the pints. "Not a happy camper, our Rosie, pal."

Calla taking the pint off Willy – "Aye well, it's her own fault. I was choking for a tin when I woke up," as he takes a massive drink and winces. "Gan get a tin of lemonade, man, Willy, that tastes like rats' piss."

Willy sits down putting his foot up. "Fuck off, you get it – my ankle's sore."

Calla pots 4 stripes and puts a twenty down. "Your shot, get your money down," as he walks to the bar.

Willy stands up evaluating the table. "No way, you cheating bastard, you wouldn't've bet if you weren't lined up."

Calla at the bar says, "Can of lemonade, huffy, and I had my earphones in so we are both in the wrong."

Rosie putting lemonade on the bar – "Whatever, Calla, just a smile when you walked in would've done."

Calla taking the lemonade and smiling – "There, happy now, grumpy hole?"

Calla walks back over as the door opens and Hags walks in. "Alrighty chaps."

Willy, potting a ball, exclaims, "Fuck me, the oldest swinger in town is back."

Hags walks over shaking the lads' hands. "Where's your lass?"

Calla asks.

"She's away home. I got a text off big Jim asking if I would stay on and cover the lads' leave, take mine after so I sent her packing for a month."

Willy laughs. "How was your holiday?"

Hags takes off his jacket sporting a grey cardigan and fastened black shirt, Calla laughing. "Fucking hell, man, Hags, you look like a right chav, 50 going on 15."

Hags looks down at his outfit. "Armani, boys, Armani! Ay holiday was nice, cruise around the Caribbean you know?"

Willy shakes his head. "You're one posing cunt, what dya want to drink!? Suppose a cocktail is it?"

Hags laughs. "Aye, why not,carry on as I left off: pina colada, William! Is that right about the bonus?" Hags sits on his stool next to the bar.

"Ay one month's in this week's packet."

Hags rubs his hands together. "I'll be telling the missus captain fuck and major all aboot that bastard by the way, you ken."

Willy agreeing – "Aye understood, roger and affirmative. Rosie, pint of shite for this man, pet."

Hags turns around. "Never mind sadsack – pina colada, pet, ta."

Calla potting the black – "next" – as he marches over "and 2 more pints, Rosie."

Willy limping away – "Aye, forgot, your round, ballbag."

Calla putting his hand in his pocket – "Nee bother to me, Willy, you got one Hags?" as Rosie puts the big fancy glass down with an umbrella in, Hags picking it up. "Aye, cheers son."

Calla looking at Hags – "Haweh!? You're fucking kidding me?"

Rosie laughing – "No, that's an extra 18 bucks please."

Calla putting a 100 dollar bill down – "Fucking Benjamin will be spinning in his box, the cunt, pina bastard coladas on a Wednesday teatime! Jesus."

Calla takes a long sip of his pint still staring at Hags' drink with disgust. Willy walking back catches Calla's eye. "What's rang wi you?"

Calla still staring – "Nowt! Might smack Hags in the mouth for drinking that though!"

Hags looks up and holds it out. "Have yourself a taste of that, man! It's lush!"

Calla leans back. "Don't wave your Club Tropicana drinks at me, ya cunt!"

Willy sniggers. "Ay the bastards wad have to be free to drink that pish!"

Calla walks back to the table behind Willy. "Free my Boro butt hole! 18 dollaroos that custard cost!"

Willy stops dead and spins round. "Get out!" pointing to the door at Hags who has stopped laughing with Rosie.

"Eh!"

"Gordon, you're from Ferguslie Park in bastard Paisley! The bairns run around the streets with nay shoes on and you're drinking semi alcoholic cocktails that cost 4 times mare than a pint of lager! You should be ashamed of yourself!" as Keegs walks in rubbing his hands – "Pint, Rosie my little lamb! Now then Hags, good to see ya... What the fuck is that?" pointing at Hags' drink.

Willy shouts over. "It's a penis collada, Keegs! What do you think about that – 18 dollars worth!"

Keegs snatches the drink out of Hags' hand and pours it in the drip tray. "Rosie, make that 2 pints please! It appears old Hags momentarily lost his mind!" throwing down the dollars and picking the pint up, Hags sat motionless.

Calla looks up. "How come you take ages to get down here, Keegs, for fuck sake! Why can't you get a wriggle on and get in with me and Willy, save a cab ride!"

"Calla son, it's 10 bucks in a cab, man, plus I like to give myself a wesh before I turn out! Even iron my shirt, you know! It's called looking after oneself!"

Calla shakes his head. "Well now it's official! All you Scotsman are queers! What with that daft cunt at the bar with an Aran cardigan on!"

Hags pipes up. "Howw hawld on there!"

Calla barking him down. "Wey man it's Orlando not Toronto you silly old bastard! Anar you're getting on but it's 20 degrees outside!"

Hags looks at Rosie. "I thought I looked areet!" Rosie patronisingly pats his arm.

Calla barking again – "Now you, Keegs, Mr I like to look after oneself!" In a girl's voice – "It's a good job for you, Willys flying the sweaty flag!"

Willy sticks his chest out at the praise as Calla leans back down. "You remind me that the Scotch are a bunch of scruffy twats!"

"Howw!" Willy spins and looks at Calla who is sticking his tongue out while taking his shot.

The bar doors bounce open and Mally and Jimmy walk in, Davey walking behind whistling away. "Hags! Bonny lad! How's your lass?" squeezing Hags' leg.

"She is away back, Davey! Had a smashing time!"

Jimmy orders the pints then turns around. "Gordon Haggerty the only man cheeky enough to ask for holidays when he is working away!" Hags laughs.

"How smelly, what did do you do for deodorant, Hing a pig on your back as an air freshener?" Davey shouts over to Calla who sniffs his arm pit.

"Au Natural, the scent of a man, Davey!" Calla takes out his phone and thumbs through a text smiling. "They can't resist the Boro charm!"

"Wee is it, hen?" Willy quizzes.

"Wor Sharon! Tells me she is looking forward to seeing me the morn!"

"Must've been squeezing her some shite over the phone, Calla! Bet she has a fanny like an Eskimo's glove! Them stewardesses are all on the Atkins diet! Meat only! Fucking love the boawby!"

Willy laughing nudges Keegs. "What was the name of that one you were podging?"

Keegs scratches his chin deep in thought. "Oh you mean etcha sketch! Ay she was a belter, she loved it!"

Calla leans on his cue. "Etcha sketch!"

Willy and Keegs laugh. "Keegs said she loved her nipples being played with! Used to twist them both to draw a smile on her face!" All the lads are rolling.

"Where's cat weasel Mark at?" Keegs lining up his shot. "Don't know. Calla he said he might not come out!"

"I sometimes wonder about that boy!" smashing the balls on the table.

Jimmy walks over with a tray of beers. "Right lads, make this your last and we will get down to the steak house a bit sharp! They do 2 pinters in frosted glasses and they are the bollocks!" handing the beers out.

"Hawld on, Jim, we are still waiting on Jonny, Kris and Mark!"

"And Chuck and big Tom!" Willy chirps in.

"Tom and Chuck said they will meet us down there as they are getting their laundry done and Kris, Jonny and Mark should be here! They knew the time! You 3 hang around here and follow us down, nee bother – the table's not booked until 8!"

"Fuck them I'm following the open bar, me!" Calla pours half the pint back. "I'll give Jonny a bell, had on!"

Jimmy and Davey pull up their usual pews at the bar next to Mally, "Mally, you ever take them fucking sambas off?" Hags pointing down at Mally's trainers.

"I did once to change my socks." Mally laughing out of his weasel nose to himself as Calla walks over.

"Mally, why did you not want me and Willy in your squadalero?"

Mally leaning back takes a hard think. "Er, that's cos you're a pair of twats I think."

Jimmy and Davey laugh as Calla turns on his heels. "The pair of twats that has made your bonus this trip, ya cunt."

Mally folds his arms. "Thanks for that, bigun." Davey nudges Mally. "Behave winding the fucker up, Mally, you know what he's like"

Mally shouts over to Calla. "Calla, one day you will be as good as me, son."

Kris, Mark and Jonny walk in all dressed up. "Here they are, the 3 bum bandits from Sunderland," Calla laughing as Mark gives him the finger and Kris corrects him. "Newcastle, bonny lad, I'd rather be a Jock than a mackam–" as Willy chirps in: "I'd rather be a ginger with buck teeth than an Englishman."

Calla stubs his ankle with the butt of his cue as Willy yelps. "Shut up, you little jockroach! And by the way you are a fucking ginger with buck teeth!" Willy frowning rubbing his front teeth "Anyway benders, where you three been? Topshop skip?"

Jonny walks over. "Says he, dressed like a fucking sailor's mate." Sarchastically saluting Calla.

Calla takes his money off the table and downs his pint. "Right, haweh, there's away."

All the lads are drinking up as Mark comes back with the drinks. "Wait on, man, we have just got these in."

Calla – "Nor, you were told the time, leave here at half 6." Sticking his finger in one of the drinks flicking the froth at Mark.

Kris chirping in – "That's balls, there was no time."

Willy stands up. "Aye well, we just made that up, get a taxi down after us."

Jimmy walks over. "Relax, finish your drink, lads, the driver isn't there yet."

Calla's at the bar chatting to Rosie. "So are you round tonight then? This is your last chance, mind."

Rosie smiles. "I can't tonight, Calla, I have to stock take – but if I get finished I'll be round! Keep your phone on!"

Calla smiles. "Rite he ho. Davey, you're looking smart the neet."

Davey looks down at his white shirt and black slacks. "Aye son, made an effort tonight, it's a meal like."

Calla laughs. "I bet the front of that fucking shirt looks like the map of Africa with gravy stains later on."

Hags standing up puts his long crombie on. "Haweh, Don Cheech, get the nanny goat on, we are expecting a bit of cold weather! Fucking silly owld cunt." Calla winding Hags up as Hags smiles.

"Bit of class, son, you wouldn't know it if it bit your arse."

Willy walks over. "Fuck me, Calla, is Edward Woodward allowed out with us?"

Davey laughing loudly – "Remember that cunt, the equalizer? He was some man for one man! Sunday neet, half 8 lying on the settee, full of sprouts and stones."

Calla pulls a face. "It's the fucking neutralizer your lass would have been looking for with the smell of your arse."

Jimmy shouts over to the lads: "Right he ho, minibus is there, lads."

Mark and Kris pour the last of their pint down then catch the rest of the group up, Calla squeezing Jonny's neck as they walk out. "Give owa, man Calla, that knacks."

Calla then holds the door open for 2 girls walking in, rolling out his hand. "Evening, ladies! Breaks my heart, I'm leaving!" Both of them smiling as Calla shouts behind. "Andrew Callaghan – you will find me in the phone book in the fun section!"

Mark letting the girls enter the inner door – "That's right, lasses, clown for hire!"

Mark walks out. "Canny funny for you that like, little Marky!" Letting the door slam in his face Calla leaps on the bus and shakes the back of Davey's chair. "EEEEE, you excited for ya free coo piglet features!"

Davey shaking with the momentum of the chair – "Will you pack it in, man!"

Calla rubs his hair then sits back flicking Keegs' lug. "What the fucks got into you, ya balloon! You're like bastard zebbedy! Just settle yourself down!"

The door slides shut and the minivan quietens down and eases out onto the main road. "YAAAK!" Calla screams at the top of his lungs, all the lads jumping out of their skin, Davey muttering 'the young'ns not right in the head' Willy leaning forward tapping Calla on the shoulder. "Hey, big man, any chance of a sub?"

Calla spins round. "No William, you're already overdrawn! I'm not a bank y'nar! Anyhow the neet is FOC!"

Willy is about to reply as Mally shouts back: "Willy, I've just got an email there from Martin Watson, the site H&S, about an incident leaving site!"

Calla pipes in. "WORK RELATED! We will discuss it while on the clock tomorrow, Mally!"

Mally turns around. "This is serious, dipshit! The security guard is claiming you delibrately tried to run him over! Logged a complaint and Martin has it down as a near miss! This isn't good this, Jimmy," handing Jim the Blackberry, eyes squinting he reads the email.

"Mally, that isn't what happened at all! I, he was..."

Calla cutting in – "Mally, man, tell him to fuck off and Willy, don't admit a thing! Lie, lie and if you can't lie, deny! Turn round, Mally, we will talk about this tomorrow! We can say Willy didn't see him! Perfectly believable seeing as though his lamp oils have been turned down since he was 8! Go on now, turn around, you're making me feel sick!"

Mally stares at Calla. "Don't get smart, Calla and remember who is in charge!"

Calla pulls a face. "Not you! And not in a minivan after we have clocked out so turn round and button it!"

Mally is going to reply, but – "Calla son, shut your fucking hole and Mally, we'll sort it out tomorrow! Dinnit spoil the neet!" Jimmy hands the phone back to Mally.

"Was only a joke, Mally, the lad didn't try and run him over!" Davey trying to calm the situation. "If he really pushes the incident, save Willy getting any hassle on his last day, he can pull a shift the morn and we will say he has already flown back! I'll sit with Martin and take a bollocking. I'll assure him nothing like that will happen again! We're not coming back anyhoo, everything will be ok!"

Jimmy leans across Mally, preventing him replying. "Hawld on, Davey! Look, it was me that was really driving the bus, not Willy!" shouts Calla.

"Bollocks Calla, it was me, you lying bastard!" Keegs snaps.

"Keegs, you haven't got a licence – it was me!" Jonny shouts, the lads all laughing, Jimmy and Davey sniggering in the front with Mally sulking. Jimmy digging him – "Smile man, you miserable faced twat!"

The van pulls up, the lads all getting off. Willy gives Calla a dig. "See Mally's face, the fucking baby! Hate that cunt, like. I hope someone mistakes his arsehole for a pipe end's and sticks a fucking purge on the cunt!"

Calla laughs. "I might just have 10 frosted pints and purge his nose with my fist!"

Both the lads laugh. Davey pulls Calla to one side, the rest of the lads walking in. "Calla son, behave yourself with Mally – regardless to your thoughts he still has Jimmy's ear and few others and he could scupper your Shanghai plans."

Calla shrugs him off. "Davey man, let the skinny vommit try, he can't cut these bastards off!"

Holding both his hands up, Davey lights a tab. "Wey anar, but just use your loaf!"

Calla walks off holding his thumb up.

Inside the restaraunt Chuck and Tom are sat at the bar with 2 large beers in front of them. "Hey y'all, this is what we're talking about!" Holding the glasses up, Jimmy waves the barman over and shouts the round in. Calla leans across Mally without a word, taking a beer off the bar.

"Excuse me!" Mally snaps.

"Nee botha, Mally!"

Calla winking gives him a toothy grin and walks towards Willy. "Table for 11, bonny lass!"

"Davey – he having a tab?" Jimmy asking Calla as they sit down.

"And letting his belt out!"

"Nice sweater, Hags!" Tom nodding to Hags who gives himself a look.

"Cheers, Tom! See, someone with taste!"

"You win that at a raffle?" replies Tom, Chuck laughing.

"What you laughing at, Chuck? What's that shirt doing!"

Chuck looking down and shrugging – "Covering my body I guess!" totally unfazed, taking a drink. "Word is this is a free evening! Don't like rolling down in my good wear!"

Davey flops beside Jimmy rubbing his belly. "Bring it on! Where's the menu?" Jimmy hands Davey the menu, Davey whistling in exitement with what's on offer.

Willy shouts over. "Here, Davey, 2kg of prime Angus and 3 sides here! They reckon it's on the house if you can finish it! 20 bob says you can't."

"Ay Davey, we'll all give you 20 bucks if you can do it in!"

Davey squints at the menu reading the offer then looking up. "Am I still allowed a starter?"

The lads laugh. "Bollocks to that, man, lads, that's a lot of cow! I've still a gallon of ale to hem in round the sides ont! I'll take it easy and just have the short ribs for starters and the 500g New York strip well done with a jacket chetty and some of that coleslar!" putting the menu down rubbing his hands. "Tom, Chuck! You two areet?"

Tom looking down at the menu – "Mm, we looked already, Dave, we are going to have the 300g sirloin with all the trimmings."

The rest of the lads are pointed to a large private table in the corner of a very swanky looking steakhouse. "We'll have a beer at the bar first, thanks, if you could, a few bottles of red and white sent over to the table, thanks." Jimmy then moving to the bar putting his card in the hand of the manager with a wink.

"It's a nice place this, mind, isn't it, son?" Davey nudging Calla.

"Not bad, Dave, got an even better feel to it cos it's for nowt! I love owt for nowt me."

Willy takes a pint off the bar and passes it along the line. "Keep them coming, Jamjars." Calla taking it off Willy, passing them along to the lads.

"Ah fuck, we invited the Yanks down the bar tonight, we'll have to go in on the way home and take the piss out of them." Mark takes a swig out of his pint.

"They won't be there for more than 2 pints unless there's a football game on tonight," Keegs putting him wide.

"You won't be going back there tonight after this feed lad." Davey making his way over to the table as Calla points. "Watch, I bet he sits nearest the bogs, back to wall," as Davey looks around spotting the toilets then sitting down back to wall nearest the toilet, stretching his legs, raising his glass at the lads. "Haweh man, lads, as fucking clamming here."

All the lads get their pints and wander over, sitting down as the waitresses bring over the wines and put them on the table, Willy picking one up, squinting at the back. "How many volts?" passing it to Calla.

"Give me a look, fucking hell man, Willy, it's like being out with Mr Maggoo! 14, it'll dee," nodding to the waitress who uncorks it and pours a drop into Calla's glass. Calla smelling the drop of wine, circles the wine around the glass as the lads watch him. He drinks it and gargles in his throat loudly before swallowing and offering his glass back – "fucking horrible that, fill it up."

Willy laughs as he offers his glass as well. "Same here, hen, fill her up."

Jimmy shakes his head at Dave. "We better call 911 at 11 o'clock, marra."

Davey chuckles and looks at the wine bottle. "I'll not be drinking any of this piss man, fucking bunch of fannies."

Jimmy taking the bottle off him – "Bit of etiquette, Davey. I told you on the plane as you drank it out the bottle, red wine is an acquired taste."

Davey taking a big swally out of his pint – "Ay, you're reet, son, although I don't mind it with a bit chocolate on a Sunday neet after the Equalizer."

Keegs laughs. "Do you not mean cheese, Davey?"

Davey looks over at Keegs in disgust. "Nightmare? On a neet? Why lad, I've enough to be fucking scared of lying next to our lass without chowing a bit cheese afor I sleep."

All the lads laugh as the orders are taken. No starters, just mains for everybody as the waitress walks off, shaking her head.

"There must be a trailer-load of steak on that order." Chuck laughs as he tops his glass up with wine.

"Half of the bastard is Davey's 500g steak – fucking hell, man, Davey, you'd eat two more tatties than a pig." Calla looking at Davey in disgust as Davey signals more pints and shouts back, "and what did you get son? A 200 gram mignon fillet from Argentina? I've had bigger bits of meat stuck in me frigin teeth!"

Willy flicks Calla's ear. "Get the vino topped up, captain sadsack, it's free, remember!!"

"Howw, Tramp features, what did you get?" Calla quizzing Hags.

"Chicken breast on a bed of mushroom tagliatelle!" smiling back.

"Hawld on, you got chicken in a steak house? Keegs, smack him for me!"

Keegs leans in digging him in the arm.

"Howw, man, red meat plays havoc with my guts!"

"Penis coladas and chicken breasts! Cardigans! I think your lass flew home early coz you're gay!" all the lads offering a limp wrist to Hags who shakes his head. The waitress puts the basket of bread on the table, Davey lurching forward snatching a piece and the butter. "Any jam love?"

"Erm?" the waitress is thinking.

"I'm only kidding, bonny lass. Tell the chef to mak sharp with the coo though and fetch a basket of breed for the rest of the lads!"

"You know when you have that heart attack, Davey? Can I have your caravan at Haggerston Castle?"

Davey looks up at Calla. "Fucking dark cloud owa there! Nar, ya can't have the caravan!"

Calla rocks back. "You're one selfish man, Davey! I was going to suggest to your lass to bury you in the cunt the size of you! Could just tow it to the cemetry!"

The lads all laugh. "Well ok, another request – you don't ask for me to carry your coffin!"

Davey laughs. "I could just see that! I'd be rowling oot afor I got to the hole with you, ya clumsy bastard! You watch what you're saying anyhoo, I might come back and haunt you!" waving his bread at him.

"Nowt would change there then, you have haunted me for the last 15 years!"

Davey nods. "Ay mare like carried your slack arse for 15 years!"

"Mare wine man!" Willy waving his glass at Calla.

"For fuck sake! The last twice you have spoke has been orders to fill your glass! I'm not the frigging waiter!" tipping the bottle upright as the wine glugs into Willy's glass. "Well that's handy! Look at the clip of that man!" looking at the red wine all over the table cloth. "Excuse me, can we have the same again please!"

Most of the lads looking down at their half empty glasses, Calla shouting, "Look, my name's Willy! Och noch fuckn jock! Haha," holding two empty wine glasses to his eyes like goggles, the lads all laughing, Davey taking a photo.

The waitress returns with a trolley and begins handing across the plates, Davey whipping the napkin out and stuffing it in his shirt as a bib. "Nom nom nom nom!" rubbing his hands. "Hey love, where's my fighting irons?"

The waitress hands across the steak knives, Calla peering over to Hags' meal. "Looks areet that actually," then at Willy's – "howw, yours looks bigger than mine!" stabbing his fork in it and swapping it with his before Willy can stop him.

"Calla, you bastard!" as Calla licks it! Jimmy and Mally look at each other, the lads all tucking into the veg and chips that are in the middle of the table, passing each other the jugs of peppercorn sauce.

"What's that, Chuck?"

"Blue cheese sauce, Keegs, try?" offering it to Keegs who waves it away. "No way, you dirty bastard!"

Tom nods. "Only thing I have with my meat is more meat!"

Davey holds his thumb up to that, Calla lashing the mustard and spicy steak sauce on his. "Mustard, man Tom, it's all about the mustard!"

Tom disagrees then looks over to Willy who is slapping the bottom of the ketchup bottle, his glasses dropping an inch down his nose with every slap. Thumbing them back into place he looks up, the lads watching him. "Wha?"

"Good bit of meat of that!" Jimmy tucking well in, splashing it back with a sip of red! "Try that, Davey!" – who follows, ramming a mouthful of chips in and washing it back with Jimmy's red, shaking his head – "Shite!" then swigging back his pint.

"After the meat, bird brain!"

Davey takes a piece of meat and again washes it over with red wine. "Still shite!" rolling the red wine and steak round his mouth.

"Why do I bother!" Jimmy takes the glass back from him, Keegs carving off all the fat from his steak making a small pile on his plate.

"Stop fannying on and get it down you, man!" Calla pointing over with his knife.

"It's all fat, Calla man!" pushing it to the edge of the plate. "Fussy bastard! How is your chicken breast, gaylord?" Calla asks Hags who is sucking a length of tagliatelle up, the last piece whipping across his face splashing Mally's glasses with carbonara sauce.

"Howw, man!" Mally putting his knife and fork down taking off his glasses and wiping them on the napkin, staring at Hags.

"Lovely, Calla son!" who is smirking as Hags sprays more sauce over a slightly miffed Mally.

"It's not lovely at all, Calla!" Mally barks, wiping his face and staring back at Hags who hasn't even looked up, just continued stuffing in the chicken and pasta.

"How is the food, gentlemen?" the waitress asks.

"Lovely, pet!"

Keegs jumps in, pointing to the pile of fat that is now the same size as the piece of meat. "Actually mine is full of fat!"

The waitress walks over. "Oh I'm sorry, sir, would you like me to change it?"

Keegs lifting the plate up to her – "Thanks, if you don't mind!"

The table is a wave of tuts and sighs, the other waitress at the other side of the table uncorking two more bottles of wine, Willy leaning in. "Can you fill the table with beers as well?"

"Certainly, sir."

Calla rocks back and lets out a sigh, throwing his napkin down. Davey looks up. "Beaten, son?"

Calla looking at the half a steak on his plate – "Nope! Just a breather so give owa eyeballing my steak!"

The lads are just about finished up, even Keegs' new piece has went the journey. "Way lads, why don't we have some dessert?" Davey sitting back signalling the waiter over.

"Not for me, fatha, want room for the next hour's booze." Calla pours the wine into his glass. "Mally, you were one bone idle cunt on the tools like," Calla shouts over at an innocent looking Mally.

"You only were on one job with me, Calla, and part of a different area, so how would you know?"

Calla laughs as Davey shakes his head knowing Calla is on the wind up. "Remember Harry Carlisle? From Seaton?"

Mally takes a drink out of his wine glass. "The welder Harry?"

Calla nods. "Aye that's him. Well he was telling me you had 18 months off work when you came out of your time."

Mally shakes his head. "Lad, piss off, no chance."

All the lads snigger, Willy leaning over. "Why like, Calla – did

Harry say?"

Calla leans back. "Aye, Harry said – 18 months off! Cos he smashed his flask."

Jimmy lets out a loud chuckle as Mally stands up to go to the toilet. "Now you know why he isn't in my squad, muppet!"

Calla laughs as four more bottles of wine are dropped on the table. "Way hey, fucking more grapes."

Kris picks the bottle up and fills his and Mark's glass then passes it down to Jonny. Mally sits back down smiling, Calla not being able to resist – "Harry Carlisle also said–" as Mally butts in – "Calla, fucking shut up, you're full of shite."

Calla proceeding – "No, let me finish, when you two were working down on Seal Sands you were his pipe fitter? Correct?"

Mally drinking his wine – "Aye."

Calla laughs as big Tom jumps in. "Come on, Calla, what did he say? Mally tells us he is one of the best tradesmen to come out of Middlesboro."

Calla is still laughing. "Haweh, daftarse, spit it out," Davey shouts down the table.

"So Harry said to me this day, do you know Mally Fitzimmons? I said, the lanky streak of piss from Boro? He said, aye that's him! He is one rough cunt! We were on a job, 100% X-ray as well! Mally said to me this day, Harry how do you like your gaps? Stepped or wavy!?"

All the lads burst out laughing as Mally stands up. "I'm going for a fucking tab, can't listen to any more of this shite!!"

Jimmy stands up wiping his eyes. "I'll come with you, Mal."

Davey gets up off his chair, pushing the chair in and putting his finger up his arse. "I think I am cutting another tooth, lads, see you in a bit," as he waddles to the toilets, all the lads laughing as Mark shakes his head. "Why, oh fucking why, does he tell us every time!"

Willy stands up. "That's because he doesn't realise, the silly wold bastard. I'm away for a laugh and a joke. Calla, you get the wee ones in when rubber face comes back."

Calla looks round as the waitress comes over. "A round of your best brandy please, gorgeous."

As the girl walks away – "Fucking hell she has some acne, see the pock marks, man." Kris pulling a face.

"Aye you could lie her down on her back and empty a tin of peas on her fyass and not one would fall off!" Calla laughs.

"You're going to hell, Andrew." Chuck laughs.

"Aye that's right, straight down the devil's pole, do not pass go and do not collect 200, so fuck! And half of you counts will be there waiting."

Tom shaking his head – "Think you could be alone down there, Calla – we are all good guys."

Calla laughs. "Kris? Jonny? Mark? Good guys?! They were late tonight because they were sniffing each other's arseholes! You think God wants arsehole sniffers up there like?" Calla pointing to the sky then sniffing his finger.

"I guess not, Calla, you have weird philosophies, dude." Chuck takes the wine bottle and tops all the glasses back.

The lads come in from outside as the brandies are landed. "What the fuck are these?" Jimmy taking his glass and smelling it. "Brandy? Who ordered them?"

Calla taking a sip – "Woah, fucking strang as well, Jimmy! Willy ordered them before he went out for a tab."

Willy turning to Calla – "You lying cunt!!"

Calla is shocked. "Haweh? How am I lying, like?"

Willy takes a drink. "Your lips are moving."

Davey sits back down in his chair. "Why, fucking hell, that was like a flock of spuggies."

Mark spits his drink into his glass, laughing. "Davey man, do you have to tell us?"

Davey smelling his glass and drinking it. "Tell you what, son?"

"I'll have another one of them bastards." Calla holds his drink up in the sky, Jimmy shaking his head.

"One more, it's nearly time for home. There's an ice cream stand in the other bar if you would prefer one of them."

Willy laughs and nudges Calla. "Ice cream, my hoop, bigman. I say we head to a club after a large brandalero." Calla agrees as Chuck and Tom shake their heads. "No way, boys, we are heading to Nelly's for one and shoot a bit of pool with the guys from Tomcat's crew!"

Mally, Mark, Jonny and Kris join the lads in agreement, Mally drinking his brandy. "Make sure you are in tomorrow, Calla, remember the bonus is for the full trip."

Calla standing up leans on Davey's shoulders. "Mally, I wouldn't miss work even without the bonus. Jimmy, I will have a double brandy, hamma," as he wanders off.

"Don't ask me, he might've banged his head." Davey shrugs his shoulders looking at a confused Mally as the manager comes over. "These ones are on the house, guys."

Jimmy thanks him as the lads all smile and raise their glasses. "I'm fucking beat with that meat." Tom holds his belly as Calla comes back in with an ice cream and a chocolate flake and sticks it upside down on Mally's head. "There you go, Mally, just in case you thought I'd lost the plot being nice to you, ya lanky streak of shite."

Mally stands up. "Are you for real, Calla?"

All the lads piss themselves laughing as Willy tops Calla's drink up, shouting across the table. "No Mally, he's fucking flaking it!"

Kris stands up laughing to comfort Mally as he pushes him over the table covering the lads with all their drinks and storming off to the toilets. Davey stands up. "Righto, that's it! Home!! Youse have fucking spoilt yersels!!"

All the lads stand up as Jimmy walks to the manager and the bar. "Sorry about this, son, they are just excited about heading off."

The manager shrugs and gives Jimmy his card back. "Yeah it is sure nice to head home."

Jimmy laughing puts the card in his wallet. "No, bonny lad, they are excited about heading back to the local, Nelly's!"

The manager winces. "Is that place still open?"

Davey walks over to Jimmy at the bar, patting him on the back. "Alrite, son, is it squared off?"

Jimmy nods. "Aye, Davey, the lad here was just wondering if Nelly's was still open."

"Why aye, it is son, packed with storytales and fairytales! Have you been?"

The manager smiles. "There used to be a grudge game, once a year! It was tough but we haven't played for 2 years."

Davey leans back. "Was it baseball or summit like, youngun?"

The manager laughs. "No, it was soccer! You know, the game you guys call football."

Half the lads turn round, Jonny moving up to the bar. "Like football? Why don't we play before we head off?"

The manager smiles. "We are all Italians! And we always beat Nelly's."

Calla leans between the lads. "Football against the fucking i tiddly ey ties? Wey, count me right in!! Three year county cup winner with Hartlepool working mens' club plus playing against you spik bastards who only have one gear – reverse. It'll be like taking sweets off a bairn!!"

The manager is bewildered, not understanding anything from Calla's statement as Davey leans in. "So what's your name, son? I'll have one of the barmaids ring tomorrow to organise a pitch?" Davey nudges Jimmy excited to get all the lads playing again, knowing they all can kick a ball.

"My name is Frank. I will call Nelly's tonight and we will play tomorrow on the pitch behind Nelly's."

Frank is staring at Calla as Willy butts in. "Frank? That's some Italian name!! My name is nibbles, nibbles yer fuckin ankles tha morn, son!"

Jimmy pats Frank on the shoulder. "Thanks for a great night. We'll see you later, Frank."

Calla drinks his brandy and smiles at the manager. "To be frank, I'd be a right cunt."

All the lads pile into the minibus, the usual Davey and Jimmy in the front. "So are you two daft cunts the managers?" Kris pipes up from the back.

"Aye, and you're dropped, daftarse," Davey chuckles from the front as Jimmy leans back into the back of the bus.

"Right lads, I think this could be a serious match. Let's miss Nelly's out and head straight home like true pros."

Calla laughs as the bus pulls away. "Like true fucking amateurs you mean! Howw, driver!! Straight to Nelly's, pal, for set pieces!!"

Willy laughs, putting his arm around Calla. "Aye mine's a double voddie and coke, in off the rocks!!"

All the lads trudge into Nelly's awash with excitement about the game tomorrow. Calla marches to the bar where Rosie and Gemma are talking. "Hoy oy saveloys! Listen, we have been talking to Frank at the Grand Grill and we have agreed to have a game of football, a mean soccer with them, the morn neet!"

Rosie, blowing out a bubble with her gum, rolls her eyes and pipes up, "Calla, we haven't done that in years and they always beat us!" pointing to the wall to the right of team pictures which appear to go as far back as the early nineties.

Calla strides over. "Ha ha, look at the tash on this cunt!" pointing and waving a couple of the lads over.

"Fucking hell, always wondered who these pics where of. Hey, Rosie, these all the photos of Nelly's?" Kris shouts over, Rosie leaning on the bar.

"Yep! First team of '94 it started with the frenzy of the World Cup! Nelly's won that year and did so for 3 years off the belt and then we never won it again and 2 years ago they battered us 9-1 and then after that no-one was up for it and Frank the dooshbag has the trophy on his shelf behind his bar! He not show you it?"

Calla wandering back – "Nope! He will the morn though when we tak it off the spaghetti faced twat! Beers in now before a kill ya!" Calla toothing a grin, Rosie nodding and pullling the pints.

Davey and Jimmy are leaning on the pool table talking to Tomcat. "Where's the rest of the lads?"

Tomcat is swaying slightly with his rum and coke. "Those bastards had a couple of beers and thought you guys weren't coming, so split! I hung around playing pool for a while! Thursday night is Walmart night with the good lady! I told her this was your send off so I wasn't hurrying out of here – goddam hate that shop!"

"You play soccer, Tomcat?"

Perking up – "Where you think the cat came from! I was up at Washington in a soccer scholarship back in the day! Back before it was played by girls! Man, the game sucked back then! Played in between the sticks!"

Davey looking him up and down comparing Tomcat to a slightly taller Neville Southall. "Fuck me, Tom, when was the scholarship? You look about 90, ye cunt!" Jimmy quips.

"Well Jim, the scholarship was many steak nights ago back in 88! Don't let the fat suit fool you – I was like a string bean back then!" Davey nudges him. "When did you last play? Fancy going in goal the morra night for Nelly's?"

Tomcat swigging his rum and coke – "Nope! Never played since I busted my ankle and lost my scholarship! Man, I was way better than Keller and he ended up on your shores! Bastard!"

Jimmy nudges him. "Haweh, big man! Throw on a shirt for us! One last horah with some English boys!"

"Frank's tried every year to get me to play for him! Haha, have to say me and Daisy have had our share of meat sitting down to discuss it! Haha, he would say, come down, have some food, we will talk about it! Dude really takes it seriously! Well we would rock up! Sit down! Eat the food! Drink the wine! Talk about me playing! I'd say no! Thank him for the food and leave! Man, every year I'd get a good feed just by saying the words 'I might'!" Tomcat sways again pondering the thought.

"So you never played for him? Haweh, will be a good laugh and something to talk about at graft! Especially if we beat them! Play with the English lads."

Standing up straight – "Frank will barr my ass! And I am fond of his cuts of beef!" Davey nudges him. "And Jimmy will get your ass sacked! What about that!"

Tom leans back on the pool table. "Man, that's bribery!"

Jimmy digs him in the ribs. "See that man over there!" pointing to Jonny – "England school boy!" winking at Davey. "That man–" pointing at Kris –"Chelsea schoolboy! And that lad there, Manchester United!' pointing to Calla, smirking at Davey. "You're in the presence of once greats, Tom! And I'll lay on a free bar the morn neet if we win!"

Tomcat breathing out – "Ok! Count me in!" slamming his glass down and wandering toward the bar. Davey laughs. "The only one who was pro was Willy and you didn't even mention him! Only connection Calla has with Manchester is catching crabs off that bird in Northwich!"

"Well Tom wouldn't believe that little specky cunt was ex pro and he defo wouldn't've heard of Glasgee Rangers!"

"Reckon he was a good hand actually, Jim!"

"Aye well we will see the morn, mate. That's Tomcat in goal, Kris at centre half, young Mark at right back, Jonny and Calla centre mid, and Willy on the left with Keegs up front! Still a few short, ham!"

"Hags is areet y'nar! Getting on like but he's been about! Could've went to West Ham when he was a youngn but his father pushed him into graft! Still a great turn, was still playing for the vets back yem in the over 40s league last year!"

"Ay? Wey, we will pencil the cunt in!"

"Put me and Mally pisspot in as well! We will gan on the bench!"

Davey rocks back. "Way lad, piss off! The pair of you couldn't kick snow off a rope!"

"Hawldy hawldy! I've got a bit of weight on but I'm still areet for 10 minutes! And Mally's as fit as a lopp! He'll run all day!"

Davey shakes his head. "Reet well the management is up to me! You're having nee imput! And from now on anything pitch related I'm the gaffa!"

Jimmy laughing waves Tom and Chuck across. "Lads, game of soccer tomorrow, should be a good laugh – you 2 any good?"

"Nope!" replies Chuck. "Need a corner flag?" Chuck snides, Davey rolling his eyes. "Struggling here, Jim! Even if we play you and Mally!"

Jimmy sighs! "We'll be reet, there will be someone at graft who will get wind and have a game!"

Rosie cocking her ears while handing the lads their drinks – "You guys are overlooking the regulars who played! We have Tim, Jockey and Steve over there."

Davey glances over at the blokes around their early 30s. "Ay, Rosie, they up for a game?"

Rosie nods. "Dead sure! Steve there is a coach at the local college. I will get him to organise the college stadium for the game! Sure he may even be able to rustle up a few people down to watch!"

Davey rubs his hands. "Bingo gravy granules, love, that would be smashing! Bit short notice like. Frank suggested the pitch out the back?"

Rosie shakes her head. "It's been neglegted, Dave! The college has floodlights and facilities!"

"Gemma, you be a darling and do a bit of organising and I'll make sure everyone is back here for a few after the game! Speak with the 3 stoodges at the bar and see if they have any friends who fancy a run out – we will need another 3 or 4 – and see if they have a ref! What about kit?"

"We still have that somewhere. I'll get them all sorted, don't worry!"

Winking and walking off Jimmy punches his hand with his fist! "Class! I'll get a memo sent around the place the morn telling everyone to get their arses down to watch the football! Tell you, Davey, this could be good craic this! Hope Frank takes it seriously!"

Davey nods. "Owld Franco sounds like he takes it real serious! Probably flying the Juventus B team in as we speak!" The two blokes laugh and continue to talk about the game.

At the bar Calla and Willy are arguing about who is the best. "Willy, you're fucking shite!" barks Calla, Kris and Jonny laughing.

"Get fucked, Cal, I was at Rangers, man!"

Calla laughs. "Who? Or aye, that pack of shite owa the border! Willy man, I've played in a better standard on a Sunday morning!"

Keegs jumps in. "Haweh Calla, in fairness, mong features is fucking quality!"

Willy smiles then gives Keegs a look for the mong features remark. "Look basically I'll be the lynchpin in the centre of the park! Kris and Mark will be at the back!"

"Wingback like, Cal?" Mark butts in.

"No! And William you will be on the left frightening the fans and big Keegs up front like Drunken Ferguson!"

The lads arguing as Davey walks over. "Here, just organising the team, fatha!" Calla puts his arm around the lads.

"I'm the gaffa and I'll be choosing wee is where! Reet!"

The lads all sighing down as Calla spins and gets does a shot off the bar. "I'm el capitano though! Reet?"

Willy snaps in. “How’s that like!”

Calla wincing after the shot wiping his mouth – “I captained the club every Sunday for 6 years! And, well...I’m the best!”

Davey shakes his head. “Let’s see what the morn brings. We will have a good talk the morn. Reckon Jimmy will give us a flyer to get down to the pitch for a knock around! Rosie is having a word with that bloke over there, reckons he can get us the college pitch so we will have a 7pm kick off, eh! Under the lights all weshed and back to Nelly’s for 9:30!”

Calla whipping out his phone – “I’ll get Sharon to come down watch me! Defo bucking after she has seen me play!”

Davey laughing wanders over to the three blokes Rosie had pointed out.

“Aye ay lads, the barmaid was telling us you three play soccer? We have arranged a game against Frank’s mob tomorrow and was wondering if you fancied it?”

The lads all nod. “Count us in,” the taller of the three, Stevey, introducing the lads to Davey. “This is Tim, all left foot, played for all the semi-pro sides around here.” Tim shakes Davey’s hand as Steve continues, “and this is Jockey – he gets Jockey because of his size! Been 5 foot and 7 stone since he was born! Built like a jockey but some ball player.”

Davey smiles. “Well lads, you are looking at the one time only manager, flown in from the north east of England, to put Nelly’s back on the map!!”

Stevey raises his glass. “Well since the drumming we recieved off Frankie’s I thought the game was over for good.”

Davey shakes his head. “Tell you what lads, everywhere we work there is always a team mad to sign the lads. They can all kick a ball – one thing we need is kit!”

Stevey smiling pats Davey on the shoulder. “We have the kit off the last time – it was brand new so we had it made new for the game and I have it in the college with a load of boots too so you guys are all set!”

Davey walks back to the bar where Willy and Calla are spinning a coin. “Heads I am captain and tails you are captain.” Calla spins the coin as Davey catches it. "You are captain, Calla and Willy is on the penalties, and we have three good lads over there with all the kit we need."

Calla shrugs his shoulders. "Give me my double headed quarter back man, Davey."

Willy looks down his glasses. "You Cheating cunt."

Calla puts the coin away. "Wey that's good them lads are in cos I'd say we'll be blowing out of our arses after 10 minutes!"

Willy drinking his voddie – "aye I'm heading off after this" – necking the voddie as the lads look surprised. Willy thumps Calla in the ribs. "Heading off for a pish, sadsack, get the bevvies in – we have a big game tha morn."

Calla holds his stomach. "Little twat. Same again, Rosie, and put a rimming of tabasco round patio door face's voddie! Good lass."

Jimmy sits on the stool enjoying his pint. "Calla, remember when we played you in the charity game a few years ago, management v bears and I scored the winner? You played well, son, but Hunter rose like a salmon to nod the winner home."

Calla shakes his head. "Fucking tin of salmon, you cheating fat bastard – you stood on my foot – it was a foul all day! Never a goal."

Jimmy laughing – "That's not what the local paper said the next day."

All the lads laugh as Willy comes back, Jonny ribbing the little man. "I hope you can play, Willy, it's not as if Scotland are famed for skill."

Willy taking his vodka and coke off the bar and drinking it – "Jonny, let's get one thing right from the start: you couldn't get a fucking beachball off me in a phonebox." Calla bursts out laughing as Willy rubs his lips. "Bastard! Calla, you're some can of piss."

Davey puts his arm round them both. "Now looka here, you pair of shites. Tomorrow we will be playing for our bar as a team and hopefully another chapter in a book somewhere in years to come. Hows your ankle anyway Willy?"

Willy jumps on the stool shouting over to the three American recruits. "Lads, join us for a pre-match celebration! Spot on Davey see just sprained" The lads wander over as Willy raising his glass in a toast. "Here's to jamming it right up Frankie's arse." All the lads cheer as the 3 lads look at each other, wincing – "say what?"

Jonny and Mark wander back from the toilet. "I played for the district at school, ya daft cunt!"

Jonny laughs at Mark. "School? I'm not having that!"

"I fucking did!" Mark defends his claim.

"You're thicker than Willy's geggs, ya daft cunt – you never went to school let alone play for the district!"

"You will see the morra, Jonny, when I'm halfing their left winger.

"Left winger versus a right WHINGER!" Jonny digging Mark. "Lads, the pubeless wonder here reckons he played for the district! Reckons he was the best right back at comp!"

"Fucking hell, Mark! School? You're some bell end! I might sign for Frank's so I can embarrass you!" Calla digging Mark.

"I'm embarrassed every time you turn out, you big daft cunt!"

Calla raising an eyebrow and smirking – "Ooooo!"

"Just get the ball across to me, Willy, I'll get the bastard down ready to burst the net!" Keegs doing a chest action.

"Waye man, get fucked, Keegs, you couldn't trap a sandbag!" replies Hags.

"Fuck off, fifty pence head! You just watch and admire from the lines!"

Davey tapping Keegs – "He is going up front with you!"

Keegs spurting his drink out soaking everyone. "Cunt!" Calla snaps, wiping his face.

"I'll show you how to gan on, son!" Hags wipes his face and puts his arm around Keegs.

"But you're bastard ancient! Bet you cleaned Archie Gemmill's boots!"

"Ayy lad, megged that cunt many a time! Haha, here, son, there is still plenty life in the old dog! Never had a bad knock in all the years I played!"

Calla butting in – "Spelks in your arse off the bench don't count, eh Hags?"

Hags rolling his eyes and sipping his pint – "Keegs son, don't you worry about me. I've nee botha taking the pegs out!"

"You couldn't take the pegs out your lass's washing, you soppy old bastard!"

Hags leans into Davey. "Give me the first half, Davey! I'll silence these!" Davey tips his pint.

"Rosie, I can't believe you wear fake eyelashes for work!" Call barks, leaning across the bar, all the lads turning around.

"I do not, Andrew!"

Calla leans in closer. "Show me." Rosie leans over. "Close your eyes!" She shuts her eyes as Calla plants the lips on her, Rosie pulling back and swinging a wild slap, Calla moving back so it just misses. "Wey hey! Get the drinks in!" smiling as Rosie stands with her hands on her hips, begining to laugh. "Asshole!"

"So Jockey, did you used to play football when you were little?" jibes Calla, all the lads laughing except Jockey.

"Say what?...Oh hahah! Hey man, I just hope you run as much as you talk!"

Calla slamming his pint down – "Hawld on! Nee one said owt about running!" then laughing as Rosie puts the beers down and Tomcat bounces past. "Later y'all! Night night, Calla, you ass wipe!" then bouncing through the door moving his beer from his lips. "Haweh!"

Mally wanders over. "Where you playing me, Davey?" (Mally being one of those players that didn't stop running and his work rate made up for his lack of skill and strength.)

"Mally, you're shit!"

Mally looks over. "Cheers for that, Calla! Wasn't that when we played you in the county cup and I ran rings around you!"

"That was coz I was playing and you were sub doing laps of the pitch, ya stupid cunt!" bringing a laugh out of the boys, Mally shaking his head.

"Left or right, Mally?"

Mally rubs his head. "Right hand side, Willy!"

"Reet, Stevey tell your boy he only needs 3 corner flags – Mally's got one covered!"

Chuck chirps in. "No, just one! Me and Tom are available for flag selection!"

Tom leans in. "Sorry lads, but I think I've got a calf strain!" The lads laugh.

Jimmy flops down next to the lads, squeezing Kris's leg. "Reet o lads, bedtime for all you after this one! Into work the morra for a steady pack up then an early finish to get back and down the college!"

"Yes guys, I'll have all your boots and shirts ready!"

"How will you know our size like, flanj?"

"Flanj?" a puzzled looking Stevey at Jonny. "We have any amount of boots laying around all sizes!"

"What about paying for the pitch and facilities?"

"Don't worry, Davey. I'll take care of that! Hey Rosie, you going to let Frank know all the details?" Rosie wandering past, texting, looks up, winking.

"Well that's that, then! Roll on the morn! Reet I'm outa here! Haweh Mally, al split a cab!" Jimmy jumping up stretches his pants around his waist, the two of them trooping out followed by Chuck and Tom waving to the rest.

"Haweh then, lads, there's away!" Davey stands up ushering the rest to finish their pint, Calla spilling some. "Areet areet! Hold your horses! You're probs popping for a piss anyhow!"

"Not yit son!" squeezing out a fart. "Reckon by the time we get yem the owld lad will need a drain!"

Mark shakes his head as they all head off saying bye to the locals and barmaids.

***

At the gatehouse Calla asks Earl if he fancies a look down for the game tomorrow. "No thanks, my playing days are over, Calla."

Calla laughs. "We didn't want you to play, Earl, we just needed you to put the nets up cos we have nee ladder!"

Earl laughs. "I might pop up anyways. Goodnight, guys."

Davey opens the front door, farting. "I'm touching cloth, youngn, the coo is popping its head out me arse."

Calla wafting the smell away – "Fucking get in, you dirty bastard, afor I bring my fucker up!" Calla pushing past him, Davey laughing going up the stairs.

"Na night, son."

Calla getts his phone out pressing a few digits. "Right, that's Rosie on the way up, nice one."

Half an hour passes and Willy knocks on Calla's door. The door opens, Calla smiling then frowning. "What the fuck do you want?"

Willy stands back. "Fucking charming! Fancy a nightcap and a game of cards?"

Calla shutting the door, putting the chain on then opening it – "Fuck off, get away, you pest."

Willy turns around. "You must be bucking." He wanders off down the street as Calla rubs his hands together and gives the old lad a rub through his jeans.

Another half hour and a text from Rosie: Calla reads it then rings. "I'm here, waiting for the last hour!! Why would I? You bumped into Willy? The little bastard! Ah haway up, please! I'll pay the fucking cab! Ah fuck off." Calla puts the phone on the coffee table and puts the porn channel on as there's a knock at the door. "For fucks sake!" Calla pulling his strides up walks to the door and opens it.

Willy is standing there with a 6 pack. "Haway, bigman, one game."

Calla lets him in. "You're some cunt telling Rosie I was out."

Willy laughing opens a can, handing it to Calla. "I was just thinking about you having to hoover the sheets for tomorrow's arrival!! Least you're free."

Calla drinking the can and sitting down – "Good thinking, Poirot. However I was gonna buck her on the settee. Get the dollars out then."

Calla and Willy start playing 5 card brag for 20 dollars each with gambles up to 100 dollars, Willy sitting with 100 dollars left out of 500 and 1 can left. "Calla, you're one jammy bastard, pair of fucking threes man! I threw a pair of fives away."

Calla laughs. "Come on, daftarse, this is the hand that I take your dinner money off ya."

Willy picks his hand up with a smirk. "We'll see, bigman."

Calla drinking his can opens the cards up. “Twenty dollars, shitballs.”

Willy looks at his pile of twenties. “Go on then, I’ll go with you you, big string of shite.”

Calla looking at his hand again – “60 dollars to see my hand.”

Willy necking his can – “You’re some cunt, I’ve only got 60 left.”

Calla laughs. “Shush man, you will wake Davey. Why don’t you fold and you have 60 left for tomorrow to buy yourself a mask.”

Willy laughs out loud and puts his hand over his mouth trying to keep quiet. “How about you just lend me the one you have on?”

Calla rubs his chin. “Like the film Face Off when I take your face and look like a specky little cunt and you take mine and buck all the lasses? Lad, fuck off! Haweh, are you in or out!”

Willy puts his 60 bucks in. “Let me see them cards, bollox.”

Calla expertly lays his cards out. “10, jack, queen, king, ace of hearts – or better known as a royal flush.”

Willy stands up and pelts his cards at Calla. “Fucking cuntlips!!” kicking the empty cans over and storming out, slamming the door.

Calla picks Willy’s cards up. “King buck, canny hand but a losing hand,” as he sits back and slurps his can as a voice from upstairs bellows down. “Calla! Bed!!”

***

Up with the larks Davey and Calla are sitting on the fence as the bus pulls up. Calla climbs in with a 100 dollar bill stuck to his head. “Morning chaps.”

Kris lifts his baseball cap up. “What’s that on your head, daft lad?”

Calla quickly replying – “Oor Wullie there at the back decided to put stops to me getting my hole and forcing me into a game of cards – needless to say I took a few bob off him! The lord moves in mysterious ways y’na.”

Jimmy laughs as the boys all wave at Earl. “Does that cunt sleep?”

Jonny sits back in his chair for a 20 minute power nap.

“They give him his digs behind the hut, same as ours, so he gets his head down when it's quiet through the day and pockets a double shift. Sure he gets every Sunday off." Davey answering the question then turning round in his seat. "Right, teamsheet will be up on the wall at 11 and then it's my treat for dinner at 12 in the canteen for the big game – none of your chips and cans of pop shite! It's proper food for energy you need."

Willy sits up at the back. "So Davey we are going to eat something that is cooked there? Fucks sake man I wouldn't eat something out of there if it was given to me for nowt!"

Calla shakes his head. "It is for nowt, you stupid cunt, didn't you listen!? And it's the best offer you could wish for by the way, since I took your dinner money off you last neet, ya balloon!"

The van arrives at site and the lads slide off, Willy swiping Calla's legs as he walks off sending him tumbling over a bin, Davey barking across the orders as the lads all troop down to the fab shop. Calla stands up. "Wey I'm away to use my arse to clock in! Back in 20 lads!"

Kris digging Mark – "Listen up sosaj, I want all that gash pipe rounded up and fucked into the skip before break! I'm not lifting a glove after 11 bells!"

Mark frowns. "Doesn't sound like you're lifting a glove at all, you knob, the scrap pipe is all that needs clearing!"

The boys glance around the fab shop, Kris banging Mark on the head. "Lucky me!" then pulls up a chair and puts his feet up, thumbing through a tool catalogue. Mark opens the site box up and puts on some gloves, walking back over kicking Kris's legs away. "Haweh Kris, you lazy cunt! I'm not the only one doing owt this morning," looking at Jonny and Willy who are taking turns to pelt 10 nuts off each other's hats, Keegs and Hags marking up the drawings. "Get on with it, Mark, you have done fuck all else the whole trip!"

Mark crosses his arms. "There's fucking loads of off cuts, Kris, it will take ages!"

Kris, slightly agitated, jumps up grabbing Mark. "You're one huffy little cunt!" He gets Mark in an arm lock, Mark thrashing, "Get off you dick!" Kris shouts to Willy, "Open the site box!"

Willy bouncing the last nut off Jonny's hat swivels round and opens the site box as Kris pushes Mark inside, Willy and Kris pushing Mark in, while Jonny closes down the lid. Mark is banging and shouting as Jonny runs round and snaps the lock shut! The 3 move back, laughing at the site box shaking and a muffled "Let me out! Let me out, you bastards, I can't breathe! Ahhh let me out!" as the door swings open and Calla walks in. "What you doing?"

Kris leans back. "Mark needed some time to reflect on his back chat so we stuffed him in the site box!"

Calla laughs and jogs towards the box; picking up a large set of Stilsons he starts smacking them off the site box then laughing and smacking the site box with a crazed laugh as Keegs wanders over and grabs his shoulder. "Howw man, for fuck sake! Rap in!"

Calla throws the Stilsons down as a faint whimper from the box emerges then the banging inside again. "Haweh man, let me out, fucking let me out now!" Mark's voice crying out angrily.

"Let the cunt out now!" Keegs looks at the lads, Kris stepping forward.

"Fucking hell! Willy, pass the key!"

"I haven't got it!"

"Calla? Jonny?"

Calla holds his hands up. "I've nee key! Fucking hopeless with losing and forgetting them!"

Jonny jumps in, laughing. "Here Keegs, he is reet like. Me and dopestick here, we were working in London this time and me and him were left on site to finish off! We had to get these ends butted up for the test on the Saturday and about 2 o'clock he taps me on my shoulder, tells me he was fucking off coz it was sunny! I told him we had to get the job done and we were left to finish it! He said he didn't care and was choking for a pint! I said haweh Calla, stop and get finished off! He walked off calling me a sad cunt and that he will be in the Regent bar with a cowld Heiney if I needed him! Some cunt. I thought being the hero I am I flicked my mask down and pressed on! Had to, man, shit would've hit the fan if they weren't able to start the testing! Well me and him were in this big house that was converted into 2 flats; we were in the upstairs one. Well I got finished and wrapped everything up about 5, got home – the door had been kicked in! I thought for fuck sake we've been burgled so I ran up the stairs and into the place, the telly, my stuff, everything was there, nothing was touched – except my brand new tube of toothpaste squeezed all owa! So I spends an hour patching the door back up, finally gets it locked and down to the Regent, where this daft cunt is spangled in the corner on his own!"

Calla leans back smiling and nodding his head. "Nee cunt out yet! Everyone must've been at work!"

Jonny shaking his head – "Says to him that we have been burgled! He straightend up, looking at me! 'Eh, did the cunts tak me walkman?' All he was worried about was a poxy walkman!"

Calla all serious – "It was brand new!"

Jonny frowns. "Anyway...I shook my head at the daft cunt. 'Don't think they took owt but the door was smashed in! The silly bastard leaned back relaxing, taking a swig of his pint – he said, 'nor man, I did that, I forgot my key'!"

The lads all laughing, Calla defends himself. "Wey a had!"

Jonny waves his finger. "Used my new toothpaste and toothbrush though, you cunt!" Jonny swiping.

"Well my breath fucking honked!" Calla smiles.

"Very nice, boys! Now open the box up!!"

The 4 all hold their hands out. "Honest, we don't have the key! Mark opened the box!"

The lads laugh as Keegs leans in. "Howw dipshit, check your sky rockets – do you have the key?"

There is a silence then from the box a loud "Orrr for fuck sake!" The lads all cry with laughing, Calla banging on the top of the site box. "It's areet lad, we will wheel it down to the pitch and put it at right back! Probably make more tackles!"

"Get me out, you bastards, haweh, it's not funny!"

Hags wanders over with a grinder flicking down the safety geggs and firing it up, the grinder letting out a loud buzzing sound.

"Watch out!" Calla stops him. "Woah there, leatherface! Jimmy will go jalfrezi if we chop up the site box!"

The grinder thumbed off, the buzzing sound quietening down with the wheel slowly winding to a stop. "So we just leave fanny balls in there, do we?" thumbing the grinder back on.

Calla shouts that it would be a better idea as Hags starts at the tamper proof cover over the lock.

Mally walks past outside with Jimmy on their way to another part of the job. Jimmy stops, holding an arm across the front of Mally and cupping his other to his ear. "Listen to that, Mally son! See, last day and you thought they would be swinging it!"

Carrying on walking – "Fair's fair, they do know how to gan on at work!" wandering off whistling, Mally looking over his shoulder confused.

Hags is almost through the lock, leaning back as Calla smashes down with the Stilsons, the lock clattering off to the ground, the box flying open with a very unhappy Mark standing up. Calla walks off pointing. "What have I told you about keeping the key in your pocket?"

Mark hops out of the box. "You fucking bastards – that was shit crack that!"

Kris slaps Mark around the head. "Pipe! Skip! NOW!" Mark pulls the gloves back on and starts throwing the pipe into the skip.

Calla is writing the lads' names onto the white board in the 4-3-3 formation, his name in capitals with a C beside it. Willy strides over wiping the speckfyass off the left and writing McLaughlin, nodding at Calla. "Bastard!"

Mally wanders down on the job. "Alrite lads, remember the scrap needs to be boxed up ready for 1 o'clock."

Calla sits on the site box. "Is the cash getting shared out like?"

Mally laughs. "Ay, Calla, between management."

Willy puts another half length of 2 inch stainless in the box. "I'll get the flatback from the stores, Mally. Is the scrapyard just opposite the deli?"

Mally nods. "Ay, Willy, that'll save us waiting for the useless bastards. Fair play to you, son. Right, down here, boys, let's have it all wrapped up."

As he walks off, Calla grabs Willy. "What you up to, ya little snake?"

Willy laughs. "Pipe down, mong features and give me a hand. Get these boxes on the pallet truck and out to the loading bay."

Calla grabs the trolley and shouts at Jonny. "Howw, fannyballs, give me a hand lift this fucker."

Willy backs the van in as all the lads start loading the scrap onto the back of the van. "Get a move on, lads, for fuck's sake." Willy shouting back out of the window as Calla sticks masking tape on the back of the van reading 'honk at me, I'm a Scottish fanny'.

Willy pulls away as Kris wipes the sweat off his brow. "Fucking drink's pissing out of me."

Calla laughs. "I've spilt more down my green tie, you bell-end."

Kris shrugs. "Right o, anyway is Davey's dream team up yet?"

Mark clapping his hands together – "Oh I can't wait to fucking hump them Yanks."

Willy pulls up at the scrap yard to be greeted by some big gypsy looking fella with an Irish twang. "How's it goin, pal?"

Willy offers the bloke a tab. "Sound, mate, my fucking driving must be shite – every cunt honking their horns at me all the way up the fucking road, man."

The big lad laughs. "So what we got here?"

Willy walks to the side of the van. "Stainless. I want you to give me a bogey bill, though; the gaffer wants it all but he can go fuck himself sideways."

The fella shrugs. "Anything you like, drive her on, bud." Willy jumps in to the van, driving it onto the scales. The fella laughs as he reads the sticker on the back of the van. "That'll do, come in the office and we'll talk money."

Willy arrives back as all the lads are standing around looking at the teamsheet on the wall, Calla laughing. "Willy, you're not even on the bench."

Willy marches over seeing his name clearly on the teamsheet. "Ay thought, straight in! Anyway lads, I have 100 dollars each for us but say fuck all to Mally – the cunt was going to give us hee haw off the scrap so I skimmed a bit and daffyd the bill."

All the lads smile. "Sound as, Willy, fair fucks to you." Chuck pats Willy on the back.

"Ay, fuck that streak of lanky shite, him and fat Jimmy sharing the spoils."

Willy disagreeing – "Na, Call, I'd say Mally would love to give it to them both, but Jimmy would see us alright for a pint like."

Calla shrugs. "Fuck them anyways, we have our touch."

Mark walks over to the lads. "Lads, we have 2 more recruits, which gives us 3 subs including big Jimmy! The 2 lads are good players – Tomcat said they are Mexicans that work for the cleaning crowd but they are fucking class."

Chuck laughs. "Well at least one thing the little bastards will be able to run, what with getting over that border and dodging fucking tax!"

The lads laugh as Calla chirps in. "Aren't they called texicans because they all end up in your neck of the woods?"

Tom's face is stern. "Cocksucker."

Willy laughs. "Ay, Calla, you suck one cock, son."

Calla wraps the last of the cables up and throws them in the box. "I wouldn't be sucking yours, stumpy little toffee crisp, that's for sure. Haweh, let's get up to the canteen to see what Gordon's brother has ready for pre match bait." All the lads heading for the door.

The lads troop into the kitchen. Davey is pouring the water into all the glasses as the lads sit down. "Right lads, spag bol on the way, energy all the way."

Willy pulls up his chair. "Fucking shiteing all day as well."

Davey sits down next to Calla. "Get this into you, son. You need to show me why you were on 25 quid a game and 6 pint tokens' worth for Hartlepool club."

Calla horsing into the spaghetti – "Don't worry, fatha, I'll bang a few in for ya."

Jimmy and Mally rock up, pulling chairs up, sitting down to the feed, Jimmy chapping the side of the glass with a fork. "Mally has something to say, lads."

Mally stands up with envelopes. "Right lads, there will be 100 bucks in the envelope out of the scrap money, we'll give you it in Nelly's," as Willy coughs his spaghetti all over his plate.

"You alright, son?" Davey patting his back as Willy whispers to Davey: "I have 100 dollars for you as well, I skimmed 100 bucks a man off cos I thought the tight cunt wouldn't pay up."

Davey coughing his spaghetti out as Jimmy looks concerned. "Yous ok?"

Calla laughs. "Ay they are just choking, Jimmy, dint worry!"

The lads finish their dinner. "Get the vans, we'll head off for 1 o'clock! And no fucking Nelly's before we play for Nelly's, Calla!!!!"

Calla stretches out! "Suits me. I was thinking of a few laps around the tennis courts anyhow," winking at Willy who is frowning. "You do haemorrhage some shite, big man!" Sucking up a long bit of spaghetti sauce going everywhere. "Haweh!"

Keegs flops down next to Calla. ""Just had a word with that Michelle – she is going to come and watch us the neet. Reckon I will be balls deep in her, the neet!"

Willy laughs, spraying Kris and Mark with spaghetti. "Dream on, Casanova! You have no chance."

Kris and Mark wipe their faces. "Fuck sake, Willy!"

Keegs looks across. "What sort of crack's that like, Willy?"

Willy wipes his mouth. "The truth, Trev! It's called the truth!"

Keegs looks at Calla. "Seems wee Willy winky wants to lose some more dollars, Call!"

Calla fishes the last of his mince onto his fork with his thumb and jams it in his mouth. "What about Sharon's mates? Thought you were my wingman!"

Keegs, wiping some of the sauce Calla sprayed on him, replies: "Thought about that, Call! I've seen Michelle and she is smart! Sharon's mates might be all fat muntas!"

Mark sits down with his tray of spaghetti and piece of garlic bread, Calla swiping it. "Howw, fanny chops, where did you get the garlic breed like!" wiping the last of his sauce up and jamming it in his mouth before Mark couldn't even react.

"Call, you're some pig faced cunt!" He stands up and wanders back to the counter.

"Fetch us all some!" shouts Willy, Calla pouring the rest of his coke down his neck and letting out a huge burp then standing up. "Boileralls off, boots in bin and yem for a pre match wank!"

"Just what I need when I'm eating my dinner, the thought of you huffing and puffing with your stinking cock out!" Kris shakes his head, Calla walking off laughing.

Inside the cabin Willy finds Calla's overalls stuffed with polythene and sat upright with blown up gloves for hands and a hard hat on top of a bucket for a head, with a big daft smiley face on it. Calla laughs and points at it. "That's probably more useful than you, Calla, you berk!"

Calla pelts a washer at Willy. "Let's get the fuck out of here!" Willy shimmying out of his suit then kicking his boot off, knocking the head off Calla's man. "Howw, you specky faced twat! That's shit crack!" putting its head back on, Willy laughing.

"Reet there's away!" The rest of the lads all ready for the off, Davey popping his head in. "Or, fucking haweh, lads, man!" looking at the pile of overalls, high vis vests and hard hats scattered everywhere. "Tidy the place up a bit!" then shaking his head at Calla's man sat leg crossed at the desk. Willy uses his foot to drag the pile of overalls under a desk, Davey walking out.

The lads join him, looking at the van. "Who has done that cunt?" Looking at the masking tape along the side with Nelly's team bus written along it and a red cross made out of sparky tape, Calla's smile beaming.

"Need you ask, Davey!" Willy snides, tearing the red cross off.

"Howw, man!" Calla snaps as the lads hop in.

The bus is sat barbling away when Tomcat leans into the window. "What time at the college, Davey?"

Davey lighting his tab on the cigarette lighter – "Doon for 5, son, so we can have a knock around!"

Tomcat bangs on the roof as Tom and Chuck jump in, Willy barking: "Haweh, Davoid, set the cunt away!"

Davey frowns into the mirror as he eases the van out, tooting the horn at the Americans, Calla spinning around flicking a waving Tomcat the finger. "Ay wey! That's the last I'll be seeing of this place! Just another peg on the monopoly board!"

Willy is puzzled. "What you on about?"

Calla crosses his arms. "Wey y'nar!"

Willy shakes his head.

"Davey, call into the 7-11 man, I want to get me mother a present."

Davey rolling his eyes – "What you gonna get her like, a tin of biscuits?"

Calla sitting forward – "Did you see tins of biscuits like?" The lads laugh. "You need to give owa spoiling that lass!"

Outside the 7-11 Davey is sat with the horn on looking into the shop where Calla is giving him a daft wave pointing to his Tampa Bay Buccaneers hat on! "Fuck sake, he has been in there 10 minutes, what's the simpleton deein?"

"Getting left is what if doesn't mak sharp!"

Calla wanders to the van struggling to open a bag of jelly beans with the 2 bags in his hands. "Areet!"

The lads all sigh. "What's in the bags like?" Willy trying to peek, Calla snatching them back.

"Fuck off and mind your business, dick breath!"

Davey looking at Calla through the mirror – "Manage to get your mother a tin of biscuits?"

Calla smiles. "A did, worzel gummidge, now get ya foot down!" The van gets going and Calla starts pulling foam beer coolers out of his underpants. "Here are lads, i nicked these for you! Pass them around!" the lads all looking at the Buccaneer football shirt beer cooler design. "You're all heart, Calla!"

Keegs raising an eyebrow – "So now you have become a thief?"

Calla shakes his head. "I've always been a thief! In the Boro we steal our shoes before we can walk!"

The lads laugh as Keegs hands Tom his beer cooler. "Say it's pretty warm for a beer cooler?"

Keegs pats him on the shoulder. "That's coz it's just come of out of Calla's pants!" Tom drops the window and glides it out, rubbing his hands on Chuck's shirt, Calla calling him an ungrateful cunt from the back of the bus.

The van heads into the estate, the van pulling up outside. Davey jumps out and Tom slides across to take the wheel. "Back here before 5, ok?" Tom salutes Davey and pulls off. Calla is only halfway out of the door falling onto his hands on the grass as the lads all cheer. Willy sticks his head out of the window. "That will learn you, dipshit!" sliding the door back shut.

Calla stands up looking at the mud on his palms. "Some cunt!"

Davey is at the door with a tab. "Get your arse in there – let's have an hour! Am busting for a shite!" Calla thanks Davey for the info as he walks past and into the toilet to wash his hands. Davey puts the kettle on. "Brew lad?"

Calla shakes his head and lifts the bags onto the bench. "No thanks. Stick these in the fridge, Davey!" Pulling the bags down revealing 12 cans, Davey squeezing the bag. "None of them before the match mind, daft arse!"

Calla nods. "Anar anar, for fuck sake!" He flops on the sofa and flicks through the channels. Davey loads the fridge up then sits in the chair opposite and drinks his tea, glancing up. "Thought you were busting for a shite?"

Davey sipping his hot tea – "Ay son, am just baking it 5 minutes!"

Calla pulls a face. "You're a horrible man sometimes, Davey! There's fuck all on this telly! Am bored! I'm gannen upstairs for a tug!" Davey pulls the face.

***

Jimmy and Michelle are going through the accounts. "All bonuses on top of the salaries?" Michelle nodding. "All flights booked and taken out of costs?" Again Michelle nodding. "When does it clear in their banks?"

Michelle punching the laptop –"Friday 3pm."

Jimmy smiles. "That's perfect. What's the average for the lads?" Michelle hands Jimmy a printed piece of paper. "Smashing, pet, thanks. So you coming to watch the game I hear?"

Michelle smiling – "I am, want to see you guys in shorts."

Jimmy laughs. "You won't be disappointed. Listen, do you fancy staying at the hotel in town tonight – save you the drive?"

Michelle blushes. "Is that a proposal, Jim?"

Jimmy blushes as well. "In 18 months I haven't once, way out of my league, pet – it was an offer on the company."

Michelle puts her hand on Jimmy's. "I'll say yes if it's a double room and we head off to collect my things and go down there now?"

Jimmy's jaw drops as he stands up, kisses her on the cheek and chuckles. "I'll get the car and follow you."

Mally is outside having a tab as Jimmy walks out. "You getting the car, Jimbo?"

Jimmy stutters, "Erm I but, you get one of the locals to run you. I have to go and erm pick up some bits and bobs for the match." Jimmy marches round the corner as Mally flicks his tab away – "Fucking charming."

Jimmy pulls up in the car park after spending the last half hour following Michelle with a smile on his face, into the wallet making sure the ever reliable packet of blobs were there and taking a Viagra out, snapping a bit off and wincing as he swallows the bitterness with a mouthful of warm flat coke.

Jimmy and Michelle check into the plush hotel 15 minutes from the college and are all smiles as they open the room into a very nice suite. "I know you have a game later so you lie back on the bed and make yourself comfortable as I unpack." Jimmy flicks his shoes off, opening his shirt buttons and lying back on the bed. Ten minutes later Michelle is on top of him. "I think I am dreaming" – Jimmy's big hands unloosening anything that needs unloosening.

Davey is sound asleep on the sofa as Calla sneaks out of the front door and walks up to Willy's digs. "Haweh daftarse, we only have an hour."

Willy walking out turns back. "Ay Hags, for fuck's sake you sound like me fatha."

Calla and Willy walk into the empty bar as Rosie is cleaning. "2 pints of your finest and 2 voddy cokes."

Rosie is shocked. "Boys, you have a big game later?!"

Willy laughs. "Aye that's why we are only having 4 – 4 each."

Rosie shakes her head. "No way! I am not serving you," as Calla grabs her off the stool. "Get off me, Calla, you big fucker."

Calla marches over to the disabled toilets and kicks the door open. Shutting it he pushes a chair angled against the handle as she shouts, "Let me out, you bastard."

Calla walks back rubbing his hands. "Right fannychops, vodka was it?"

Calla stands under the optics as Calla's phone goes. Putting the glass down he looks at the phone. "Fuck it's Davey!"

"Fuck that, don't answer it!" Willy waving back towards the vodka.

Calla lets out a sigh. "Nar, he will butcher me if I don't answer. Hallo Davey. Eh, she did?...Anar but we were going to...just the...you couldn't!..you wouldn't...Reet! Ta ra!" Calla walks out from behind the bar.

"Well, big man, what he say?" Calla looks up. "Yeh, he said if we have a drink he will drop us and if we are dropped we can kiss the bonus goodbye!"

Willy leaning over the bar lifts up the vodka. "Bollocks! He cannae do that!" lifting the soda gun and adding a splash of coke, lifting it to his mouth.

"Maybe! But he also said we can kiss our railings goodbye as well!"

Willy stopping the drink at his lips – "He that serious?" Calla nodding his head and wandering to the disabled toilet, Willy putting the glass down and rubbing his teeth. "Ay well fuck that, it's taken me ages to grow these bastards!" rubbing his teeth.

Calla lets Rosie out of the toilet. She digs him in the chest. "Bastard!"

Calla lets out a small groan, Willy hopping up off the stool and putting his arms over his chest protecting himself from Rosie who goes straight behind the bar. "Davey call you?" she spits, Calla straightening up. "Did you ring him!"

Rosie smiles. "Yep!' Told Gemma last night to keep our eyes peeled and not to serve you two."

"Grass," Willy mutters under his tongue, Calla standing up straight.

"Arr, weyy, a suppose we can do without a slurp just this once!" Leaning onto the bar – "Still have some time tho', sexychops, any chance of a quick shag?"

"No chance!" crossing her arms, Calla leaning back.

"Fair enough!" moving back, tapping the bar. "Haway, speckled hen, let's make like bananas and split!" the two trudging towards the door.

"Fuck sake, big man, I'm fucking gasping for a drink!"

Calla nudges Willy. "Haweh, I have an idea!" as they wander across the road to the shop.

In the hotel Michelle is back in the bathroom, Jimmy all sweaty, sitting upright, breathing heavily with a tab quivering on his bottom lip. "Fucking hell!" he sighs to himself smiling. Lifting the quilts he sees the old fella still stood to attention as Michelle comes out of the bathroom and slides onto the bed kissing him on the neck. Leaning over he stubs out the cigarette and rolls her onto her back, shouting "Round two! ding ding!" Michelle giggles.

Stood at the fridges inside the shop, looking at the stacks of booze, Willy pecks Calla on the cheek. "Genius! We cannae go back to mine though, Hags will gan mad!" looking at Calla.

"Wey, Davey will punch out our teeth!"

The two laugh. "It's like being 12 again!"

Calla looks up. "I think that was the last time I was sober?"

Willy nods. "Ay me too! Can't remember what it's like to have a firm shit! Had the squirts since my first tin of Tennents!"

Calla lifts out two 6 packs of Sky vodka lemonade, Willy struggling to hold them both. "What's this! Man, get the lager!"

Calla closes the fridge and taps on Willy's head. "Haweh, man! Use your basin – you can't smell vodka!"

Willy nods in approval. "I like it!" putting the two packets onto the counter and 3 packets of extra strong Chiclets chewing gum. Calla glances over to Willy. "Just in case!"

The two lads wander out of the shop. "Your bus stop or mine?" both lads laughing.

Davey is all dressed and waiting outside, having a tab when the bus swings by. Hopping in Kris asks, "Where's Calla and Willy?"

Davey pulls on his seatbelt. "They went to Nelly's for a pre match pint!"

Kris leans forward. "Or for fuck sake!"

Davey smiles. "Settle down. I knew they would so I told Gemma not to serve them! Rosie rang me from the disabled bog and told me the pair of piss heads had shown up!"

"Disabled toilet?" Tom quizzes.

"Run through walls them two for a pint!"

"So where are they now?"

Davey shakes his head. "I rang Calla back and he said they said they were making their way down to the college!"

The van pulls up to Earl's cabin. He holds 2 fingers up and ushers them into the back. "What's he doing?" as the gate lifts and he wanders out with his jacket, sliding open the door and climbing in, the whole van shaking with the size of him. "Hey y'all!"

All the lads look at him puzzled. "What you doing, Earl?"

"Watching you guys play soccer!"

Davey turns around. "Who will watch the gate?"

"It's open! Don't need no watching!" Davey nods. "Ok then! Foot down to Wembley Way, Tom!" all the lads cheering.

***

"Pass me one of them bottles, Calla!"

Handing Willy the bottle – "Y'nar, Willy, life doesn't get much better than this!"

Willy frowns, looking around. "Eh! We are sat on a park bench supping alcopops like bairns?"

Calla throws his bottle into the bin and opens another one. "Nar man, I mean sitting here Friday afternoon, few quiet scoops afor a footy match! Back to the boozer for a good piss up then back yem at the weekend with a hip full of cash! Haweh, couldn't be better!"

Willy squints at the back of the Sky bottle then takes a swig. "How many do we have left?"

Calla looks down at the bag. "Two each!"

Willy looks across. "So, let me get this straight, Socrates..."

Calla jumping in – "Socrates?"

Willy laughs. "Monty Python, the philosophers' song..." Willy leaning back starts singing: "Immanuel Kant was a real pissant Who was very rarely stable. David Hume could out-consume Wilhelm Freidrich Hegel, And Wittgenstein was a beery swine Who was just as schloshed as Schlegel. There's nothing Nietzsche couldn't teach ya' 'Bout the raising of the wrist..." Calla laughing puts his arm around Willy, joining in – "SOCRATES, HIMSELF, WAS PERMANENTLY PISSED!" both laughing, swigging the Sky back.

"So Socrates, as I was saying, we are on a park bench drinking potent panda pop –only 2 left by the way! – we are about to run our pods off for 90 minutes and when Davey sees us blowing out our arses or you chucking up we will be without our teeth? I can think of a million things that could be better!"

Calla hands Willy another bottle. "Shit there's the lads there!" pointing through the trees to the van pulling up outside the College, watching them all climb out and start stretching. Calla laughs. "Willy son, what you like playing football with a drink onboard?"

Willy looks at Calla in disgust. "Put it this way! I don't know what it's like to play sober!"

The lads stand up and take the last gulp of the bottle, Willy throwing the rubbish in the bin. Calla holds the last two bottles, handing one to Willy. "Chin chin!" as the two clink bottles and neck them over in one, slamming them into the bin. "Say that shite is just like pop!" Calla stuffing a full pack of Chiclets into his mouth, passing Willy a pack. "Get them into you, dog breath!"

The two lads then jog down the hill towards the van, spitting the chewy out, Calla putting his hands on his knees. "I beat ya, Willy, ya slow cunt! Hand owa the 100 bucks!"

Willy looks at Calla. "Eh!"

Calla standing up, stretching in front of the lads, holds his hand out. "Haweh, Willy, don't make a SPECTACLE, a bet's a bet, hand it owa!"

Davey chirping in – "You two ran here?"

"Wayaye! Willy bet he could beat me! Now haweh, Willy, hand it owa!" staring at Willy who begrudgingly takes out his wallet, holding out the hundred. "Che ching!" Calla snatches it. "Reet haweh, let's get a warm on ready for these bastards."

Davey grabs his arms. "Let me smell your breath!" Calla breathes out into Davey's face. Davey leans back. "Owa minty son! Where have you been?"

"Rosie wouldn't serve us so we ran here!"

Davey squints at Willy who is holding his sides, breathing heavily. "Wey, a dinnit believe ya! Haweh, let's get changed!" He walks off.

Calla digs Willy. "Fucking ace acting that pal! Reckon he believed us!" walking after Davey.

"Eh? I really have a stitch!" Willy lets out a sigh, catching them back up. Inside and Stevey and the two others are there with all the kits hung up and a bag full of boots. Calla, like a dog digging a hole, barges forward raking through the bags, boots flying behind him, coming out with a pair of Copa Mundials size 11. "Wey hey, them bastards will do!" looking at Stevey. "These are bumped by the way, bonny lad!"

Stevey pulls a puzzled face "Bumped?" still holding out the bag as the rest of the lads all scrat around coming out with all sorts of boots, Willy the last finding the only size 7s, a pair of yellow Nike mercurials looking them up and down. "These are fucking shite!"

Calla laughing and pointing – "Think the bairn has bigger feet than you! Bananaman!" The lads laugh. "Howw, Davey, you don't need a pair – pass them here!"

Davey holds them tight. "They are for Jimmy, daft arse and they wouldn't fit you anyway!"

Willy tips the bag out looking at what's left – the nearest to his size a size 9.

"Fucking hell, man!" Stevie laughs. "You want me to get the girls' boots?"

"Arr fuck it, I can play in sandals – these jallopys will do!" sitting down next to Calla who has already grabbed the no 7 shirt leaning in whispering, "Haweh Cal, giz the hundred back!"

Calla jumping up and taking off his t-shirt – "NO WAY, WILLIAM! I WON FAIR AND SQUARE! YOU CAN'T HAVE YOUR MONEY BACK!"

Willy shakes his head, looking back down at his boots. "Cunt!"

Jimmy wanders whistling into the changing room which is now awash with the hustle and bustle of the lads changing and the smell of deep heat, Davey handing him the boots. "You areet?"

"Weyaye! Never better!" squeezing Davey's shoulder and taking a seat.

A puzzled Davey shouts to the lads. "Reet everyone, quieten down...Howw lads, haweh, be quiet...SHUT THE FUCK UP!" the lads all freezing, the place falling silent. "Right, plan of attack: we finish getting ready and wander up to the pitch; we will run a few drills and get warmed up! Then we will head back down and I'll run through the team! Jimmy, where is Mally and Tomcat!" Davey still smiling away.

"Mally will see us down here and I haven't a clue about Tomcat?" as Tomcat slaps Davey on the back and sits down. "This is Raoul, Jose and John!" introducing the 3 Mexican looking guys, all the lads pulling puzzled faces towards John who catches Davey's stare – "Brooklyn, New York! Third generation immigrant!" casting a toothy smile and sitting down, the lads sniggering and Davey rubbing his hands. "Right lads, get dressed and head up to the pitch," turning to walk out as Mally bumps into him. "Howw, stretch, get ya arse in and changed!" Mally nods and walks in.

Davey gets out up onto the pitch admiring the set-up. Hearing a whistle, he turns around to see Tom and Chuck, the only two in the stand, both waving cans of beer at him. Waving back he then wanders out onto the pitch rubbing his hands as Kris, Mark, Hags and the rest start to troop out with footballs. Davey lets them all get onto the pitch. "Right o lads, let's get warmed up." Calla blasts a ball a mile over the bar; Willy laughs then lets rip himself, the ball crashing off the post and flying in. "That's how, bonny lad, that's how!"

Davey waves them in. "Give owa man, everyone in!"

Calla scoops a ball, it bouncing straight off Mally's head, laughing. "Now Willy, that's how! I say that's how!"

Mally and Davey are both unimpressed. "Right knackers, get in here!" Davey barking to Calla who slowly wanders across.

All the 14 lads huddle round Davey as he gives out the orders. "Right lads, this is how we work – you listen to me and only me. I've watched more football than some of you cunts have had hot dinners."

Calla chirping in – "Bet you haven't, fatha, the reckon these Mexicans have chilli con carne hot enough to set fire to your undercrackers, man!"

All the lads laugh as Davey pipes in again, "Shut up, dafthole.

Right, so you listen to me and only me. So we get in a nice straight line and Calla will take you through a few warm up drills and stretches. Ten minutes, Calla, off you go."

Calla jogs over to the touchline. "Right lads, nice and easy, jog across the park, flicking them up at the back halfway."

The lads all jog across the pitch as Jimmy is at the door waving at Davey, Davey waving back. Jimmy shakes his head and ushers him over as Davey tappy lappies over. "What's up, bigun, nee jersey to fit ya?" Jimmy points down at the tentpole in his shorts. "Fucking hell, lad, I knew you were excited to play but that bastard will have your eye out."

Jimmy laughs. "I nar that, I'll skip the warm up and get a tracky on, tell you about it later on, reet."

Davey chuckling walks back to the lads who are now stretching off. "Right lads, in a circle, two touch follow your pass." Davey kicks the ball over to Jonny who brings it down, the ball bouncing off his shin. "Fucking hell, Jonny you have a touch like a fucking rapist" the three Yanks laughing as the lads get on with the drill. Davey looks into the stand to see Michelle and a few others gathering. "Some crowd, lads, so let's put on a show! Righto let's have you in," Davey ordering the lads over.

As they get to the dressing room, Frankie's team run out, all Italian looking fit lads apart from 2 heavy lads with thick tashes. "Fucking hell, the Chuckle Brothers are playing for these cunts." Willy nudging calla who is still blowing out of his arse.

"Don't make me laugh, Willy man, I can hardly fucking see." Willy tripping him up as he flies into the bin right in front of Davey. "How do you think I feel now, you blind cunt."

Davey picking Calla up out of the bin – "Are you sure you haven't been drinking? Do you want to go sub?"

Calla gets up. "No I haven't and no I dont!"

Davey kicking him up the arse – "Wey get in there then, you balloon."

Jimmy is sitting in the changing room as the lads walk in, Mally smoking a tab. "How come you get to skip warm up, Jimmy?"

Jimmy smiles. "Cos I am the boss. Give me a couple off that

bastard, I'm gasping." Mally hands him the tab as the lads sit down.

"Fucking hell, Mally, that's the first time I have seen you wearing summit different than sambas since 1986," Calla pointing at Mally's copas.

"They look the same though," Mally admiring them

"Bet the cunts don't smell the same," Calla replying as Davey shuts the door.

"Right lads, the shirts are all on the pegs and I want you all to wear them as I call them out so if you have one on take the bastard off." Willy smiles as Calla takes the 7 off and pelts it at him.

"Right, listen up: one, tomcat;

two,Kris leftback;

three, Jose rightback;

four, Jonny sweeper;

five, Stevey centre half;

six, Mally centre mid holding;

seven, Jockey right side..."

Willy pelts the shirt back at Calla who is smiling like a cheshire cat then his smile disappearing – "Eh? Fucking Jockey 5 foot Wilson, the cunt."

Davey pushes his glasses up his nose. "Aye!

"eight, Rauol in the centre with Mally;

nine, Calla up top;

ten, Willy floating between Calla and midfield;

eleven, Tim on the left, son. I have heard off Stevey you are top banana and all left foot so try and get the ball into Calla so he can hopefully bang a few in! And listen, lads, I know these bastards beat Nelly's a few years back so let's show them what a beating is eh? And aparently the ref is a bit of a stickler for swearing so keep it down, lads! Plenty of drinks onboard and let's get out there. Calla, you're captain!! That leaves Jimmy, Hags, John, Jose and Keegs on the bench the bench when he gets here.

Hags stands up. "What's my number? Can't believe it's not number 9 like!!"

Davey looks at his pad. "Well I have you down next to 13, being Mr dark cloud."

Willy laughs. "Stop your wimping, grandad and put the cunt on – you will be on after 10 cos Calla is still fucked from the warm up!"

All the lads stand up giving each other a hug or a handshake just as the ref comes in. "Warwick Hunt?! Ah fucking haweh."

Warwick stands there smiling. "Yes, Calla, that's me! So listen up! Swearing: I will tolerate it at ground level; anything at me or loud you are in the book! Let's have you, boys," as he walks out.

"Jimmy, what the fuck's that in your tracky? A traffic cone!?" Willy points down as Jimmy goes red.

"It's my lad, lad! Bit excited you know."

All the lads hurry out, laughing. The lads are stroking the ball about as Walter walks over to the sideline. "Hey, Calla, come here, son."

Calla jogs over. "Areet Walter, thought you escaped only once a week?"

Walter laughs. "Wouldn't miss this for the world. See the number 5 there for Frankie's? He's Frankie's son – he is one dirty bastard! You put him in his place and I'll buy you a beer."

Calla looks over at the big centre half. "No bother, Walter, I'll burst him."

Calla walks over to the centre circle and is talking to the ref and the big fat chuckle brother spins a coin as a good few hundred people are now gathered watching. Willy walks behind Calla and pulls Calla's shorts and his underpants down, all the lads bad laughing as Calla leaves them there checking to see if it is a head or a tail.

"Pull your shorts up, son." Warwick points at Calla's balls as he bends over.

"Sorry ref."

Davey shuffles into the dugout. "Some set up, lads, eh?"

Jose, John and Jimmy sit next to Davey as big Earl leans round the dugout. "Hey guys, any trouble – I have your backs."

Davey laughs. "Sound, Earl son; do you not fancy getting kitted out, like?"

Earl climbs in. "You have four subs, Davey – isn't that enough?"

Davey shakes his head. "We have another lad on the way mate but could always use more, man, gan on, get kitted out."

A massive smile on Earl's face as he heads off towards the changing room and Jose pipes up, "Boss, I play midfield, I am like a Mexican Diego Maradona"

Jimmy, leaning by Davey, looks at him. "Jose you look like a Mexican dago never mind the Maradona."

Davey leans out of the dugout. "Chuck, give us one of them beers, man"

Jimmy laughs. "Get me some ice for me underpants," the two Mexicans shuffling away.

Calla and Willy are standing in the centre circle. "20 bucks a nutmeg, 20 dollars a goal?" Willy stands, smiling at Calla.

"Deal, twinkletoes," Calla replies, pushing the ball to Willy, Davey nudging Jimmy swigging his bottle of beer. "Look at them two shaking hands, team bonding already."

"Be some dodgy bet more like, pair of bastards." Jimmy leans back, pushing his tackle to one side as Davey leans round the dugout. "Chuck, any scran, son – hotdog or owt will dee?"

The ref blows his whistle as Frankie's centre forward comes running in as Willy pushes the ball forward to Calla who fakes the pass back and megs the centre forward, running round him then clipping the ball wide to Jock. "20 bucks, hamma" as Willy runs past him – "Cunt!"

The ball goes forward with Jockey skinning 2 players and laying into the screaming Calla who gets flattened by the defender, the defender pumping the ball down the pitch then offering his hand as Calla climbs to his feet himself. "You fucking dirty wop cunt."

The centre half speaking perfect English – "I was raised in the USA, you fucking English imbecile."

Calla jogging off past him – "Aye, by wop cunts."

Big Stevey brings the ball down expertly and lays it to Jonny who cooly takes it past a player and gives it to Kris who plays it first time to Mally. Mally turns and pushes the ball through to Willy who is onto it like a flash!

Davey stands up banging his heed in the dugout. "Ah, ya bastard, go on, Willy son!"

Willy shapes to shoot as the big centre half comes in from behind and brings him down. "Penalty!!" Davey shouts, standing up, banging his head again. "Ow, ya cunt," rubbing his head as the ref points to the spot.

Willy holds his ankle as Calla picks the ball up. "Get up, ya daft cunt, out the way so I can take this."

Willy waves over to the bench for some help as Davey looks over squinting then gives Jose the dregs of his lager and a bucket and sponge. "Here son, run on and give Willy these – it's owa far for me."

Willy sits up as the young Mexican arrives. "What the fuck's that?"

The Mexican shrugs his shoulders looking at the quarter of beer label. "B.u.d. l.i.g.h.t," as Willy snatches it and takes a gob-full, handing him it back as Calla walks over and empties the bucket on Willy's head; taking the bottle off Jose, he swigs it and puts it in the empty bucket. "Are you going to get up, ya diving cunt, or what?"

Willy standing up as the ref gives the number 5 a yellow card. "You give him the Weetabix card!!? Fucking hell, ref, that was a Ready Brek card if ever I seen one!!!" Warwick not having a clue what the Scotsman was on about.

"I'm on penas, Calla," Willy trying to grab the ball.

"Are you fuck, you're on fucking get owa there and watch Calla score." Calla walks forward as Frankie's goalie standing 6 feet 4 is on the penalty spot intimidating Calla as Calla holds the ball in front of his eyes. "See this, cunt? Yes, bigman, it's a ball! And the very next time you touch the bastard it will be when you pick it out of the net!"

The ref orders the goalie to his line as Calla runs up and lashes the ball into the bottom right corner, remaining to be standing on the spot as the goalie picks the ball up and turns. "See! I fucking told you."

All the lads congratulate him as they run back to the restart, Walter right at the front. "Yeah pick that out, you motherfucker."

Davey leaning round the box – "Haweh, Walter son, there's bairns in the crowd."

Frankie walks down the touchline screaming at his team to go forward. "Frankie, your lace is loose," Jimmy shouts over as Frankie lifts his trousers and looks at his foot. "No, the other one," as he repeats the same, Davey and Jimmy then singing the circus song, "da da da da da da da da da olay!"

Frankie is fuming. "Hey, fuck you two," as the bench pisses themselves laughing. Kris slides in, taking the ball off one of the forwards and lays it into Jockey, who flicks the ball past the midfielder and scampers down the line, floating the ball into Willy who turns and switches the ball with accuracy to Calla who nods it down to the on running Raoul who spoons it with his right foot wide.

"He's getting the shepherd's hook at half time, that cunt, Raoul, my fucking hoop."

Davey nudges Jimmy as he looks at his pointy cock poking through his tracky. "Don't look at me, this bastard's still full force."

Davey looks at the three other lads – "Get warmed up." John and Jose jump out and run up the line. "And you, Hags."

Hags looks at Davey as if he is mad. "I am warm, Davey, I have a tracksuit on and I'm sipping coffee."

Davey shaking his head looks on the pitch. "Mally!! Push the ball wide," as the ball lands wide to Jockey who whips it in first time, no one in the box for Nelly's.

"Where the fuck's Calla?!" Mark shouts over at the bench as Davey gets out and wanders up the line spotting Calla running back out of the bushes onto the pitch. "Where have you been, you fucking idiot!"

Calla looks over. "I went for a shit, I was bursting! Had to throw me skiddies away tho, Calvin n all."

Davey storms back into the dugout as Jimmy and Hags are crying laughing. "He's been for a fucking shite, a fucking shite, y'nar!"

Nelly's get a corner and Kris cracks the ball over as Stevey connects with a bullet of a header which crashes off the bar landing at Willy's feet who fakes the shot and the keeper dives and the defender slides in, Willy squaring the ball to Calla. "There you go, bigun."

Calla going to shoot then squares it back to Willy. "Nar, dint fancy it," as Willy slots it in. "Pick that cunt out, lerch," as the lads get mobbed!

Frankie throws his bottle of water onto the pitch as Davey shouts over: “Toys out of the cot, son?”

Chuck and Tom walk down to the dugout handing the lads hotdogs and beers, all the lads tucking in. “Hags, give me that here, you daft cunt – you’re going on second half,” Davey snatching a hotdog off him.

“Davey, I’m hank marvin, man.”

Davey takes a massive bite and hands it back, trying to speak through a bun. “There yar, greedy cunt.”

Hags looks at the the bit of hotdog left and eats it, shaking his head.

“Davey, I have cramp – I’ll be off at halftime.” Mally holds his leg as Calla runs over pulling Mally’s shorts down and pushing him over, Mally holding his balls and screaming, at the same time holding his leg and ballbag. Frankie’s push forward and the ball falls to one of the Chuckle Brothers who blasts it at goal, the ball smashing into Tomcat’s face and going wide for a corner as Willy and Calla roll about laughing. “Haha fucking Billy the fish, man.” Calla wiping away the tears as Willy points. “He saved the cunt with is face! Fucking brilliant”

“Davey, he’s knocked out!” Mark shouts over, Davey giving Jose the bucket and half a bud light .”Jose, gan on son, get yourself owa.”

Jose runs over to Tomcat who is covered in blood, getting to his feet. “What the fuck’s that, dude?” Tomcat looking at the Mexican who looks at the bottle. “B.u.d. l.i.g.h.t,” as Tomcat pushes him to one side, taking the sponge out of the bucket as Jose looks at the bottle and shrugs, taking a massive mouthful, then spitting it out.

“Earl, get on, get Tomcat’s gloves – you’re going on for Tomcat the goalie.”

Earl just about to sit down then walks onto the pitch, all the lads staring at the size of him as he takes Tomcat’s gloves. “Is this vest ok, ref?”

Warwick not daring to say no – “Yeah, that’s fine, bright yellow, no problem, son.”

The corner comes in and Earl punches the ball which lands at Calla’s feet on the halfway line who turns the centre half while crunching his elbow into his mouth, the ball being played wide as the ref waves play on.

Tim sprinting away down the line cuts inside then, before squaring it to Jonny who has sprinted all the way from the back to smash it into the top corner, running on, slapping Calla. "Bit of mackem magic, son!!!"

Davey jumps up. "Get in, ya bastard, ah ya cunt!" banging his head off the dugout again. The ball goes back to the middle as the ref blows for halftime, Jimmy standing up out of the dugout as Michelle waves over, Davey catching the glimpse. "Ya owld fox, Jimmy, right lads in here!!" All the lads trotting in as Davey opens the cooler box. "Bud lights for the lads and Corona for the Mexicans." All the lads tuck in, laughing, as Davey puts his arm around Jonny. "It's like watching fucking Newcastle in 96, lads."

Jonny looks at Davey. "We'll probably conceed a few then, if that's the case!!"

"Right lads, a few changes. Mally is coming off, his cock's owa small and Raoul, son, you come off as well – if you're a footballer my cock's a kipper! John, you go centre mid," ushering the Mexican sub to get his tracky off. "Willy, you drop into the middle of the park and Hags will come on up front," all the lads letting out a cheer as Hags takes a bow. Davey nods. "Aye the old ledge, and Keegs just texted me saying he's finished on site and will be here the last 20 minutes so he'll be getting a run out. Earl, you ok, son?"

Earl nods as Calla opens his second bottle. "I feel fucking champion now. Haweh lads, let's get another few past these wop bastards."

Walter puts his arm around Calla. "Hey, Calla, the number 5 is away to hospital – that means I owe you 2 beers!"

Calla laughs. "Aye I think one of his molars is stuck in my fucking elbow."

The lads all get ready, Willy having a tab, over talking to Michelle and the barmaids. "Haweh, Willy, game on!" Davey shouts over to Willy as the teams wander on for the second half.

There is a good number of people in the crowd, Davey noticing this for the first time as he holds a thumb out to the old storeman from site sitting down. "Fuck me, there is a canny turn out!"

Jimmy splashes some beer on his crotch. "Ay there's mare here than a fucking MLS game!"

The lads are all into position as Frankie is making his last orders to his men. "Haweh, Frankie ya cunt!" Calla shouts over, Warwick casting a hard stare. "Match hasn't started yet, ref!" smiles Calla.

Warwick walks over. "Look Andrew, we spoke about this; please keep the language down!"

Calla crosses his arms. "Well REF! Would you send me off if I called you a cunt?"

"Yes I would!"

"Well what about if I thought you were a cunt! Would you then?"

Warwick shakes his head. "Of course I couldn't send you off for thinking I was a cunt, Andrew!"

Calla patting him on the shoulder and jogging off – "Well Warwick, son, I think you're a cunt!"

Warwick is about to respond as Davey shouts over, "Haweh ref, Frankie fucking 4 fingers is ready!"

Warwick looks around and blows the whistle. Frankie's men are a lot more organised after the first half battering, keeping the ball in small triangles, Davey's lot running after it like headless chickens. "That's it, boys, let's run this lot down!"

Frankie's men playing some lovely one touch, Calla flying in missing, Willy and John doing the same. "Bastards!" Davey shouts to a look from Warwick, Frankie grinning as a ball is lifted out of the middle into the run of his left back who takes it down. Davey's lot, who have been drawn out, scamper backwards as the ball is whipped into the box and their number 9 heads it to the left of Earl (GOAL). Frankie jumps up and down as the lad goes to get the ball out of the net and begins to run back with it as Earl grabs it. "Let go!" the number 9 shouts, Earl heaving it back – "NO WAY!"

The number 9 and Earl pull the ball back and forth just as the 9 is over-stretched pulling the ball, Earl leaves go sending the lad flying onto his back, everyone laughing. Frankie holds his hands out to Davey. "What's that about, Davey! Man, was only going to celebrate!"

Davey shrugs. "Get yasel owa and tell Earl!"

The ball is put on the spot, Calla rolls his foot over the ball, then flicks it wide to Jockey who is closed down straight away and Frankie's lot break playing neat one twos through Davey's lot. "Get a bastard foot in!" Davey is out of the dugout screaming at the lads, half hearted tackles missing as the ball is smashed low and hard past Earl. Davey throwing his bottle off the grass – "For fuck sake!"

Frankie is out clapping his hands with a daft smile on his face. "Fuck me, he was like a fucking whippet!"

Calla's hands on his knees – "Anar! Fuck! I need a tab!"

Willy breathes heavily as the match resumes. "Keep it simple, lads, let the ball do the work!"

Nelly's knocking some decent passes around as one of Frankie's men crashes the ball. Willy sees his opportunity and dives to the ground holding his leg. A fracas starts up with Warwick in the middle trying to calm it down. "Never touched him, ref!" as Warwick waves on the Mexican with the bucket, Davey squinting. "Weez that Willy?" barking after the Mexican. "Had on, son, had on!" the lad turning and jogging back. Davey takes a cigerette out and a lighter – "Tak this!" The mexican frowning looks down at the tab then jogs back to Willy, kneeling down. "Where does it hurt?"

Willy leaves go of the ankle. "My chest!" snatching the fag and lighting it up, lying on his back letting out a pleasurable sigh.

Calla is over at the touchline taking a lashing from Davey. "These bastards are all over us!"

Calla spitting the water out agrees. "That cunt's give them all a wrap of speed a think!" flicking the finger at Frankie.

"Look son, we need to sort that number 8 out – he is fucking brilliant!"

Calla looks across at the lad adjusting his hair with his head band, Calla nodding. "Reet, al see how fast the cunt can limp!"

Willy is back on his feet as Warwick blows the whistle, Davey slapping Calla's back. "Haweh son!"

"Number 7 you're shit!" a scream from Frankie's dugout as Calla spins around. "So is your lass in the sack!" grabbing his balls and throwing the v sign.

"Areet Davey, what's the score?" Keegs sits down already stripped with a tracky top on, Davey's face brightening up. "Trevor, son, just what we need. We are 3-2 up but these cunts are doing us with fitness now! Gan get warmed up and al fetch you on. I want you to sit in the middle and tighten things up!"

Keegs nods then jogs along the line to the far corner stretching. Frankie's number 8 takes a short pass to his feet, does a step over and knocks it through Calla's legs.

"That's the 20 cancelled, Calla!" Willy laughs as the number 8 spins on the ball, past Jonny to the left, knocking the ball and running onto it as young Mark flies in 2 footed getting the ball but carrying on straight through the lad who flies up and lands screaming, the lads all cheering, Frankie out of his box screaming at the ref.

"That's a fucking excellent tackle!" a faint cry from Keegs who is at the opposite end of the pitch doing stretches.

"How do you know! Do you have a pair of binoculars in your pocket?" Frankie shouts, Keegs turning slowly. "No you daft cunt! A fist!"

Frankie shying away, Warwick waves the Mexican on as Calla grabs the no.8's arms and drags him off the pitch. "Get off, you soft cunt." The lad is clearly in pain as he rolls his sock down and throws his shin pad looking up confused at the Mexican with the bottle of beer. Frankie's captain is signalling to the bench that the lad's fucked, Frankie sending on a sub. "That was real dirty, Davey! Should've been sent off!"

"Sit down, man, he won the ball!"

Warwick is still surrounded by Frankie's men who are calling for action, Warwick waving Mark across and dismissing the Frankies lot, Calla pulling one away. "Fuck off!"

The lad squares up to Calla. "Say what?"

Calla leaning back and wafting his hand over his nose. "I said fuck off! And now I'm saying fuck off and brush your teeth, your breath stinks!"

The lad leans in, biting his teeth and glaring at Calla who is laughing. "See you at 90 minutes, mate!" the two walking off still staring.

"Look, Mark that was a needless, reckless challenge and you see both feet were off the ground and it was two footed and you were the last man so I'm afraid I will have to send you off!" as the ball bounces off Warwicks head Mark holds out his hands as Warwick holds the red card up looking round to see who threw the ball to the applause of Frankie's men and the shouts of Nelly's.

Frankie is out of the dugout. "Well done, referee! Well done!" looking at Davey.

"Frankie, until now I only thought you were a cunt! Now i think you're a RIGHT CUNT!" Frankie, puzzled, sits down as Davey stands up. "Steve, haweh, off son, you're playing shite! KEEGS!" Keegs runs up taking his jacket off. "Sit yourself in front of the defence, lad, and steady the ship. We will drop to 3 at the back. Tell Kris to move across to right back and them other 2 Yank cunts to move in!" slapping Keegs on the back. "Ref, sub!" as Mark trudges off with Steve. Davey puts his arm around Mark. "Well done, son! That bastard was whopping us!"

Mark sits down with a bottle of water, old Walter leaning around the dugout. "Nice one, son! Taking out the queer with the headband!"

Frankie's sub is stood over the ball waving at his men who are jockeying for position in the box; Calla, Willy and Hags in the wall, Calla squeezing Willy's arse. "Cut the cunt out, Calla!"

"We jumping or standing, lads?" quizzes Hags.

"Jump if it's high, stand if it's low, you stupid cunt!"

"Fuck off, Calla!" As the ball loops straight over the top of them and on towards the top corner, everyone is gazing as big Earl half dives and stretches out an arm, palming the ball away and landing with a crash on the deck, everyone cheering. "Told you to jump!" barks Calla at Hags as they run back, the ball landing at Frankie's number 9, who squares it to their number 4, who dinks it to the other side of Earl who can't move across in time. Frankie jumps out of his dugout. "Get the fuck in! Get in! YEH YEH YEH!" jumping up and down. "In your face, Davey! YEH YEH!" Frankie doing a dance as a hot dog hits him on the head to a cheer from the crowd. Davey is pissing himself as Frankie stops, wiping the mustard from his hair looking at Mark. "That's your fault, knackers!"

"EH? Haweh!" Mark holds out his hands.

Walter leaning around – "You silly little boy!" Mark crosses his arms in a huff.

Davey is out of the dugout waving orders to the lads as Nelly's kick off, Calla stroking a ball to Jockey who fires down the wing, cutting inside and shoots the keeper, tipping it around the post for a corner. "That's it lads!" Davey screams.

"You're gonna have a heart attack you, ya cunt!" Jimmy pulling Davey to sit down. Willy wanders to the corner with the ball, placing it down, squinting through his geggs. Calla's inside the box pushing and shoving with one of Frankie's lot. "Howw, man, get your fucking hands off me, dogshit breath!" as he pushes back just as Willy knocks in the ball. Calla stamps on the guy's foot and jumps, flicking the ball to the back post where Hags blasts it wide.

"You fucking useless cunt, Hags!" Calla turns around and holds out a hand to the bloke writhing in agony. "Fuck off you dirty bastard!" Calla walks off rubbing his eyes mimicking a baby crying as the play is resumed and goal kick is lumped up field and taken down and laid out wide, then back inside Frankie's men, passing the ball well again as a bullet of a shot flies from the midfield and crashes off the bar and back into play where the no. 6 volleys it back goalward, an outstretched Earl unbelievably tipping it from the top corner over the bar for a corner.

Calla helps Earl to his feet. "That was a fucking excellent save, Earl – where did that bastard come from!"

Earl dusting himself down – "Fuck off, Calla, I was still going for the first shot!"

Frankie's getting the ball taken off them by Keegs who leans in and forces the guy off the ball, steadying himself, standing with a foot on the ball as a tackle flies in, rolling the ball back then forward, not even looking down the tackle totally missing him, he knocks a ball straight over the top towards Hags who brings it down dead, tapping it to the side and smashing it straight at goal for an easy save.

Frankie's shouting across to his captain who is holding out his hand in protest, Calla jogging backwards with his tongue out to Frankie. Looking up he spots Sharon with her friends waving. Doing a small curtsy he shouts towards Keegs, "12 o'clock, sunshine! Nee munters among them!"

Keegs smiles. "Well hellooo, ladies! Get me the ball, Calla, I want to score!"

Calla nods. "Fuck off, Keegs, dinnit gan spoiling yourself now!"

The ball's kicked off and Keegs storms in with a tackle, the lad lying like he has been run over, Keegs laying the ball out wide and jogging backwards into position looking down. "Trevor Keegan, son! Autographs at the end of the game!"

The lad sits up getting his breath as the ball is skipped across to Willy who beats one player and does a double step over and drags the ball back from the line and spins through two other defenders with an "Arrr! see ya!"

The ball is centred and straight away they are on the attack again, the winger cutting inside who squares it to the centre forwards as Keegs pokes it off his toe and sets away. "I'll have that, chucklevision," clipping it through the middle. Jockey is onto it like a flash, backheeling it to Willy who recieves it and megs the midfielder "40!" as the midfielder comes back he sweeps his foot round and megs him again "60!" as Calla slides in and crunches Willy taking the ball.

Davey is out of the dugout. "The daft cunt's tackling his own men, Jimmy, he's not piss wise."

Jimmy climbing out of the dugout – "think I'm ready now" – smiling over at Michelle then looking down – "ah, for fuck's sake maybe not."

The ball lands at Mexican John's feet, he dummies his opponent and plays it through for Kris who has bombed on. Kris pulls it back only to see the defender punt it up field right over the back 3 and the Frankie's centre forward is onto it. Taking the ball round Earl he slots it home as Keegs slides in stopping it on the line. Standing up he sprays it out wide to Tim. "Fucking class, Keegs! It's like watching Prince Phillipe Albert!!" Davey shouts as he asks Walter how long left.

"20 I'd say, Davey, you need to get some fresh legs on I would say."

Jimmy is taking his tracksuit off as Davey looks down. "Perfect timing. Right, Jimmy, you know the dance right up top! Fuck them, tell Keegs we are going long everytime!! You, Hags and Calla, Willy behind you! Shit or bust!"

Jimmy runs on to the cheer of the crowd. "Keegs, go long every time," as Keegs bangs the ball down field, Jimmy flicking it on as Hags runs onto it. Just before he can get there Calla beats him to it, lobbing the keeper, the ball dropping into the net as Calla runs away taking his shirt off, swinging it round his head. The ref runs up to him taking out the yellow card. "You're fucking kidding me? You have been watching too much TV, Warwick!" Calla protesting as he puts his shirt back on.

"So have you, son," as Warwick walks away. Davey doing the Morecambe and Wise past a furious Frankie. "Chuck, get me a beer, son, I feel a celebration coming on."

Frankie's kick off as Rosie wanders down to the sideline. "Calla, score me another and I'll come back to yours to wash your back after the game."

Calla looking over – "Give me a gobfull of that beer and I'll think about it."

"Gan on, bananaman!" Calla shouts, Willy skipping into the box before dropping a shoulder and running round the goalkeeper with an open goal just about to pull the trigger – only to be 2 footed by one of the Chuckle Brothers, Willy collapsing with a yelp. "Ahh, me ankle!"

Calla runs straight over, getting him by the throat. "You fucking dirty cunt!" the bloke grabbing Callas hand pushing him away. Both teams pushing and shoving each other, Davey on his feet kicks a water bottle into Frankie's dug out. "You bunch of horrible bastards!" Red faced, waving his fists, everyone in the dugout shitting their pants.

Warwick blows his whistle as the Chuckle Brother and Calla are still tangling, people trying to prise them apart, Willy being helped up by Kris and the Mexican waving to Davey it's game over. "Frankie, you tell your man not to walk this way when he gets sent off! I'm telling you that was fucking horrendous – could've broke his fucking leg!" pointing the finger as John and Mally pull him back.

Warwick waves the red card at the Frankie's player and a second yellow at Calla who storms over and pushes Warwick who stumbles over. "You fucking nobhead Warwick!"

Warwick is on his back blowing his whistle out his mouth, his voice shaking. "Off! OFF OFF! Get off!"

Calla throws a handfull of mud at him and storms off after the Frankie's player, barking over his shoulder to Warwick, "Willy bucked your lass!"

The Frankie's player takes off his shirt to some whistles and some boos from the crowd, the bloke around 6'3" and built like a brick house, storming off full of hell.

Davey straight over as he walks past: "You dirty twat!" the bloke not breaking stride, pushing Davey away and throwing his shirt towards the bench. "Fuck off, old man!"

Calla catches up and swipes his ankles, the guy stumbling and turning round. "Come on, you daft cunt, let's go!" Calla throws his shirt on the ground, Frankie going in the middle pushing them apart as Davey pulls the guy's arm, pointing in his face. "That was out of order that tackle! Out of bastard order!"

The bloke waving Davey's finger out of the way – "Don't point at me, old man!" The 2 benches up, all pulling their team mates back. "I'll be seeing you later on! Mark my words!"

The Chuckle Brother laughs and walks towards the dressing room. "Look forward to it!"

Walter shouts over. "Hey you!"

The guy glances up. "Get fucked!" hitting his arm, sending up a fist.

Everything is calming down as Calla sits on the bench next to Davey. "I'm gonna kill that cunt!"

"You will have to bring the bastard back to life then, coz I'm going to kill him first!" Davey foaming and shaking with rage as the game resumes. Davey orders the Mexican to drop the bucket and get on. "Up front, son! Keep it simple, good lad!"

Keegs steps up to the penalty and smashes it straight down the middle. Davey leaning out the dugout – "Get yourself back to the restaurant for that trophy!" Frankie leans back grunting.

The ball seems to go in slow motion as it rips down the back of the net –"and that, my son, is 5-3!" Hags shouting as Willy jumps on him. "Get in, you old bastard," as Davey does the moonwalk past Frankie – "Pick it out, frankyo."

Frankie's unable to keep the ball as the heads have dropped, as Keegs goes 50/50 with the centre half, winning the ball and sliding it through to the Mexican who squares it to Hags who taps it in, all the lads hugging as the crowd cheer. "Haven't lost it!" winking to Keegs, Davey leaning back. "This cunt's give me heartburn, pass a bottle" taking a bottle from Mally who is shaking his head. "So much for a friendly!"

Davey lets out a burp. "Who said it would be friendly like, Mally, you soft shite," pointing his bottle towards the pitch where Jimmy has flown in with a wild tackle catching the ankles of the guy who had already played the ball. Warwick running over. "Yellow card, Jimmy, come on!"

Jimmy is running backwards. "I got there as fast as I could, ref!"

Nelly's now just keeping the ball with easy passing, Frankie's men lightly jogging showing little interest knowing with 5 minutes to go they are well beaten. Frankie is out of the dugout. "Fucking come on! It's not over yet!"

"Sit down, man, you idiot, you've been beaten like a ginger stepchild!" Calla sits sipping his beer as Frankie walks back to the dugout, shaking his head, the ball slipped through to Jimmy who taps it into the the goal, clearly ecstatic, running towards the corner flag pretending to ride it like a horse as the lads all pat him on the back and wander back into position. Jimmy clutching his groin wanders off the pitch as his cock is now hard again.

"Or for fuck sake!" Davey standing up cups his hands to his mouth. "HAWEH, JIMMY, YA SILLY SOD! WE ONLY HAVE 9!"

Jimmy trying to force his cock down, waves back as the game is re-started and Frankie's men string together a move that catches Earl napping, scoring a goal, Keegs unable to get back in time. "Haway, Earl, man!"

Earl shakes his head, stubbing the spliff out on the post. "Hey, sorry y'all!"

Frankie is screaming, "Come on!"

Davey shakes his head. "Are you for real?"

Frankie unzips his top. "Ref, sub!"

Davey pisses himself. "You're going on?"

Frankie nods as Warwick waves him on, Frankie stood waiting for the off doing a series of stretches, twists and jumping up and down on the spot. The ball is kicked and Warwick blows for full time. Frankie looks over. "Ref, there's got to be time to add on?"

Warwick is monkey sick of the shit he has put up with –"90 minutes is long enough, Frankie! You were never gonna win! Come on!"

Frankie walks towards him, still arguing. Davey and the rest are bad with laughter as all the players troop off the pitch, no-one shaking hands, Davey hugging each of his players as they walk off, congratulating them as Frankie walks past. "Good game, Frankie!" rocking back with laughter as Frankie sulks past, the last in, Davey laughing. "Did you tap that cunt in with your cock or what?"

Jimmy is still shuffling his rod. "Bastard won't pissing go down!" Davey holds out a hand. "Well I'll shake your hand – you're not getting a hug!" the two walking off towards the changing room.

All the lads are singing in the chainging room as Davey walks in. "Fucking excellent show, lads! I'm proud of you all!" rubbing the young Mexican's head. "Willy son, how is your ankle? The dirty bastard!"

Willy takes a long puff on his tab and squints as the smoke fills his glasses. "Like a fucking puddin, Davey! Need mesel some lager to tak the swelling down!"

Davey looks at his ankle, which is twice its normal size as Warwick sticks his head in. "Well done, boys! Calla, I'm sorry but you lifted your hands and you were on a yellow!"

Calla nods. "Ay Warwick, ya areet! Sorry for pushing you!"

Warwick smiles. "And the remark about my girlfriend?"

Calla looking up – "Ay, you're reet!" looking across to Willy – "Sorry for dropping you in the shite!"

Willy holding up a thumb, Warwick's face straightening and he is about to speak when Keegs pushes his head out, slamming the door. "Ya don't miss a slice out of a cut loaf, Warwick, son! Now piss off –

players only!"

The lads are continuing their celebrations as Tom and Chuck wander in with a cool box. "Well done, boys! Yo dude, cut it out!" Tom nodding to Calla who is wiggling side to side letting his knob slap on either leg. Davey opens the coolbox and throws the cans to the lads. "Ay everyone did us proud! Tomcat, how's your nose!"

Tomcat leans forward with 2 black eyes beginning to show. "I'll live!"

Calla grabbing big Earl – "I reckon big Earl is man of the match! Dare say he kept a good few goals out!" Earl leans back, smiling, eyes glossy still slightly stoned. "Thanks man!"

Davey looks around. "Well I'd say Keegs changed the game and set the tempo! But ay, Earl did us reet for a quater back!"

Keegs nodding as the rest of the team cheer Earl's name. There is a knock at the door as Davey opens it. "What do you want ya cunt?"

Frankie steps in with his hand out. "Well done, Davey – no hard feelings!"

Davey leans back contemplating shaking his hand then shoving Frankie out the door, Frankie with his hand still outstretched. "Ay, nee hard feelings, now fuck off and get the trophy!" slamming the door but re-opening it. Frankie's still stood there. "And tell that daft cunt who nearly broke Willy's foot I want to see him!" slamming the door again.

All the lads are getting ready."Right, back to the digs, quick cobble and out!" Davey giving the orders.

"Yeah guys, see you there." The American lads head out followed by the Mexican lads.

"Haweh Willy, get up son, I'll help you with your ankle." Calla and Keegs help him up as the lads head out to see another big fracas by the cars. "Jesus, what now?"

Davey and Jimmy walk over, pushing the crowd to one side, seeing the big Chuckle Brother lying on his back coming round from an apparent punch. Frankie turns to Davey. "Think you need to have a word with your players." Pointing at Earl.

Davey looking at the size of the lump swelling on the lad's cheek – "Ay, will do. I'll thank him for saving me a job!"

Earl is leaning against the van as the lads walk over. "Nice one, Earl."

Earl winks. "My pleasure, Davey."

Jimmy gets in his car. "I'll see you in the bar, lads" shutting the door and spinning off before anyone could say anything

"Queer cunt, isn't he? Why the sharp exit, like?" Calla climbs in beside Davey.

"He's got to see a man about a dog."

Calla shrugs. "I've got to see a bit blurta about her blurt so get a fucking wriggle on, fatha."

Kris is squirming in the back. "What's up with you, sadsack?" Willy nudging him.

"Me gonads are on fire, man."

Calla laughs. "That'll be the ralgex I put in your undies, ya balloon."

All the lads are singing we are the champions as the bus turns round in the carpark, Hags pressing his big hairy arse against the window as it passes Frankie's bus, half their team on board. "Get it up, ya wop bastards!" Hags shouts, opening and closing his arse cheeks with his hands!

Davey pulls the bus to a halt, one of Frankie's lads laughing. "His arse reminds me of a 3D film – it almost jumps out of the window it's so motherfucking ugly," as the sliding door opens. "You like 3D? Well try this motherfucking hotdog – does it feel almost real?" as Chuck splats a full hotdog into the bloke's face.

"Drive on Davey," Willy shouts as the bus pulls away, all the lads singing 'we are the champions' again.

***

Davey is on the phone to Marj as he walks in Calla's bedroom, Calla standing there balls naked as Davey doesn't batter an eyelid. "Here, son, tell our lass about today."

Calla covers his bits up with a shoe as he takes the phone. "Hiya Marj, did he say? Aye he's been hailed the new Bobby Robson! Is he? Ok I'll look after him! Ta ra," Calla handing Davey the phone back as he walks out, shutting the door.

Hags is on the sofa having a tab when Willy hobbles out of his room. "Ey hen, dee me a favour and bandage up my ankle!"

Hags flicking his cigerette ash into the cup on the table squints at Willy's foot. "You need to get that checked oot, boy!"

Willy hops to the sofa and flops down waving the bandage. "Just get it strapped and I'll get it looked at the morn."

Hags stabs out the cigarette and stands up and wanders into the kitchen, soaking a tea towel and emptying the ice tray into it, Willy turning around. "Howw, fill that back up, man, we may fancy a few voddies when we get in!"

Hags shakes his head and fills the ice tray up. "In fact, hen, pour me one now while you're up!" Hags salutes Willy and splashes the vodka into two glasses and walks back. Handing Willy the two vodkas and a can of coke, he bends over lifting Willy's foot onto the table.

"Owww, ya cunt!"

"I'm being as careful as I can, you soft shite," pressing the towel full of ice onto his ankle.

Willy leans forward. "Or man, get me some ice out of there!"

"For fuck sake, Willy man, your ankle is more important!"

"Isn't!" Willy snaps, taking some of the ice cubes from Hags and dropping them in his glass and leaning back taking a big gulp, offering the other one to Hags who is sitting back on the sofa lighting another tab, still looking at Willy's foot. "I bet something's broke!" shaking his head.

"No man, I've dodgy ankles – they always come up like that! And it's still sprained from Miami! Hags, can I borrow a shirt!"

Shaking his head – "Ay lad, which one?"

Willy thumbs his glasses up. "I like that white Armani one you have!"

"No way! It will be returned like an Afghan soldier's! You can borrow the black or the navy one!"

Willy pondering – "Reet, the navy one, but I'll need them dark replay jeans and your tan belt, or and them tan loafers you have!"

"Fucking stroll on – you need some undies and socks as well?"

Willy puts the half empy glass down, rubbing his hands. "Fucking champion! I wore my last clean pair the day and they have a grass stain up one side!"

Hags stands up and wanders to his room. "Some cunt, William McLaughlin! Some cunt!"

Kris is in the shower as the doorbell goes, Mark answering it to a delivery guy. "Hello sir, large pepperoni pizza with green peppers and a portion of fries? 18 bucks!"

Mark hands over a 20. "Keep it!" kicking the door shut with his foot, a muffled shout from the bathroom – "Is that the scran?"

Mark is already on the sofa shovelling a slice in. "Ay!"

"Don't eat it all, I'm starving!" the muffled shout returns. Mark is well on, lifting another slice up as Kris carries on singing Wet Wet Wet's Wishing I was lucky.

Calla wanders out of his room pulling his belt through his belt holes while trying to text. Davey, already dressed, looks over. "See you've made an effort, son, that's a canny shart!"

Calla half turns. "Ay D&G, forgot I fetched it, only found it there at the bottom of my case!"

Davey stands up. "Gist!"

Calla stops. "Eh!"

Davey holds out his hand. "Haweh son, you can't go out with it like that – it needs a warm iron owa it! Give me it, al dee it!"

Calla wriggles out of the shirt and hands it to Davey then wanders to the fridge as Davey walks across and switches the iron on. "Tin like, Davey?"

Looking up – "Is my arsehole rotten?" Calla frowns. "Yes! Leave it on the bench. I'll get it in a minute!"

Calla's arm is already swinging – "Catch!"

Davey, straightening out the sleeve, ducks slightly and catches the can, which was heading for his face. "You silly sod!" tapping the top and bottom of the can before opening it, the beer still fizzing. Davey slaps his lips around it, Calla laughing and sitting down. "Y'nar, Davey, I'm not sure about this Shanghai trip!"

Davey steaming the shirt sleeve – "Ay son! What's rang like!"

Calla all serious – "Well them Chinese lasses, man! I'm not sure they dee owt for me!"

Rolling his eyes – "Calla, son, you would fuck a frog if you could stop it hopping!"

Calla sniggers and pulls on his socks.

"Howw, man, they are mine! Davey snaps while turning the shirt over, Will you give owa wearing my candys!" Calla pulls on the second one as Davey catches a glimpse of Calla's scabby feet. "Have you athlete's foot?" Davey quizzes a smug Calla.

"Wayaye, did you not see me out there the day?"

"The infection, dopey hole!"

Calla taking a sock off and inspecting his foot – "Or aye, looks like it.

"Arr well, you can keep them then son! I'm not catching that bastard!"

Calla pulling the sock back on – "You can't catch athlete's foot, dipshit! You get it from not drying your feet properly!"

Davey holding out the shirt and switching off the iron – "Course you can!"

Calla pulling the shirt on wanders into his room already on the internet with his phone checking the facts on athletes foot. A few minutes later he walks back past Davey who is having a tab in the kitchen. "What does Google know!" Davey laughs drinking the last of his can.

Willy and Hags are all showered and ready with Hags knelt down strapping Willy's foot up. "It's fucking massive this, man!"

"Oww, ya cunt!" Willy yelps. "Just wrap it up and hurry up, you rough cunt!"

Hags sticks the pin into the finished bandage – "Done! How you gonna get a shoe on that?"

Willy bends owa and pulls a sock on, then forces Hags' loafers on, smashing down the back of it. Hags grabs his arm. "Howw man, they are Hugo Boss!"

Willy finally gets his foot in. "Wey, now they are Hugo Bosst!" Letting out a loud laugh, standing up. "See, fucking champion!" looking in the mirror adjusting his hair. "Hags, a reckon I look better than you in this gear anyway! Think I'm deffo shagging the neet!"

Kris finishes off drying himself and wraps the towel round his waist and wanders out to see Mark pushing the last slice into his mouth. "You greedy little cunt! You have eaten a large pizza?"

Mark throwing half the slice back into the box clearly stuffed – "Fuck! Soz Kris, but I was starving and it was lush!"

"And what am I gonna eat?"

Mark holding his finger out – "I haven't touched the chips!"

A red faced Kris lurches forward grabbing Mark by the neck as the pair wrestle onto the floor from the sofa both towels falling off as Jonny walks in. "Am I interrupting something here, lads?"

The two stop and clutch their towels. "This greedy little cunt has scranned all the pizza!"

Mark stands up and heads for his room. "I left the chips!"

"Hurry up the pair of you, man, will ya!" Jonny barks, Kris picking the box of chips up, wandering into his room, Jonny helping himself to a can from the fridge.

Tom, Chuck, Hags and Willy are all at Davey's door when Kris, Jonny and Mark troop along. "We waiting for Calla?"

Willy nods as Calla barges out of the door flicking him on the nose. "Reeto ladies, let's rock!" striding off towards Earl's gatehouse leaving the group and Davey locking the door.

Earl is standing at the gate with a grey suit on and a white shirt. "Where you going, bigman – to a christening?"

Earl laughs. "I don't get out much so I like to make an effort, Calla, you should try it."

Calla laughs. "I get out too much so I don't have to, hamma."

Nelly's is packed as the lads troop in to see Frankie at the bar with a big bottle of champagne and a huge trophy. Walking over to Davey he hands them both over – "Here you go, Davey, well deserved, a famous win for Nelly's."

Davey takes the cup as Calla takes the champagne and shakes it as Frankie tries to intervene –"that's a 1980..."

Bang! as the cork goes and Calla sprays everyone, eventually getting his mouth around the bottle as Davey grabs it and pours it into the cup. "Give me it here, you fucking balloon."

The lads push their way through to see a roped off section in the lounge with buffet tables and champagne on every table. Jimmy and

Mally are sitting there with Brad and Walter. "Come through lads, the bar have put a spread on for the champions." All the lads walk through, shaking hands.

"Drinks are on the house until 9pm, boys." Brad stands up taking the cup off Davey and filling it up some more. "Back where it belongs! Pops will be down soon for a drink"

Calla smiles. "Aye, with my fucking winnings. What time's the race, Walter?"

Walter stands up taking the cup off Mark, slurping the champagne. "8 o'clock Calla. We will watch it on the screen above the bar – look."

Calla looks at the racing already on. "We will that, Walter, I'll be riding the bastard home, I hope."

Davey walks over and sits down with his mouth full of bread, crumbs all down his shirt. "Is the buffet open yet?" spraying Jimmy with crumbs. "It fucking looks like it, Davey, fucking stroll on."

Willy hobbles in and sits down on a bar stool with Hags, Earl and Keegs following suit. "What time's them slappers coming in, Calla?"

Calla looks at his watch. "About 8 o'clock I think, you four eyed git."

Willy takes his pre-poured vodka off the bar. "Who got this?" as he takes a big drink.

"On the house until 9 o'clock, courtesy of the owner," Brad shouts over holding his glass up to Willy.

"Is there a man of the match award?" Keegs taking his pint off the bar.

"Well if there was, son you would get it," Davey shouting over as Calla disagrees – "Fuck off man Davey, you talk some shit. Only been managing five bastard minutes and you think you're Alex jock-faced Ferguson."

Keegs handing the drinks to Chuck and Tom – "Who would get it then?"

Hags laughs adding to Calla's argument and continuing: "Firstly, I am very fucking surprised Davey never give it to Tom or Chuck for feeding him hotdogs for 80 minutes! Secondly I thought Willy had a very good game and he would get it."

Calla shakes his head. “Aye well you would, being jockanese like.”

Frankie walks into the area as all the lads boo, Frankie laughing. “C’mon boys, no hard feelings, it’s only a game.”

Willy points at his ankle. “Aye maybe, but this could be broke and I could be suing that fat Terry McDermott-faced cunt tomorrow.”

Frankie puts his arm around Willy and faces him to the bar. “William, I know this was very bad.”

Willy lifts his jeans up. “I am off work now, fucked! So tell the Chucklebrother I think he’s a cunt.”

Frankie puts a crisp wedge under Willy’s hand without anyone seeing. “Marco is my brother; he doesn’t think sometimes. Please take this as an apology and I will make sure the bar stays free until closing, but please relax about things.”

Willy flicks through the 100 dollar bills, turning round with his arm around Frankie. “Lads, let’s join forces tonight and get pissed. Frankie is right – it’s only a game!! Let’s raise our glasses to Frankie’s team.”

All the lads raise their glasses as Frankie picks up a glass. “The bar bill is on me, gentlemen,” as an even bigger cheer goes up.

“Go on, Frankie! Some man.” Davey salutes him.

“He’s a cocksucker.” Davey is surprised as Walter leans back and drinks his Jack and coke.

“What was that Walter?” Davey leaning in.

“He’s a cocksucker, he sucks dick.”

Davey laughing. “He can suck what the fuck he likes as long as he’s paying for my jungle juice, youngn.”

All the lads are in the lounge end when Sharon and her four pals all walk in to the greeting of wolf whistles and “git up” from the lads. Calla wanders over. “Well you scrub up canny Sharon. I’m Andrew,” offering his hand to the other girls as Rosie gives Calla a stare.

The lads introduce themselves to the girls as Michelle walks in. “Well done, boys, you were brilliant.” All the lads saluting her looking stunning as she sits down next to Hags sending Jimmy a wink.

“You’re some cunt Hunter.” Davey nudges Jimmy.

“Aye, Davey, couldn’t believe my luck, all this time she was mad for a bit of Newcastle sarsage.”

Davey laughs. "Bet you gave her the best five minutes she ever had!"

Jimmy laughs. "She'll be getting another five minutes later as well. Say nowt though – don't want the lads finding out."

Davey shouts over to Calla. "Here are, son, here's the race."

Calla, mid-conversation, turns away from the girls. "Rosie, turn this up, will ya?"

Rosie slides the remote along the bar. "Turn it up yourself, Romeo."

Calla looking at her then Willy – "Haweh!"

Walter joins Calla and nudges him. "I hear you have went big on Bonustime Boys, son, good chance although you should've maybe went each-way! Castlerock is a tough favourite."

Calla nods. "I know, that cunt has 500 spuds on him."

Keegs winking at Calla – "Number one Calla, just so you know."

Calla looking at the telly – "Number seven just so you know who we have as well, fuckface."

Brad wanders over. "So what's the split, Calla, for your two thousand?"

Calla taking a drink of his pint – "Grand for me and Willy, five hundred for Tom, hundred for Davey and Jimmy and kris and Mark have hundred and fifty a piece on."

Keegs laughs. "You have to catch the favourite first, daft arse."

All the horses are loading up when Sharon's sister pipes up. "I used to ride horses."

Willy dropping his glasses down his nose responds, "Play your cards right hen and I might let you ride this stallion round the block later."

Calla spitting his drink out – "A fucking shetland pony more like, ya daft cunt."

Davey tells the lads to be quiet as the stalls open and all horses bolt out. "Fucking die, ya bastard." Calla is first to shout as Keegs' horse goes straight to the front by about three lengths.

"Watch and learn, Calla son, this horse is a fucking belter! Unbeaten."

Willy gives Keegs a dig. "It's not even into the second furlong yet man" as the bar is quiet except for the commentary on the TV. The lads start to get a bit rowdy, Castlerock going even further clear with about seven furlongs to go. "Go on, my son!" Keegs cheering for his horse as they go round the corner into the final three furlongs.

Walter gives Calla a nudge. "Vesqeuz will push the button two out as he stays further than this."

Calla looks at Walter. "How the fuck do you know, thought it was a tip?"

Walter winks at him as Castlerock goes over two lengths clear as they hit two out. "Go on then, you fucking Mexican bastard, hit the cunt!" Calla putting his pint down makes a whipping imitation towards the telly as the bright yellow jacket of Bonustime Boys is switched outside. Passing four in front it hits the final furlong half a length behind Castlerock whose jockey is at full force trying to get as much out of the favourite as possible. Willy jumps on Calla's back, Calla grabbing his legs as Willy rides Calla. "Hit the bastard, fucking hit the bastard."

Davey and Jimmy laugh. "Them two aren't piss wise."

Jimmy agrees. "I would love it to win for the crack though Davey."

Davey is glued to the telly. "Fuck the crack, we have a fortune coming back man, if it wins. It's coming, son, it's coming."

Both horses flash past the post on a head bobber. "Get in, you fucking bastard." Keegs punching the air as the TV calls a photograph. "Hold on there, bud, think it's closer than you think." Tom leans over to Keegs as Walter walks out of the door.

"Where's he gone?" Willy climbs down off Calla.

"Probably gone because his tip didn't win." Keegs rubs his hands together as the photo is up on the TV.

"I think the one closest won," Sharon is telling Calla as he turns his back on the telly.

"Fuck it, if it has then the drinks are on me; if it hasn't the drinks are on you, Keegs."

Davey wanders up to the bar whispering in Calla's ear, "That's won by the way."

Calla shakes his head. "I don't know, fatha, looked close."

Davey smiles. "Walter was gone like a flash, man, he's not daft."

The commentator comes over – "We have a result at Arlington in the eight o'clock race, a real head bobber but Castlerock has been beaten for the first time this season by a 16-1 shot, Bonustime Boys!" the screen shot showing the closest of finishes as Calla picks Willy up. "The fucking milky bars are on us the neet, hamma!!"

All the lads are going mental – "Champagne all round, Rosie."

"It's free anyway, Calla, you dooshbag," Rosie shouts over as the music comes on: Queen – We are the champions. The lads belt the chorus out as Frankie joins in with his arms round Calla and Davey, Keegs shaking his head. "Fucking robbed!"

The door opens and Walter comes in with a big smile, followed by Brad's old man. "Well done, boys. A good day all round – congratulations," shaking everyone's hand as Brad introduces him to Calla. "Pops, this is Calla; Calla, this is my old man Bob"

The well dressed bloke shakes hands with Calla. "Hope we haven't put you out of business, Bob." Calla smiles.

"No son, I had that bet laid off to cover more than my ass."

Calla takes a drink. "That's good to hear, when do I see my green?"

Bob taking a glass of champagne off the bar – "I'll drop it in to you tomorrow, don't worry, it's safe."

Walter joing the conversation – "He's good for it, Calla, don't worry about that."

"And where did you head off to before the result?"

Walter laughs. "Went to phone my man at the track."

Keegs leaning in to the circle – "Which man, the bastard who took the photo?"

Walter laughing – "No, son, the trainer, he's my horse you see."

Keegs stepping back – "Fucking bastard, I wish I'd knew that!!"

"So how much did you win, Andrew?" Sharon asks Calla.

"Ah a few dollars, nothing major – maybe about 34 thousand of the bastards."

Sharon looks shocked. "Well done! Where are you taking me tonight!" as Willy whispers into Calla's ear, "Up the arse, more than likely."

Calla laughs. "Anywhere you like, love."

Mark and Kris are getting on famously with the girls, while Calla and Willy plot their escape. “Rosie has the right hump, Willy. We are going to have to go into town with these in a bit before she stabs me.”

Willy winks as Calla joins the girls. “Leave it to me, bigman.”

Willy nudging Brad – “You fancy taking us into town, young Brad?”

Brad getting off his stool – “No problem. I’ll get the wheels around to the door – there in 10.”

Willy taps Hags on the shoulder. “Come on, Hags, we are headed into town; Calla,– do you fancy it?”

Calla smiles. “Aye Willy, ok, we’ll finish these drinks and head in.”

Frankie offers the lads a lift. “No thanks, Frank, we have took enough off you today, the trophy, your pride and your champagne,” Calla ribbing the Italian.

“We are staying here, lad,” Davey shouting over, him, Mally and Jimmy happy with the free drink.

“Suit yourselfs, don’t wait up!” Calla drinks his pint as Willy puts his arm around the girls. “Are you girls joining us for a few in town?”

Sharon smiles. “Of course. We have a car so we will follow you, if that’s ok.”

Tom and Chuck are sitting at the end of the bar eating ribs, covered in the sauce. “Count us out, we are happy eating pig.”

Calla shaking his head – “Davey mustn’t’ve seen them otherwise you would be eating fuck all by the way.”

Michelle walks over to where Davey and the lads are sitting. “I’ll join you guys. The boys are heading off.”

Davey shakes his head, standing up. “Aye, bonny lass, you can sit there. I’m cutting another tooth,” as he wanders off leaving her looking confused.

“See you later, Rosie.” Calla waves at Rosie as she smiles and gives him the finger. “Eh, bit harsh.”

Calla looks at Willy who is bad laughing. “Harsh but fair, bigman. There will be no cheap drinks for you anymore.”

Calla shrugs. “So fuck, I only have one more session in here tomorrow and that’s it! Plus it’s packy bags neet tomorrow and a few bob off Bob, yee haa!” All the lads climb in Brad’s motor as the door slams and the wheels spin off towards the bright lights.

Brad drives the big suburban downtown as he passes through a traffic light, the bright flashing lights of a police car speeds up behind him. "Ah fuck." Brad putting his signal on and slowing down.

"Will I dump this bag of charlie or what?" Calla nudges Brad.

"Oh fuck, Calla, please tell me you haven't got any drugs, man."

Willy laughing in the back – "He better not have, the cunt, or I'll be having words for a sniff of the bastard."

Mark looks over his shoulder. "Here he comes, deputy dog."

Brad shaking his head – "You have to sit tight until the cocksuckers get up to you. Please guys, no funny shit."

The window is tapped by the officer as Brad lets it down and hands his licence over as the cop shines his torch in. "Some kind of roadtrip, boys?"

Calla leans over. "What's the problem, pal?"

The officer shines the torch at Calla. "Breaking a red light at 43mph."

Brad shakes his head. "It was amber, sir."

The cop is having none of it. "Step out of the vehicle, son. Have you been drinking?"

Brad shakes his head even though he has had 2 bottles. "Walk the line, then blow into this please." All the lads looking as the cop looks at the machine. "It's borderline. I am going to have to ask you to come to the station. Is there anyone else that could drive the vehicle?" as Calla gets out – "I can drive."

The cop looks at him. "Ok, blow into the machine," as the machine light goes red straight away. "You are 2 times over the limit!" The cop is not very happy as Calla looks at him.

"You asked if I could drive, not if I was pissed, man." All the lads laughing as Calla climbs back in. "Wey, he fucking did, the balloon."

Kris laughs. "You're the balloon, you balloon."

Just then Brad's old man pulls up in front and walks over to the cop; they shake hands and walk off. Standing laughing for about 5 minutes as the cop walks back and gives Brad the ok to jump back in the motor. "No more drink, Bradley. Didn't realise when I seen your surname you were Bob's boy."

Brad smiles as his dad sticks his head in. "Always hauling your ass out of trouble, boy. Meet me down at Roxy's – I'll square a tab off and I'll be in the usual spot. See you and the boys there."

Brad puts on his belt and sighs as Calla laughs. "Fucking lucky there. I thought the plod asked you for a line not to walk the line! I was thinking it's like bein' back in Redcar and nearly give him one."

Keegs pulls up alongside the lads at the traffic lights in the girls' car. "Alreety, we parked up back there laughing at you daft cunts getting pulled by the rozzers. What did they say?"

Calla leaning his head out of the window – "They said they were on the look out for a tight ugly jock twat! Don't worry tho Keegs we said we hadn't seen you." Leaning back laughing as Keegs flicks Calla the finger.

In the door of another swanky looking bar full of decent clientele and trendy music as the lads troop in. "Fucking areet in here, how," Kris shouts as him and Mark make their way to the bar as Brad moves them away. "My dad normally has the corner section for himself and my stepmother and her friends on a Thursday" as the lads walk towards another roped off area.

"Fuck me, we have been in more VIPs and drank more free drink this trip than George Best." Calla marches over nudging Willy. "Look at the fucking blurt in here, by the way."

Willy laughing – "Aye, bigman, and you are on the arm once Sharon gets in."

Calla scowling – "Ah, fucking bollocks, I'll get rid if I see owt any better."

"That'll not be hard – look at Brad's stepmother. Fuck me pink!" Willy pointing as they walk in and are introduced by Brad's old man to a 6 foot blonde stunner.

"Boys, this is my wife Kim – she is from Norway."

All the lads introduce themselves. "I worked in Bergen – do you know it?" Willy speaking like a robot to the girl as Calla laughs. "She's from Norway not the moon, you daft jock cunt," Calla nudging him as everyone laughs, and takes a drink off the waiter's tray just as Keegs and the three girls walk into the bar.Keegs nodding at Kim and shaking Bob's hand.

"Yes, I know Bergen; however, I've lived in the US for 18 years!"

Kim replies.

"All your life then!" Willy smiles, nudging Calla.

Kim laughs. "Thanks! Work transferred me here for 6 months and I've been here since!"

Willy leaning on the bar – "And what line of work are you in?"

"I'm a lawyer."

"She has a brain, hammer, that's you fucked!" Calla whispering in Willy's ear.

Willy leaning forward – "I considered studying law, perhaps contractual law!"

Calla bursts out laughing. "Only thing you know about the law, ya geggy bastard, is how to break it!"

Willy nudges Calla as Kim laughs and leans towards Bob. "Can you get me a mojito, darling. I'm off to powder my nose!" Bob not even looking up as he he lights his cigar and waves out the match.

Willy smiles as Kim walks past. "Hear that, Cal? She is off to powder her nose! I bet she's eyebrow deep in the Columbian dancing dust! Ooo I would love to knock one in her!"

Calla leaning with both elbows on the bar – "William McSpaz breath, allow me to highlight a few key points: firstly she has eyeballs and secondly sense!"

Willy nudging Calla – "Fuck off man!"

Calla sniggering – "No, seriously, she is married plus if she wanted something little and ugly I'm sure Donald Trump over there would buy her a chihuahua!"

Willy necking his pint and wiping his mouth – "I'm not trying to pull her – I'm just being nice!"

Hags rolls his eyes and wanders towards Kris and Jonny, leaving Calla shaking his head, stepping forward, putting his arm around Sharon. "Here she is! Hello beautiful! What you drinking?" ushering her towards the bar. Willy on his own lifts his next drink towards Bob who is puffing on his cigar looking back without reaction.

Back at Nelly's the old hands are all laughing. "I bet silly bollocks loses our winnings to Brad's father in a coin toss!"

Davey putting his head in hands – "A wadn't be surprised, Jim!"

Tom, wiping his mouth and licking his fingers, wanders over to the booth. "You boys wanna shoot some pool?"

Davey stands up. "Haweh then Tom lad, I've no shame taking your money off you!"

The pair wander to the table, Mally looking at Michelle's hand resting on Jimmy's lap. "Something I should know about you two?" Michelle whipping her hand back, blushing.

Jimmy swigging the last of his beer off – "Ay Mally, we are leaving coz your crack is shite!" Jimmy easing Michelle out and waving to Davey who winks back exaggerating the chalking of his cue.

"Speak to you all tomorrow! Haweh, Michelle, you can drop me off – you look dog tired!" as Michelle waves and the pair disappear out of the bar.

Mally sits back. "Well isn't that bastard charming!"

Tomcat swigging the last of his beer – "Fair shout, Mally! Your crack is shit!" Standing up he wanders out of the bar waving to the lads.

Davey nudges Tom. "Look at skinny bollocks there sat on his own!"

Tom laughs. "He isn't! Chuck's there!" pointing to Chuck who is sat staring at the wall oblivious to Mally talking to him.

"Chuck's all picture and no sound there, Davey! He has been switched off a while!"

Frank walks over. "Say boys, what about we round a few up and head into town – show the young bucks we can still swing!"

Davey nodding towards a few of his lads and the Chuckle Brother with the shiner – "I'd say that would be the perfect way to get everyone locked up! I'm staying put!" Leaning over and smashing the balls – "Say, where did big Earl go?"

Big Earl is at the end of the bar getting his ear roasted off Rosie. "He went into town with them fly ins!"

Earl shakes his head. "He is entitled to after the big game celebrations – give the guy a break – and anyway they head off tomorrow I think."

Rosie's jaw drops as Earl realises he's dropped Calla in the shite. "The dirty rotten bastard told me he was here for 2 years, back after 2 weeks in the UK, he said to me a few weeks back!"

Earl smiling shakes his head as she storms over to get her phone.

"Hey Davey, I think Calla told Rosie he is here for 2 years – she doesn't look happy."

Davey taking his shot and laughing – "She'll not be the first bit skirt that's heard that off him, Earl."

Earl takes a big drink of his Jack and coke. "Well tell him to watch out, Davey and make sure you call into see me tomorrow before you leave."

Davey smiles. "Will do, Earl son, I'll be heading to the cot myself in half an hour – got to pack the bags for blighty!"

Calla is doing shots with the girls as Bob walks over. "Hey, Calla, do you have a minute?"

Calla turns around. "If it's about the thousands of dollars you owe me I will have it in large notes."

Joe shakes his head. "It's not about that – that is safe and sound. I was wondering if you would mind keeping an eye on the missus for an hour. I am heading home and Brad over there looks to be getting wasted! Maybe she could join the girls here?"

Calla shrugs. "Aye, send her over. What time will I see you tomorrow?"

Bob looks at his watch. "Say 12 o'clock in Nelly's?"

Calla rubs his hands together. "Sounds like a plan Bob, don't be late mind!" as Bob wanders off to speak to his missus.

Mark and Kris are drinking with Brad. "Look at Calla surrounded by lasses – he's some gobshite." Kris racking another round up. "Aye, roll on tomorrow though. Did Davey say anything about a plan? We haven't got tickets yet?"

Mark shrugs his shoulders lobbing a bit of ice off the back of Calla's head then signalling him over.

"Arite sadsacks, what d'ya want?" Calla signalling the barman

"Have we tickets for tomorrow or what?"

Calla shrugs. "Four tequilas, mate. Er last I heard we fly at 7 o'clock, leave Nelly's around 3."

Kris lifts his drink. "Perfect: lie in until 2, shit, shave and shower, pack the bags and yemski."

Mark and Kris are both leaning round the back of Calla. "Willy, Keegs and Hags are like flies round shite. We'll wait until they buy all the drinks then crash and burn, then we'll come over."

Calla laughing – "Good lads, I like your plymouth argyle," Calla knocking his tequila back, walking back over to the gang squeezing Sharon's arse.

"He's one mad motherfucker." Brad swallowing his drink.

"Yep, he'll probably try and buck your stepmother later as well."

Brad laughs and shrugs his shoulders. "He's welcome to her. The old man is probably away screwing her best friend as we speak."

The lads are both shocked as Willy hobbles past. "Howw, ya bunch of fannies, you won't get your Nat King Cole standing at the bar like a set of plums talking shite."

Kris laughing – "You wont get your hole, full stop, ya mong."

Willy arrives back after the toilet into the crowd as Calla is looking at his phone. "You're some cunt'!? Bit harsh," as he shows Willy the text off Rosie.

Willy laughs. "She isn't as stupid as she is cabbage looking!"

Calla shakes his head. "Or wey, doesn't look like I'm bucking that now!" flicking the phone shut, Brad's stepmom cocking her ears. "Bucking?"

Calla putting his arm around her – "I said looking! Looking AT that!" winking to Willy who is pretending to lick one of the girl's ears behind her back.

Kim sniggers. "You guys are crazy and Andrew, my glass is empty!" shaking her glass at Calla.

"Now then, you on a mission, Kim?"

She hands Calla the glass and sighs. "Well my beloved husband is probably off screwing one of his secretaries so I may aswel have some fun!"

Shocked – "I'm sure he isn't doing that!" Calla glances across at Willy who is slowly sliding over.

"Ay Kim, bastards, men, the lot of them! I bet he is doing just that! You see, if you were on my arm, my eyes and attention would be nowhere but with you!" putting his arm around a blushing Kim who pecks him on the cheek. "How sweet, William."

Calla walks off to the bar gesturing his fingers down his throat to Willy who winks back and gives Calla the sly thumbs up.

Brad stumbling around with another tequila – "Calla, my man!" giving Calla a big hug.

Calla easing him off – "Haway man, Brad, ya pissed cunt!"

Brad throws the tequila down and stots the lime off the barman's shoe. "What's Willy saying to Kim the slut!"

Calla spinning – "Haweh lad, she's your step mother!"

Brad leaning on his elbows staring across at her – "She's a gold digger, Cal! I tell you, she is with my pa for nothing but his money! Look at her, Chanel this, Gucci that! He even paid for her tit job and she had one of them designer vaginas put in!"

Calla laughs back spraying his tequila. "A designer fanny put in! Haha, it's not a fucking fireplace!"

Brad nudging Calla – "Man, I bet it was like a ripped out fireplace before the op!" He leans forward and spews all over the floor.

Calla leaps back. "Or haweh, ya dirty bastard, man! These are me new glorias!"

The doorman wanders over placing a hand on Brad's shoulder, Brad squaring up. "Hands off!" the doorman stepping back. "Ok Brad, now come on, time to go!"

Brad jerks his head forward towards the doorman. "Leave when I say so, asshole!" clicking his fingers at the barman. "Yo, yo, clean this up!"

The doorman looks at Calla who shrugs. "Cut it, though, Brad, we have other people in here!"

"Get fucked!" waving the doorman off. "Yo barman, more tequilas ASAP!" winking at a frowning Calla who picks his drinks up and wanders back to Willy. "Here Kim, Willy," handing them the drinks.

"What's going on over there?" Willy takes his drink.

"Too many shandies, William, you know how it is! Bastard splashed me trainers though! Cunt."

Kim sighs. "That boy is a liability! Always getting drunk and fighting, Bob always pulling him out of trouble!" looking across to Brad who flicks her the finger.

Calla sniggers. "See, he likes you!"

"Yeh, he thinks I am spending all his father's money! Truth is Bob wouldn't have a cent if it was not for me! He is another one who would gamble his last dollar!"

Calla nods "Aha, so, Kim! Interesting topic came up! Designer Vagina? Heard you have had one installed, care to share?"

Willy sprays Calla with his beer, Calla wiping his face laughing. Kim tilting her head. "Share my vagina or what I've had done!" Willy thumbs his glasses back onto his head smiling. "Yes I've had some work done, boys, no big deal!"

Calla raises his glass. "Well, Kim, here's a toast to your honesty and time travelling blurt!" chinking Willy's glasses. With a wink and a nod towards Kim he wanders toward Sharon. "Helloo, sweet cheeks!" kissing her on the side of the head. "And what's these dopey shites been saying?"

"Nothing much. See you have quite an eye for the girls, Calla?" Sharon keeping her distance as her friends give him a look that would kill.

"I was being friendly to a friend's mother to be honest – and her new designer vagina!" as all the lads burst out laughing, Hags wiping the tears away.

"Oh aye, Calla, hello Mrs, me and your son are pals – how's the new fanny these days? Fancy a scone and a cup of tea!"

Calla shaking his head – "Design this phrase into something, you daft old cunt, off fuck."

Sharon is still not the happiest as a bottle smashes and a large spray of water soaks the party, Brad smacking the bouncer as one more grabs him from behind. "Back in a minute." Calla steps over the stool cracking the bouncer holding Brad, knocking him into Willy's direction.

"Ah, for fuck's sake, man." Willy, soaked, limps over, kneeing the bouncer in the face as the music stops and the lights come on.

***

Sitting in Hardee's, Calla, Willy, Kris, Mark, Hags and Keegs are stuffing their faces with burgers and chips, Hags laughing.

"What you laughing at, daft arse?" Calla throws a chip at Hags. "You daft cunts, how quickly things can go wrong! Women at your mercy, some of them with magic fannies and a free bar and here we are eating burger and chips, lucky not to be in jail! Same old shit with you cunts."

Willy shakes his head. "Fuck off, Hags, you depressing cunt. As if you would've pulled anything!"

Mark chirps in, "Aye Hags, last thing you pulled was probably a wrinkly old cock."

"Bit like your mam, Mark." Calla laughs as Mark gives Calla the finger.

"Look lads, we are away tomorrow so let's get a taxi and fuck off – it's only 2 o'clock." Kris stands up looking out for taxis.

"We might get a lock-in at Nelly's?" Willy rubs his hands together as Calla jumps up. "Fucking good shout, fannyballs, last chance ranch here we come!"

Two taxis pull up outside of Nelly's and the lads climb out falling into each other. "Fucking happy days, lights are still on." Calla smiles, marching towards the door. "Bastard, it's locked!"

Just then Chuck bursts out of the fire exit, spewing into a plant pot. "Good man, Chuck, I knew those ribs would come in handy, ya greedy fucker." Calla marches past him as all the lads walk past Chuck, patting him on the back as he continues to spew his guts into the plants.

"Six pints of your finest, gorgeous." Calla marching up to the bar as Rosie looks surprised to see him.

"Your girlfriend dump you!?"

Calla shakes his head. "Nope, firstly because I haven't got one and secondly I wouldn't want one."

Davey wanders up to the bar, pissed. "Alrite youngn?"

Calla turning back to hug Davey – "Alritey, father. Get him one as well, Rosie – and who else is in?"

Rosie stands with her hands on her hips. "Mally's asleep through there, and you lot! Do I get some kind of apology?"

Calla laughs. "I am sorry I have had to wait more than 2 minutes for a pint."

Rosie shakes her head and starts pulling pints. "You're one asshole of an asshole, Calla. Be pleased to see the back of you."

Calla, putting a hundred dollars on the bar, responds, "Aye, thanks, so am I."

"Rosie, the jukebox is off, pet," Hags shouts over as Keegs sets the pool balls up. "That's because it's nearly 3 o'clock, Gordon!"

Hags squints through his glasses looking at his watch. "She's fucking correct! And you win a prize! So will you put the cunting thing on or not?" as the sound of Rod Stewart's 'I am sailing' starts blasting through the bar, all the lads standing around Hags, putting their arms around each other, as Rosie sighs and starts lining the shots up, necking one back, looking at Calla raising her glass as the lads belt the words out: "We are sailing, we are sailing, home again across the sea, We are sailing stormy waters, to be near you, to be free!!"

Just then Mally walks in to the bar after his cat-nap in the lounge, standing with a look of disgust at the lads, wiping his eyes. "What a fucking crew," as Calla shouts over: "Same crew that pays your wages in, you lanky streak of poodle shit."

The lads all gather around the bar and raise the glasses of Mickey Finn, Willy standing on the stool. "To packy bags neet." All the lads laugh as Keegs shouts them in again and the sun starts to shine through the gaps in the curtains.

Calla throwing his shot down his neck and wiping the film of hot saliva from his mouth – "Rosie! One question, that's all!" steadying himself as Rosie rolls her eyes and leans in.

"Yes Andrew?"

Calla leaning in too – "What's the chances I'm bucking you tonight?" as Rosie straightens up slapping Calla across the face.

"Not in a million years!" She spins and storms off, Calla rubbing his stinging face and raising an eyebrow shouting, "Hmm, so you say there's a chance?" laughing.

Davey throws his arm around Tom. "Big man! Brilliant day and class night!"

Tom swaying slightly – "Yep, y'all know how to throw a good knees up!"

Chuck wanders past rubbing his eyes. "I can't see!" Davey and Tom both laughing.

Mally wandering to the bar shouts to Rosie. "Last round, sweetheart, then close the bill!"

Calla waving Mally's hand down – "No need, Mally, you got them in last Christmas!"

Mally shaking his head steadies himself, eyes all bleary. "Calla...ya nar what! Fuck off!"

"Wey hey!" Putting his arm around Mally – "You know what I like about you Malcolm?"

Squaring himself, locked by Calla's hug – "What's that?"

"Fuck all!" Calla pushes Malla away who loses his balance and goes head first over the bar stool clattering into the table, letting out a whimper. "Arr me ankle!"

The lads are pissing themselves as Mally sits up square with his glasses twisted on his face.

Davey pats Tom on the back. "And on that note I'm away yem!"

Tom nods. "Let me walk you, David!"

Davey smiles. "Well thank you, Thomas!" as the two wander to the bar and give both barmaids a kiss and thank them. Davey shouting to Calla who is dragging a very drunk Mally up off the floor, "Calla son, na neet!"

Calla looking over at Davey leaving lets go of Mally who thumps to the floor in a slump murmuring, "Cunt!"

"Hawld on, Davey man, ya can't go yet! Haweh, it's the last neet!" Davey, ignoring his plea, disapears through the fire exit.

"Actually I'm calling it a night as well!" Hags says wandering past.

"Me too!" Keegs and Kris stand up and wave everyone off, the rest of the lads shuffling out of the door, Mark helping Mally up and following behind, Calla shouting "Pufters!" flopping onto the bar stool next to Willy and Chuck who is asleep. "Areet, Will?"

Willy squints at Calla, covering one of his eyes to help focus. "I've shit myself!"

Leaning in for a sniff – "By you have anarl!" Calla jumping up. Willy laughs. "Rosie Gemma! Threesome?" Both barmaids shaking their heads. "Lezzaz!" Calla barks throwing Chuck over his shoulders and heading for the door. "Haweh, shitty pants, there's away!"

Willy easing out from the stool and carefully standing up before waddling 2 steps to the door swivelling around one hand over his eye and the other over his arse. "Nee pursuading yaye hens back for a nightcap?" Both girls shake their heads. Willy shrugs and waddles out the door.

***

"Calla, haweh son, get up, I've made you egg bread and beans."

Calla is lying on the couch bollock naked under Davey's jacket, the remnants of his shirt, shoes then socks then jeans and underpants trailing towards where he is lying. "Fucking hell, Davey, it looks like I was vaporized last neet – look!"

Davey, putting Calla's breakfast down on the table with a cup of tea – "Has the pissed the settee?"

Davey flops back in the armchair with a big fried egg sandwich as Calla looks under the jacket smelling. "No, but I might have shit in your pocket – something definitely isn't reet down there."

Davey quaffing half a loaf into his gob mumbles, "It's probably your bell-end, son."

The two lads are all showered and packed as Jimmy knocks on the door.

"Rent?! Spent!" Davey shouts as he waddles up and opens the door, Jimmy with a bag of envelopes. "Morning, hamma, here you go – two tickets and two payslips."

Calla takes the envelope off Davey, ripping it open with his teeth, reading his wage slip. "Champion and the bonus, when does the wages go in like?"

Davey laughs. "Fucking hell, I wish I had a pound for every time I have heard that in my life!"

Jimmy smiles. "This afternoon, Calla."

Davey opening his – "What time do we fly?" – as Calla walks past him putting his jacket on. "Fuck, flying! What time does that cunt turn up with my winnings first! It's nearly 12 o'clock! Fatha, fetch me bag to Nelly's."

Calla is crossing the road as he sees Bob pulling up in the car with a big hard looking bloke. "Afternoon Calla, did you see the result from last night?"

Calla looks puzzled. "Aye and so did you – my 16-1 winner won!"

Joe shakes his head. "After we turned it off the race went to a steward's enquiry, Castlerock was awarded the win and Bonustime second!"

Calla drops his cigarette and clenches his fist.

"Just before you get a little crazy that's why I left last night early as I had a lot of money riding on the horse to cover your bet and obviously with my own expected winnings as me and old Walter have a few ventures together."

Calla swaying with anger – "Have you got the paper?"

The big henchman passing the paper to Calla with the result as Calla reads it. "Still don't fucking believe that! Where's Walter!?"

Bob takes out an envelope. "Here, it's all there – 2 grand. To be honest that's still a good bet, money back on a loser. Old Walter did alright yesterday and didn't want you boys looking for trouble! Believe me I wish I was paying you for the win – and look here is Keegan's 1200 bucks also, as his actually did win."

Calla flicks through the hundreds. "So is this what you brought this big daft cunt for?" Calla pointing at the big fella, who turns to Bob seeking permission to have a go.

Bob laughs holding his arm across him. "We have some business with last night's doormen – nothing to do with you, son."

Calla shakes his head. "Fucking can't believe this like, thousands down! What the fuck was the steward's enquiry for anyway?"

Bob climbs into his car. "Cut across Castlerock's path, the third horse made no difference to the result but you know the saying, boy – you win some, you lose some?"

Calla nods. "Suppose you're right, stroker. Thanks anyway, ya jipping cunt"

Bob buzzes the window down. "No, thank you – for looking after Bradley's back last night," as the car pulls off.

Calla bumps into Keegs walking into the bar. "Pint, bigman?" Calla smiles at Keegs.

"Why not? Be rude not to, Calla," as the lads go into the bar.

Sitting on the bar with two shandies, Calla turns to Keegs. "You know that race last night, well your fucking horse was awarded the win! Mine allegedly hampered it in the final furlong! Here, look at the paper."

Keegs laughing, rubbing his hands together, reads the paper. "Belter, so did the bookie give you my winnings or what!?"

Calla shakes his head. "No he said he was going to drop them in to you at five bells."

Keegs looking at his watch – "We fucking fly at 7, Calla!"

Calla shrugs his shoulders. "Not my fault. Do you a deal I'll give you a grand and I'll take the chance on Bob showing up with the winnings?"

Keegs thinking – "Righto, fucking deal! I am not missing that flight."

Calla counts the 10 hundred dollar bills out. "There you go, sadsack," as Willy marches in. "Show me the green, cuntlips!"

"Well for starters, you misey Scottish cunt! I was halfys out the goodness of my heart and as I just told Keegs the hoss turned out to be a cheating bastard like you, so it finished second – so no winnings! Just a little softener from old man Walt!"

Willy shrugging, noses closer. "So what's the softener, big man?" again rubbing his hands.

Calla puts his hand into the envelope swiftly pulling out his middle finger. "None for you, ya cunt! We are still down!"

Willy flops down on the stool signalling a pint to the barmaid.

The rest of the lads troop in, Davey holding his hand out to Calla doing a little jog. "Where's my money!"

Calla turning sideways bats the newspaper across Davey's chest. "What's this?" taking out his reading glasses and flicking open the paper, reading the article, muttering under his tongue then looking up – "The fucking bastard!"

Calla handing him a pint – "Anar, it's been a sad end to a fab trip!" handing Davey his stake back.

Willy squeezes out a fart, Davey stepping back. "Woah there, treacle, you safe to be doing that?"

Willy wafting the fart towards Kris – "Ay Davey, the old nipsy was a touch slack last neet but it's all mended now!" shuffling his underpants from the back of his trousers.

Calla shakes his head. "You make me sick! You're one horrible little man!"

Kris is starting to gag. "Byy! If that's not shit it will dee till we get some!" snides Davey wandering away.

Mark is into his bag pulling out his deodorant, Willy sat laughing. "What's rang? I'm sorry it's nee Estee Lauder like!"

"The rest of them heading down?" Davey cocking his own arse letting out a squeaky fart.

"Think they will be after Jimmy's been with the post! That cunt was looking worn oot this morning!" Mark leans further back squirting towards Davey.

"He was off bucking that Michelle, man!" Calla jabs Keegs, looking up.

"Bollocks!"

Davey nods. "True, lad, he's been riding her the last few nights!"

Keegs stands up. "Not having that like!" Then burping in Calla's face! He walks off laughing.

"Smells like cum!" Calla wafting his face.

Jimmy and Mally walk into the bar with the rest of the lads in toe. "Right, lads, there's a bus taking you to the airport leaving here at 4, gives you 3 hours to get pissed," Jimmy explains to the troops.

Hags is in behind Jimmy looking like he is going to a wedding with a blue suit on and white shirt.

"Fucking hell, Hags, have you upgraded yourself to business?" Keegs having a go.

"No Trevor, I have upgraded myself from pipefitter to suave and sophisticated."

Willy laughs. "Your fucking set ups are far from sophisticated, you blind cunt."

Hags laughs. "I wouldn't set up what I couldn't weld myself, William. By the way you owe me a new pair of trousers"

Willy laughing "Hags man it was only a bit shit!"

Davey shuffles over to Jimmy. "You having a pint, son? Why can't you drop us to the airport like?"

Jimmy shakes his head. "I've a mountain of paperwork to get done, Davey and I'm not a lover of farewells!"

Davey offers his hand. "Well, thanks for the trip and the craic, lad."

Jimmy shakes Davey's hand. "Always a pleasure to get the best in the business over and I'll see you in Shanghai, man."

All the lads get up to give Jimmy a handshake. "Safe travels, Calla, you big mad bastard, make sure you put some of those wages in the right direction."

Calla laughs. "Aye cheers, I'm sure I'll be pink lint by the time we go to Shanghai!"

Jimmy wanders over to the bar as Mally has a word with Willy. "Willy, do you fancy staying on for two weeks, bit of testing? A doddle to be fair."

Willy shakes his head. "Mal, I have done a 16 week trip – our lass is going fucking apeshit. The dog will probably gan for me when I get in it's been that long."

Mally nods as Calla chirps up: "You mean the dog will go for you cos you have a fyass like a pork chop, ya cunt!? Mally, I'll stay if you pay for my expenses; I'm a bit skint!"

All the lads laugh, Mark standing up. "Calla, it would be cheaper to fly another crew over than pay your bar bill!!"

Mally agrees. "I think I'll pass on that one and get a couple of Yanks to help me out, Calla."

Calla picks up his pint. "Good cos I was just going to tell you to fuck right off anyway!"

Jimmy gets to the door and shouts Mally over. "Right lads, you all got tickets and wage slips? There's 500 dollars behind the bar! See you all in a bit," as him and Mally walk out into the car park. Chuck and Tom call in just as the lads are ordering the drinks.

"There's a familiar sight, you guys at the bar." Big Tom walks over to the lads.

"Aye there's another sight, you fucking Yanks coming in last when all the hard work is done." Kris patting Tom on the back.

"We are back to Texas, boys. No doubt still on the road when your plane lands in the United Britain."

Willy laughs. "United Kingdom, you numb bastard."

Chuck laughs. "All the one really, just a pool of mental cases waiting to get out and terrorise the world."

Keegs shouts over to the lads. "You two gays having a beer or what?" both lads shaking their head – "soft cunts."

Calla is drinking his pint as Kim walks in looking rather tasty. "Well hello Kim and her designer vagina." Calla gives her a peck on the cheek as she laughs with all the lads drooling, Willy coming back from the toilets with two eyes drawn on his balls and them both squeezed through his fist – "look lads, ET!"

Calla nudges Kim. "You're right, it is a ringer for ET! By the way your jacobs are hanging out!"

Willy spotting Kim tucks his balls away, zipping them up and wandering over, offering his hand. Kim looks at his hand keeping hers well back. "Hello William!"

Calla slaps Willy's hand. "Put ya minging cheesy trotter away, pervert!"

Willy digging Calla – "Drink, Kim?"

"No thanks, Joe said you guys were leaving and I was passing so I thought I would call in to say good bye!"

Willy eases past. "Suit yourself!"

Kim grabs Calla's arm. "What time you guys leaving?"

Calla looks across at the lads making their way to the pool table sipping their beers. "4 bells the venga bus arrives!"

"Venga bus?" Kim asks Calla leaning in.

"Look let's skip the chit chat, bonny lass! I know you have came in here looking for a good bucking so why don't you say your goodbyes, wander outside and I will follow you out in 2, then a quick hop up to my digs for possibly the best sex in your life, then you can drop me back down before my pint gets warm?" leaning back showing all his teeth and winking.

Kim gasps, shocked, leaving go of his hand, shaking her head. "And what sort of woman do you think I am?"

Calla glances over her shoulder watching Willy walk past, sniggering, giving him the v sign, leaning in again and pointing below her waist. "Look it doesn't eat grass, bonny lass, it needs a bit of top quality meat now and again." He sips his pint, leaning on the bar as she smiles, whispering, "See you outside in two," before she turns and waves to the rest of the lads who all wave back. Willy blows her a kiss as she leaves. Calla is propped against the bar. "Hoy hoy there, bet you asked her for a shag and bet she fucked you right off?" Willy shouts over.

"Aye William that's exactly what she said, I'm away for a piss get the pints in–" standing up and wandering off the other way to the toilet area and out the fire exit and straight into Kim's Range Rover. "I can drive with this if you want," Calla pointing to the tent pole in his jeans and signalling up the road towards the accommodation. "You keep this to yourself, Andrew, I'm warning you!" waving her finger.

"Chill your beans, man, I'm outta here at 16 hundred hours, never to be seen again!"

Kim is all smiles as big Earl stops the car. Calla leaning out the passenger window – "Alreet, Earl, I forgot my wallet!"

Big Earl winks as they drive through the gate.

Kim eases off towards Calla's place. "I can't believe I'm doing this! God, I have never done this sort of thing before!"

Calla rolling his eyes at the statement – "By the way, bet Bob's hung like a sewing machine and goes like a donkey!!" as he jumps out of the motor and runs up the path, kicking the front door in.

"Did you not keep your key?" Kim steps in past the broken door, Calla kicking the door shut.

"I did, gorgeous," pointing at his right Adidas Samba then planting the lips on her and dragging her jacket off. Calla drags Kim onto the stairs, she pulls his jeans off, grabbing through his Calvins as he hitches her skirt up admiring the view. "It does look like it has had short back and sides, by the way."

Kim giggles, dragging him forward, unbuttoning her top. "Go on, try it for size, Andrew."

Calla smiles as he whips a blob out of his pocket, ripping it open and putting it on his knob like an assassin putting his rifle together and is just about to slide it in when the broken door swings open.

"Areet son, I think I've left me glasses on the bedside cabinet."

Calla and Kim jump up as Davey walks in. "Calla, what ya dein, have we been burgled? The door's hingin off?"

Calla stands up, pulling his jeans up. "For fuck sake, Davey you could've knocked, erm on the window or summit."

Kim straightens herself up, mortified, as Davey pushes past them going up the stairs. "Areet, bonny lass."

Calla shrugs his shoulders, as Kim awkwardly steps over the door frame, putting her jacket on and pulling her skirt back down as she makes a hasty exit. Calla walks into the kitchen and opens the fridge, pings the last of two cans of lager as he hears the Range Rover start up and drive off. "Fuck it, fatha, do you want a can?" Calla shouts up the stairs as Davey walks down into the kitchen with his glasses on.

"Aye, gan on, these are me crossword glegs, Calla, I'd be fucked without them." Davey sips the can. "What was that all about on the stairs?" rubbing his belly. "By, something has got my guts fucked!" squeezing out a watery fart.

Calla pulls his head away and pulling his t-shirt over his nose – "Davey man, that fucking stinks, here, nearly had my Redcar missile in a designer fucking vagina before you came back!"

Davey laughs, clashing cans with Calla. "Why lad, fuck off, I bet it's just like our Marj's!"

Calla spitting his drink all over the kitchen – "I'll tak your word for it, Davey!!"

Back up at Nelly's, Willy is ribbing Mally about not getting the Shanghai trip. "I say you will get bagged after this gig here finishes, Mally, you're about as much use as one-legged man in an arse kicking competition." Mally shakes his head. "Willy, if only you could graft as well as you spoke shite!" thumbing his glasses back to the bridge of his nose just as Keegs gives him a playful nudge in the back sending them wonky again. "Take no notice, Mally."

Davey and Calla wander up to Earl's little office and tap on the window. The big man comes out all smiles as usual. "Here, pal, there's 200 dollars for the damaged door. We forgot our key, if you know what I mean."

Earl laughs as he waves the money away. "No need, Cal, I will have maintenance look after it. Consider it a farewell gift."

Calla shakes his head. "Ok, we'll go halfs – how about that?"

Davey shakes Earl's hand. "You coming for a beer?"

Earl shakes his head. "A beer? Like one? With you guys!? I'll wave you off when you leave, thanks all the same." The taxi driver blasting the horn in front of Earl's gate gesturing the lads to hurry up, Calla giving Earl a big hug. "You guys should try walking to Nelly's." Calla stops at the cab door – "And waste precious drinking time?" before jumping into the back behind Davey, squeezing his shoulders. "Back to wor Nelly's!" – shuffling his pants down, pressing his bare arse against the window at Earl walking back into his hut laughing.

Back at the bar Keegs is leaning over the table about to play his shot when Willy pulls his pants down, Keegs potting the stripe unfazed pulling his strides back up shaking his head at Willy as Calla walks in. Kris looks up sucking the froth off his beer. "Hey fannyballs, where have you been?" Calla striding past blows the top of the pint sending the froth into Kris's eyes – "Helping Davey find his goggles, blister brains. Rosie, pint please!" Willy limps over whispering into Calla's ear, "Buck Kim, big man?" Calla sipping his pint shakes his head. "Well, William McCocksplash, a gentleman never tells!" Willy laughing – "That's a no then, hopeless hole!" Davey flipping the beermat off the top of his half finished beer and taking a huge mouthful lets out a loud burp. "By Christ!" digging himself in the chest. "Well, Rosie, great bar by the way and one we will all have good memories over! No doubt your takings will drop." Rosie laughs. "By eighty percent I would say. We will miss you lot, even that asshole!" pointing at Calla who sticks his tongue out as Davey leans across the bar giving her a kiss on the cheek.

The lads are all having the last pints as the bus pulls up, necking the last dregs and throwing their kit bags over their shoulders. The lads all troop out giving the barmaids a kiss on the cheek as they actually shed a tear. All the lads get a team photo outside Nelly's, big smiles and laughs as they pile into the bus, Calla sliding the door shut. "Davey, lend me your phone, man."

Willy, lighting up a tab, opens the window. "What you doing, fannychops, phoning the ex, telling her to slide down the banister and warm your tea up, you're on your way home!?"

Calla holding the phone up – "No pal, I'm just seeing if the wages are in."

THE END

Lightning Source UK Ltd.
Milton Keynes UK
UKHW010802150822
407319UK00002B/418

9 781910 223116